AF566594

Interfungal Parasitic Relationships

CAB INTERNATIONAL is an intergovernmental organization providing services worldwide to agriculture, forestry, human health and the management of natural resources.

The information services maintain a computerized database containing over 2.7 million abstracts on agricultural and related research with 150,000 records added each year. This information is disseminated in 47 abstract journals, and also on CD-ROM and online. Other services include supporting development and training projects, and publishing a wide range of academic titles.

The four scientific institutes are centres of excellence for research and identification of organisms of agricultural and economic importance: they provide annual identifications of over 30,000 insect and microorganism specimens to scientists worldwide, and conduct international biological control projects.

International Mycological Institute

An Institute of CAB INTERNATIONAL

The Institute, founded in 1920, and with a staff of 72:

- Is the largest mycological centre in the world, and is housed in a complex specially designed to support its requirements.
- Carries out research on a wide range of systematic and applied problems involving fungi (including lichens and yeasts), bacteria, and also on the preservation of fungi, the biochemical, physiological and molecular characterization of strains, and on crop protection, environmental and industrial mycology, food spoilage, public health, biodeterioration and biodegradation.
- Provides an authoritative identification service, especially for microfungi of economic and environmental importance (other than certain human and animal pathogens), and for plant pathogenic bacteria and spoilage yeasts.
- Undertakes extensive computerized and indexing bibliographic work, including the preparation of 7 serial publications.
- Supplies advice and undertakes a wide variety of project and culture work on crop protection, environmental, food spoilage and industrial topics.
- Offers training in pure and applied aspects of mycology and bacteriology, both at the Institute and overseas.

The Institute's dried reference collection numbers in excess of 355,000 specimens representing about 31,500 different species, and a genetic resource collection holds more than 17,000 living isolates by a variety of the most modern methods. The library receives about 600 current journals, and has extensive book and reprint holdings reflecting the Institute's interests.

International Mycological Institute
Bakeham Lane
Egham
Surrey TW20 9TY
UK

Tel: (0784) 470111
Telex: 9312102252
Telecom Gold/Dialcom: 84:CAU009
Fax: (0784) 470909
E-mail: CABI-IMI@CGNET.COM

For full details of the services provided, please contact the Director.

Interfungal Parasitic Relationships

P. Jeffries

Biological Laboratory
University of Kent at Canterbury

and

T.W.K. Young

Division of Life Sciences
King's College
University of London

CAB INTERNATIONAL

CAB INTERNATIONAL
Wallingford
Oxon OX10 8DE
UK

Tel: Wallingford (0491) 832111
Telex: 847964 (COMAGG G)
Telecom Gold/Dialcom: 84: CAU001
Fax: (0491) 833508

A catalogue record for this book is available from the British Library.

ISBN 0 85198 670 6

Typeset by Solidus (Bristol) Limited, England
Printed and bound in the UK at the University Press, Cambridge

CONTENTS

LIST OF FIGURES

List of Tables

Foreword

While fungi are well-known as parasites of plants and animals, their role as parasites of other fungi has received scant attention. Yet interfungal parasitic associations are far from uncommon, encompass a fascinating array of strategies at the structural and biochemical level, and assume economic importance as in diseases of edible mushrooms and the biocontrol of plant pathogens.

This wide-ranging and authoritative review of the mycoparasitic fungi is the first major synthesis of current knowledge on these little-known but potentially much more exploitable fungi.

In focusing on fungi in which mycoparasitism is defined in a strict sense, it is pertinent to emphasize that there is an immense number of reports of fungi growing on other fungi where almost nothing is known of their biology. No overall estimate exists, but there are not less than 1200 conidial fungi occurring on other fungi (fungicolous fungi), and perhaps 2000 fungi of all groups growing only on lichens (lichenicolous fungi). The proportion that is mycoparasitic as opposed to commensalistic or saprobic is unclear, but it is inescapable that many of these associations occur in the tropics, and represent an enormous resource for assays for the biocontrol of other fungi. The elucidation of the nature of their interactions at the physiological and ultrastructural level is challenging. The elegant results from well-studied cases documented in this volume, especially in the zygomycetes, show the degree to which this is possible. At the same time, patterns of host specificity have the potential to generate datasets against which postulated phylogenies can be compared.

This survey can be commended to mycologists as a review of what is known of the range of mycoparasitic interactions from the biological to the ultrastructural and physiological level, the ecological and disease significance of the interactions, and of progress in their utilization in biocontrol.

The authors have produced a book which deserves to stimulate a renewed interest in a significant but little-known ecological group of fungi.

Professor D.L. Hawksworth
Director
International Mycological Institute

Acknowledgements

We gratefully acknowledge the help of Professor D.L. Hawksworth and the staff of the International Mycological Institute for reading and commenting on earlier drafts of the manuscript. We also thank numerous colleagues for supplying micrographs:

Dr H.C. Hoch (Figs 1.4, 4.12, 4.13), Dr P.H.J.F. van den Boogert (Figs 3.3, 3.4), Professor M.J. Powell (Figs 4.10, 4.11, 4.23, 4.24, 4.25), Dr R. Bauer and Professor Dr F. Oberwinkler (Figs 4.14, 4.15, 4.16, 4.19, 4.20, 4.21); Dr A. Tsuneda (Figs 3.7, 3.8); Dr Y. Persson (Figs 3.1, 3.2); R. Phillips (Figs 2.2, 2.3); and Dr J.T. Fletcher (Figs 6.3, 6.4, 6.5, 6.6, 6.7). Fig. 1.4 was reproduced with permission from Springer-Verlag. Fig. 4.20 was reprinted with permission from Mycologia, 76, 1039–1048. Copyright 1984. The New York Botanical Garden, and Fig. 6.8 was reproduced with permission from Intercept Press.

We also thank Sue Davies and Anna Sherriff for their patience and understanding as they frequently retyped new versions of our text.

Finally, we thank our wives, Jackie and Pam, for their help and support throughout this work.

Note Regarding Classification Scheme

Modern advances in techniques of molecular biology have resulted in revised systems of classification for the fungi. Many systematists now favour the division of the Eumycota (*sensu* Ainsworth, 1971) into more than one kingdom such that the old concept of the fungi will no longer exist. In this book we have used the term 'fungi' in the traditional sense of 'organisms studied by mycologists' as suggested by Hawksworth (1991). We have, however, tried to anticipate the new systematic scheme that will appear in the forthcoming eighth edition of *Ainsworth and Bisby's Dictionary of the Fungi* due to be published by CAB INTERNATIONAL in mid-1994 (Hawksworth *et al.*, 1994). In this system, the fungi and fungus-like organisms are classified into three different kingdoms, such that the chytrid-like organisms and the Oomycetes are separated from the kingdom Fungi. This scheme is also novel in the division of these organisms into a number of phyla, and by the proposed absence of taxonomic ranks at the level of division or class. The basic scheme is outlined below, and many of the existing orders of fungi will be arranged within these phyla.

Kingdom	Phylum	Characteristics
Protoctista	Myxomycota	Amoeboid, many form aggregations, plasmodia or sporophores
	Chytridiomycota	Not amoeboid, many unicellular, chitinoid walls, flagellate zoospores
Chromista	Oomycota	Filamentous, cellulose wall, motile zoospores with two types of flagellum, sexual state oogonia
Fungi	Zygomycota	Filamentous, produce zygospores (resting spore) as sexual state
	Ascomycota[a]	Ascospores found inside an ascus, usually in specialized fruit bodies, most are filamentous, hyphal with perforate centres, walls not multi-layered, electron transparent and chitin present; or with endospores produced inside cells, lacking special fruit bodies, mostly yeast-like, septa where present not centrally perforate, chitin often absent
	Basidiomycota[a]	Basidiospores produced outside, usually non-septate basidia, clamp connections and dolipore septa usually present, multi-layered and electron-dense walls, most are filamentous; or with basidiospores produced outside, cross-septate basidia, with clamp connections, dolipore septa absent, spores germinate to produce a yeast-like phase which can predominate, walls lacking xylose

[a]Members of these phyla in which sexual (teleomorphic) states are generally absent are sometimes called deuteromycetes, 'conidial' or 'imperfect', most are anamorphs of ascomycetes.

1 Introduction

1.1 The Variety of Interfungal Relationships

All fungi are **heterotrophic** and therefore require pre-formed organic compounds as the source of energy for growth. This necessity has led to the evolution of fungi as **saprotrophs** and symbionts (de Bary, 1887) *par excellence.* Fungi which grow on dead organic matter, such as rotting tree stumps, are examples of those that feed saprotrophically. Those deriving nutrients from living hosts, on the other hand, are symbiotic as they live in close, often permanent, association with one or more unrelated species of organism. The partner in this relationship may be advantaged (**mutualistic symbiosis**), unaffected (**neutralistic** or **commensalistic** symbiosis) or compromised (**antagonistic** or **parasitic** symbiosis) by the association. Such relationships may change with time, and fungi such as the potato blight fungus, *Phytophthora infestans*, can have a saprotrophic phase on dead host plants, and a parasitic phase on a living host. Perhaps two thirds of the known species of fungi form intimate relationships with other living organisms and there can be few taxa that have not been associated with fungi in some sort of symbiosis. Not unexpectedly there are also many relationships in which different species of fungi can interact among themselves. These interspecific fungal relationships encompass a broad spectrum of interactions including such phenomena as **hyphal interference, gall formation**, and **mycoparasitism.**

There are many ways in which fungal mycelia of different species can interact such that the presence of one in some way affects the behaviour of the other. At one extreme, one living fungus serves directly as the nutrient source for another; this is mycoparasitism, the major concern in this book, particularly in relation to **necrotrophic** (Chapter 3) and **biotrophic** (Chapter 4) associations. At the other extreme, the two mycelia may simply be growing together in the exploitation of a particular environmental resource. If the two organisms do not apparently affect one another this is termed neutralism. This is unusual, however, as two organisms in the same ecological niche almost inevitably interact in some way or other. In between these two extremes are many other interactions in which one fungus is compromised by the presence of another, yet nutrient exchange has not been demonstrated. Interspecific interactions in this latter category have been discussed in detail elsewhere (Rayner and Webber, 1984; Pearce, 1990). There are also interactions in which one fungus derives an advantage from the presence of another but the latter does not derive reciprocal benefit although it is not harmed; this is commensalism. Where the circumstantial evidence suggests that a parasitic relationship is likely, examples will be discussed in detail in this book, but where the evidence for direct parasitism is weaker, or where commensalism may be operating, then less emphasis will be accorded. The physiological aspects of mycoparasitic relationships are discussed in more detail in Chapter 5. There are also relationships between fungi that alter over time and an association that began in a mutualistic fashion can develop into an antagonistic interaction as physiological conditions change. In nature parasitic relationships between fungi probably play an important role in the development of community structure; ecological aspects of interactions of fungi are dealt with in Chapter 6. The ability of antagonistic fungi to inhibit the development of other **pathogenic** fungi can be exploited in **biocontrol**, and this has received particular attention ever since the phenomenon of interfungal parasitic relationships was discovered (Chapter 7). We have excluded from this book consideration of those antagonistic interactions that can occur between mycelia of the same species. The phenomenon of fungal individualism whereby different clones of fungi can recognize and respond to others has been well-described (Rayner, 1991). Similarly we have not discussed in detail the intraspecific mycelial interactions that involve the cytoplasmic transmission of nucleic acid which codes for **hypovirulence** or includes a 'killer factor'.

There are also different degrees of specialization within interfungal parasitic relationships. The conventional concept of parasitism applies most readily to biotrophic relationships where, by definition, the parasite exploits only living host cytoplasm as a nutrient source. Owing to the limitations of transporting material across an often complex interface between the partners, it is usually only the simpler organic compounds such as sugars and amino acids that are likely to be absorbed from the host. In contrast, necrotrophic relationships involve the utilization of dead host biomass, which has been killed as a direct result of the activity of the necrotroph itself. This can involve the degradation of the major structural polymers of the host or/and the soluble components of the cytoplasm and is closely paralleled in many saprotrophic situations. It must also be realized that saprotrophy, necrotrophy and biotrophy are not necessarily mutually exclusive ways of life (Cooke and Rayner, 1984) and the relationships between fungi may switch from one form to another as the association develops. The distinction between these nutritional modes may not always be clear and it has been pointed out that the colonization of living tissues, via essentially saprotrophic behaviour involving the utilization of diffusates of dead cells, can be confused with necrotrophy and even biotrophy (Rayner *et al.*, 1985). Despite these reservations there are, nevertheless, several good illustrative examples of both necrotrophic and biotrophic mycoparasitism and these will be discussed in full in the relevant chapters.

1.2 Mycoparasitism Defined

Mycoparasitism is the parasitism of one fungus by another (Hawksworth *et al.*, 1983), and its usage among mycoparasitologists is so widespread that it remains a useful descriptive term although it has been strongly criticized on various grounds (Cooke, 1977). In rare cases the term has also been used to describe parasitism of any organism by a fungus (e.g. Ehrlich and Ehrlich, 1971), but this broader use of the term has not gained wide acceptance. Sometimes a mycoparasitic association is referred to as a **mycotrophic** association. This emphasizes the nutritional aspects of the relationship although little may be known of the mechanism of transfer of nutrients from the host to the parasite, or indeed, of the form in which the nutrients may be exchanged. The term mycotrophy, however, has also been used to refer to mycorrhizal associations, where fungi grow in close association with the roots of plants and both partners usually benefit from the association and so its use in either context should be discouraged. The term **mycophthorous** (Ainsworth, 1971) has also been used to describe the parasitism of one fungus by another fungus but appears to have been little used by mycologists. Whether the term parasitism should be used at all to describe the relationship between fungi where the host is harmed by the association has been questioned (Cooke, 1977) and the phrase 'antagonistic symbiosis' substituted. This concept has the merit of implying that the host is in some way harmed through the association, also implicit in 'parasitism'. Mycophily (Rudakov, 1978), literally fungus-loving, has been used in the description of various physiological categories of interfungal relationship although the term itself gives no clear indication of the physiological status of the relationship alluded to. Mycoparasitism has sometimes also been termed hyperparasitism (Boosalis, 1964) but this is not strictly correct as hyperparasitism refers to a situation in which one parasite is parasitic on another, and the host in a true **hyperparasitic** association is not necessarily itself a fungus.

Hawksworth (1981) recommended the use of the term **fungicolous** fungi as a neutral term to embrace the broad range of associations of two fungi living together even where the biological nature of the association is obscure. This avoids the problem of determining whether nutrient exchange occurs, and can also include wider symbiotic relationships such as commensal ones. For example, there are numerous fungi which have become associated with the external surface of another fungus as their particular ecological niche. They obligately grow in this location, but they are not necessarily parasites, and may be commensals

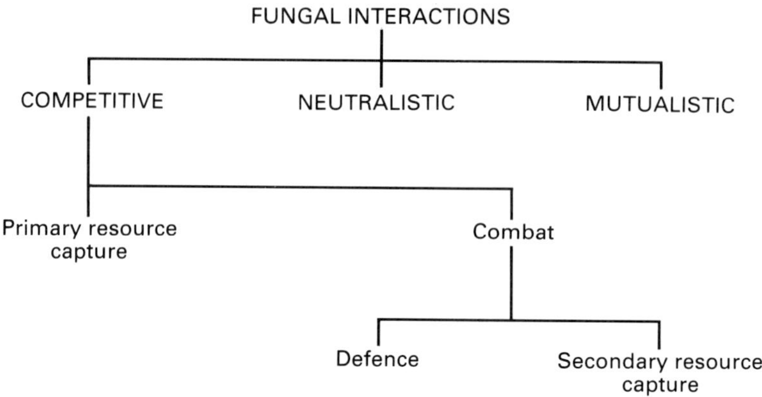

Fig. 1.1. Fungal interactions (after Cooke and Rayner, 1984).

or fungicolous saprotrophs and as such have been little investigated in terms of their diversity and biology. In some cases the fungicolous fungi are only found growing on the relatively large basidiomes of mushrooms and toadstools, but in other instances the association is between the two respective mycelia. There are also many fungi that are normally only found growing in association with lichens, the lichenicolous fungi (section 2.2).

Some of the problems in defining interfungal relationships have been addressed by Cooke and Rayner (1984). They have proposed that mycelial interactions are split according to the scheme illustrated in Fig. 1.1. The primary division is into **competitive**, neutralistic and mutualistic interactions of the two participants, depending successively on whether the outcome is detrimental to either or both, detrimental to neither but not beneficial to both, or beneficial to both. Thus in this book we are primarily concerned with those interactions termed competitive, and in particular those involving combat between species either relating to defence or **secondary resource capture** (i.e. the replacement of one species of fungus by another within a particular niche). The mechanisms which fungi use in combat with their competitors are encompassed by the term **antagonism** and are summarized in Fig. 1.2.

Competition between fungi occurs during **primary resource capture** when two or more mycelia are limited, in terms of growth rate or size, by a common dependence on a nutrient substrate or other environmental factor. The outcome of such interactions is determined by several factors, but relative **competitive saprotrophic ability** (Garrett, 1956) is, by definition, of crucial importance. The ability to antagonize a competing mycelium is an important attribute contributing to competitive saprotrophic ability as is the converse ability of being able to withstand the antagonistic action of other competing fungi. If a fungus has already colonized a primary substrate, but is subsequently replaced by a comba-

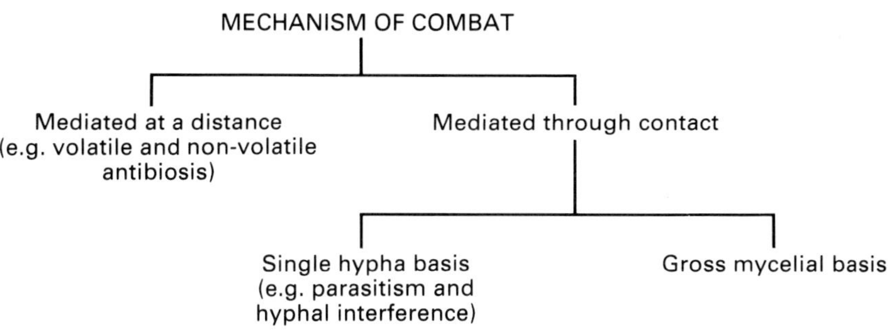

Fig. 1.2. Mechanisms of combat (after Rayner and Webber, 1984).

tive competitor, the interaction is then described as secondary resource capture. An example of this latter phenomenon occurs in mushroom compost, or on agar plates, when *Agaricus bisporus* kills the mycelium of *Scytalidium thermophilum* and grows into the domain previously colonized by the latter fungus. The antagonistic effect can be mediated at a distance, or through a cellophane membrane, and no evidence for direct hyphal interference or mycoparasitism has been observed (Op den Camp *et al.*, 1990).

In contrast, mycoparasitism of mycelium of the pioneer wood-colonizing genus *Coriolus* by the wood-decaying *Lenzites betulina* allows the latter fungus to gain selective access to wood occupied by *Coriolus* (Rayner *et al.*, 1987). *Pseudotrametes gibbosa* is similarly temporarily mycoparasitic on *Bjerkandera adusta* (= *Polyporus adustus*). *Coriolus versicolor* and *Lenzites betulina* are two lignicolous bracket fungi commonly found on, although not restricted to, birchwood. *Lenzites betulina*, in producing relatively large basidiomes, needs to occupy a proportionately large volume of the wood in order to sustain the reproductive phase. Parasitism of the mycelium of primary colonizers of the wood such as *C. versicolor*, which themselves occupy a large volume of the substratum, is believed to enable the *Lenzites* to dominate the wood. When mycelium of *Lenzites* grows into contact with the mycelium of most wood-decay fungi, recognition is usually followed by deadlock or replacement of one species by the other. If mycelium of *Lenzites*, however, encounters an opposing mycelium of *Coriolus versicolor* or *C. zonatus*, there are no signs of recognition of the former which replaces the *Coriolus* by direct mycoparasitism involving hyphal coiling and penetration. Evidence for the mycoparasitic ability of *L. betulina* is derived from the investigation of hyphal interaction, *in situ*, on malt agar plates and in experimental blocks of wood. Hyphal interactions on agar show that at a growth temperature of 25°C *L. betulina* displaces *Coriolus*, and that during this process the mycelium of *Coriolus* is entwined and penetrated by the hyphae of *L. betulina*. In wood blocks, *Coriolus* is always replaced by *L. betulina* which is clearly highly antagonistic to the former via the agency of mycoparasitism. Although mycoparasitism is usually considered as an interaction which provides a primary food source, Rayner *et al.* (1987) considered that in this case, the interaction was more important in that it gave access to the domain initially colonized by *Coriolus*, i.e. it was more important in secondary resource capture than in provision of a nutrient source. A related phenomenon is exhibited by *Phanerochaete magnoliae* (Ainsworth and Rayner, 1991), another wood-inhabiting fungus which, in this case, is able specifically to replace colonies of *Datronia mollis* as well as forming its own hymenial surfaces through the tubes of *D. mollis* basidiomes. The mode of displacement is unusual in that it does not involve dense zones of invasive mycelium, nor overt mycoparasitism. Instead, the fungus produces very sparse, rapidly extending hyphae which grow between the *Datronia* hyphae. Contact of hyphae then results in highly destructive hyphal interference (see section 3.1.2) which affects both fungi. This suggestion has also been made for antagonism by *Pythium oligandrum* towards several fungal hosts within the same ecological niche; direct exploitation of antagonized hyphae is of secondary importance to its primary role in combat between competing mycelia (Lutchmeah and Cooke, 1984). Thus mycoparasitism can be incidental to competitive success. In this way, interactions of mycelia can also affect the eventual appearance of the **sporomes**. Hyphal interference has been shown to influence the appearance of basidiomes of coprophilous fungi (Ikediugwu and Webster, 1970a), or the basidiome production of wood-inhabiting basidiomycetes as in the case of *C. versicolor* and *L. betulina* described previously.

The antagonistic ability of a fungus is determined partly by its physiological state, so that changes in physiochemical or nutritional conditions associated with resource utilization will affect the outcome of interaction of combatants (Rayner and Webber, 1984). Various mechanisms of antagonism can be distinguished, and involve, for example, antibiotic production, secretion of lytic enzymes, hyphal interference, and direct penetration of the host. It must be remembered, however, that any given fungus–fungus interaction may encompass more than one of these mechanisms either individually or simultaneously. Only those relationships in which one fungus obtains some nutrients either directly or indirectly from another fungus can be termed mycoparasitic, but the act of parasitism can involve any of the mechanisms described here. Specific examples of each of these mechanisms will be described in detail later.

It is important to realize that antagonism can be mediated either by direct contact or at a distance. These phenomena can be distinguished with use of the terms ‘direct antagonism’ and ‘indirect antagonism’. Direct antagonism involves direct physical contact between the two organisms, whereas indirect antagonism covers those situations that do not involve physical contact. Thus indirect antagonism includes those instances in which one fungus releases into the environment a material that induces a negative effect on the other; such materials include antibiotics and lytic enzymes. The boundary between indirect parasitism and competitive interactions involving primary resource capture may be very difficult to draw as parasitic individuals can benefit from the increased availability or release of nutrients consequent upon their activities, and thus alleviate competition for primary nutrients.

Barnett and Binder (1973) divided mycoparasites into two main categories depending on their mode of nutrition. Necrotrophic or **destructive mycoparasites** kill their hosts as a result of their parasitic activity, whereas biotrophic mycoparasites obtain their nutrients directly from the living mycelium of the hosts. In necrotrophic relationships the antagonistic action of the mycoparasites is strongly aggressive and the mycoparasite dominates the association. Hyphae of the parasite contact and grow in association with those of the host, sometimes coiling around them, and frequently penetrating. Secretion of hyphal wall degrading enzymes or **exotoxins** may cause the death of the cytoplasm of the host prior to hyphal contact, or alternatively cytoplasmic death may not occur until after contact has been established. Necrotrophic parasites tend to have a broad range of host fungi, and are relatively unspecialized in their mechanism of parasitism. For example, they often release toxins and lytic enzymes into the environment, are overtly destructive and usually lack specialized infection structures. In this way their behaviour parallels that of the necrotrophic fungi which parasitize plants.

In a biotrophic mycoparasitic relationship, the living host supports the growth of the parasite for an extended period of time, may not appear diseased, and its growth rate, sporulation and metabolism may appear overtly to be little affected, at least in the early stages of the relationship. The parasitic relationship is physiologically balanced and the parasite appears to be highly adapted to this mode of life. Biotrophic mycoparasites tend to have more restricted host ranges than necrotrophs, and often form specialized infection structures or **host–parasite interfaces.** Exotoxin production has not been demonstrated in any biotrophic mycoparasitic interaction. Three types of biotrophic relationship can be distinguished by the morphology of the interaction (see section 1.5).

Some necrotrophic fungal parasites, such as *Trichoderma harzianum* and *Gliocladium virens*, are frequently also able to grow well as saprotrophic competitors of other fungi. *Trichoderma* species, for example, have been studied primarily in relation to their ability to displace plant-pathogenic fungi such as *Armillaria mellea* from soils in which trees are grown for timber or crop purposes, such as tea plantations. *Trichoderma* species display strong competitive saprotrophic ability mainly due to a relatively rapid rate of mycelial growth and a tendency to recover quickly in soils treated with fumigants such as carbon disulphide. In the soil it is difficult to distinguish between the effects of direct antagonism involving hyphal coiling and penetration, or the secretion of antibiotics deleterious to the host, and the indirect antagonism of the competing fungi for the colonization of substrates in the soil and the occupation of available space. *Trichoderma* species are potential antagonists of many soil and root-inhabiting fungi and their success as competitors is believed to be due to the combination of the factors referred to above. A question relating to the necrotrophic reaction is whether the fungus is capable of growing on the living host prior to destruction of the cytoplasm. Interacting mycelia of the mycoparasite *T. hamatum* and the host *Rhizoctonia solani* (teleomorph = *Thanetophorus cucumeris*) stained with the vital dye fluorescein diacetate both fluoresce during the first 24 h of incubation, the *Rhizoctonia* mycelium losing most of its fluorescence after 2–3 days. The pattern of fluorescence is taken to indicate that the host is alive during the initial attack (Barak and Chet, 1986).

The extent to which a fungus is successful as a mycoparasite is debateable as it depends whether indirect antagonism, such as **antibiosis** for example, is included within the definition of mycoparasitism. In the narrow sense, mycoparasitism could be taken to include only direct contact of the mycoparasitic fungus with the potential host. This may involve coiling around potential host structures, possibly their penetration, the development of haustoria by biotrophs, the absorption of nutrients from the cytoplasm of the host and destruction of the cytoplasm of the host. From a practical point of view, however, the production of antagonistic metabolites which precedes the physical contact necessary to invasion of the mycelium of the potential host could reasonably be considered to constitute part of the essential reaction which leads to overt physical parasitism of the host. The morphological responses relating to the invasion of a potential host fungus undoubtedly succeed physiological reactions which lead to the observed response. In this context, the question of host recognition should be considered. How is it that a fungus may infect one host fungus but not another species, albeit a closely-related one? The answer is presumed to reside in recognition at the molecular level whereby components of the cytoplasm of the potential host, for example, may be sensitive to metabolites (e.g. antibiotics) of the invading fungus which enables the invader to gain a foothold on the host and this may be manifested visually by the so-called mycoparasitic reaction. Similarly, biochemical processes of the invading fungus may be affected by the proximity of the potential host as the former may be stimulated to produce hydrolytic enzymes capable of degrading the walls of the latter. Where physical contact between the antagonistic fungi occurs, for example during **appressorium** formation, a lectin wall-binding mechanism may be involved (see section 4.1). Mycoparasitic relationships considered in the broad sense comprise biochemical and physiological reactions which precede the microscopically visible phenomena of coiling, appressorium formation, penetration, **haustorium** formation and cytoplasmic degradation. In the many examples of biocontrol of plant-pathogenic fungi by other antagonistic fungi the relationship may involve to some extent antibiosis, wall recognition, enzymatic response and subsequent physical invasion of the potential host and it is neither possible nor necessarily desirable to divorce the biochemical and physiological aspects of the interaction from the visually observable responses.

1.3 Occurrence of Mycoparasitism

The significance of mycoparasitism in the natural environment is certainly underrated. The unequivocal demonstration of a mycoparasitic association under laboratory conditions, however, cannot be taken as proof that a similar parasitic relationship occurs in the field. Such studies are often carried out in pure mixed culture under favourable environmental conditions and nutrient excess. *In vitro* tests of the host range merely provide a strong indication of susceptible and immune fungi in the absence of natural competitors. Under field conditions the situation will be different. Microbial competition and nutrient limitation will place a premium on those characteristics that give an advantage in antagonistic interactions of fungi. Abilities that enhance the competitiveness for carbon and nitrogen substrates, or help a fungus to withstand or reduce the antagonistic behaviour of other fungi will increase the share of primary and secondary resources available to that fungus. Mycoparasitism is a powerful tool in this respect. It is difficult to make field observations of mycoparasitism *in situ*, but there are, nevertheless, some clear examples where the natural occurrence of mycoparasitism is documented. These are discussed in more detail in Chapter 6. In the case of the biotrophic haustorial mycoparasites, the highly specialized nature of their host–parasite interface, coupled with the difficulties experienced in inducing them to grow in the absence of a host, leaves little doubt that they normally grow as parasites in nature.

It has been suggested that, for **agaric** fungi at least, necrotrophic mycoparasitism may ultimately prove to be more common than is presently realized (Rayner *et al.*, 1985). At present, however, much of our knowledge of mycoparasitism among agarics is based on those which produce basidiomes on those of other fungi, notably other agarics, e.g. species of *Nyctalis* on Russulaceae (see section 2.1.2), *Armillaria mellea* inducing gasteroid basidiomes in *Entoloma abortivum* (Watling, 1974), and *Psathyrella epimyces* on *Coprinus comatus* (McDougall, 1919). Care must be taken, however, not to confuse other relationships with parasitism. For example, *Xerocomus parasiticus* and *X. astraeicola* are not parasitic on Sclerodermatales as is often thought, but simply require the presence of the latter to stimulate basidiome production (Rayner *et al.*, 1985).

1.4 Origins of Mycoparasitism

The origin of the phenomenon of mycoparasitism can only be speculative as there is no evidence from fossilized remains to indicate whether the parasitic way of life is either a primary or a secondary condition. Current opinion holds that parasitic fungi have evolved from non-parasitic ancestors through a process of increasing nutritional specialization which has resulted in various degrees of dependence upon hosts for survival. This view has been challenged (Cooke and Whipps, 1980) with respect to terrestrial plant-pathogenic fungi and arguments put forward to support the hypothesis that the main course of evolution on the land has been from dependence on hosts towards increased nutritional versatility and so to independence from them. This may also be true of mycoparasitic fungi. Obligate parasitism in the fungi may not be a belated evolutionary pathway, but a fundamental attribute of primitive groups, from which saprotrophy arose repeatedly (Savile, 1968). Necrotrophic parasites, however, being facultative saprotrophs may in some cases be derived from saprotrophs.

Many present-day fungi have the ability to behave as facultative mycoparasites in the sense that they can grow equally well in the absence of a potential host as when the host is present. Species of *Gliocladium* and *Trichoderma*, known to parasitize other genera of soil fungi, grow vigorously for indefinite periods as saprotrophs in pure culture. In soil, however, the situation is likely to be different as the mycoparasitic fungi would have to compete with other microorganisms. In a competitive situation the ability of a facultative mycoparasite to parasitize its competitors would confer a selective advantage in the struggle for colonization of soil substrates. Obligate mycoparasites, on the other hand, must by definition be unable to compete on a saprotrophic basis, possibly through the loss of some essential enzymic capability necessary to sustain a prolonged saprotrophic existence. Whether such a deficiency led them towards the parasitic mode of life, or whether a gradual transition to a parasitic mode of life caused them to lose this ability cannot be ascertained. What is clear, however, is that as a biological group the obligate mycoparasites are undoubtedly successful as they are worldwide in distribution, commonly found, and many examples are known from all of the major groups of fungi.

Many members of the Oomycota and Chytridiomycota are parasitic on other fungi, often species from the same taxa, and sometimes on algae. These organisms, although traditionally regarded as fungi, may be more closely related to the plant kingdom and are probably derived from the algae. This is reflected by our adoption of the new classification scheme outlined in the Foreword. In some members, for example *Saprolegnia* spp., it has been demonstrated that true cellulose is present in the wall of the hyphae (Bartnicki-Garcia, 1968). Nevertheless, these organisms have been studied extensively by mycologists, and are included here where mycoparasitic relationships have been described. Some Chytridiomycota, for example are intracellular parasites of Oomycota. *Woronina* species parasitize species of Saprolegniaceae when the zoospore encysts on the hypha of the host and the cytoplasmic contents of the cyst pass into the hypha via a penetration tube. It is noteworthy that filaments of the green alga *Vaucheria* are also parasitized by this fungus (Bessey, 1964) as there are morphological similarities in the sexual reproductive structures, the oogonia and antheridia, of *Saprolegnia* and *Vaucheria* species. It is often the case that a parasite has a specific host range which is probably related to recognition at the molecular level by the parasite and perhaps the presence of true cellulose in the wall of the host plays its part in this respect. Deacon (1988b) has emphasized the role of cell wall biochemistry in determining the outcome of the parasitic attack. The zoospores of species of

Pythium that are mycoparasitic are more likely to encyst on a chitinous substrate than a cellulosic one, whereas plant-pathogenic *Pythium* species encyst preferentially on a cellulosic substrate. Zoospore tactic responses, however, appear less specific and zoospores of mycoparasitic or plant-pathogenic zoospores apparently swim towards either substrate with no obvious preference. It is easy to imagine how a mutation in the recognition process could lead to the evolution of mycoparasitic pythia from plant-pathogenic species or vice-versa.

Coevolution is the evolution of one organism in conjunction with an evolving host and in this sense, stable relationships between symbiotic fungi and their hosts have exerted a major influence on their evolutionary development. The more intimate and stable a symbiosis, the greater is the probability of it being the outcome of coevolution (Pirozynski and Hawksworth, 1988). It has been suggested that mutualistic symbioses probably evolved from initially antagonistic relationships and may well be the outcome of coevolution. As many fungi are obligate and often mutualistic symbionts, mycologists have tended to take coevolution of these relationships for granted. Although some mycoparasitic relationships have been studied in detail, almost nothing is known about the biological and ecological aspects of the vast majority of fungus–fungus interactions. The acquisition of an essential growth factor from one fungus by another due to subtle alterations in membrane permeability in the contact zone, might well lead to a similar exchange of nutrient material in the reverse direction. Mutualistic equilibrium would be the outcome. For example, Barnett (1968) has suggested that the relationship between *Gonatobotryum fuscum* and a *Graphium* species was mutualistic with respect to the synthesis of reciprocally essential growth metabolites. In view of the marked differences in susceptibility of different host groups to mycoparasitism, Pirozynski and Hawksworth (1988) suggest that there is clearly a potential for the types of coevolutionary studies that have been carried out with rusts and flowering plants (e.g. Savile, 1979), and it has been recommended that in implementing any major taxonomic work on a particular group of fungi it would be worth ascertaining how the fungicolous fungi correlate with any remodelling being considered (Hawksworth, 1981). Owing to the wide range of interfungal relationships, however, ideas relating to the evolution of fungi are speculative. Detailed studies of the host range of most obligately mycoparasitic fungi have not been performed, and Hawksworth (1981) correctly points out that it is a considerable problem to estimate the numbers of potential host taxa.

In the biotrophic relationships, it is clear that considerable specialization has occurred at the host–parasite interface. Intracellular and haustorial biotrophic mycoparasites have evolved complex infection processes which avoid the triggering of host resistance responses and enable successful penetration of the hyphae of the host. In the **fusion biotrophs**, this relationship has progressed to the situation whereby parasite and host protoplasts become fused and direct cytoplasmic continuity is achieved. This situation is rarely encountered in any other parasitic relationship and suggests some close compatibility between the genomes of the interacting organisms that prevents elicitation of rejection responses. It is closely paralleled by the initial stages of intraspecific fusions between vegetative hyphae of *Coriolus versicolor*, for example, which involves the formation of a single opening which undergoes regular and symmetrical enlargement to the final diameter (Aylmore and Todd, 1984) rather than irregular breakdown. These interactions also resemble the initial stages in sexual interactions such as the fusion of zygomycete gametangia which also involve development of openings at the centre of the fusion wall which enlarge until the whole area is dissolved (Hawker and Beckett, 1971). This similarity is further emphasized in the interactions of different strains of the mucoraceous fungus *Parasitella parasitica* with other members of the Mucorales (see section 3.2). When a hyphal tip of *P. parasitica* contacts a hypha of *Mucor hiemalis*, for example, it stops extending apically, develops a septum and becomes swollen. This induces a short sidearm to develop from the contact point on the hypha of *M. hiemalis* which also becomes slightly swollen and from which short hyphal outgrowths are produced. The hyphal walls in the contact zone apparently dissolve and fusion of cytoplasm occurs. These structures have been termed **galls**, and resemble both the suspensors and zygosporangium formed in the normal sexual reaction between opposite mating strains of *P. parasitica* and also the illegitimate matings that can sometimes be induced between other dissimilar species of Mucorales. There does not seem to be any beneficial growth effect observed in *Parasitella* colonies as a result of this interaction. Burgeff (1924) reported that + and − strains of *P. parasitica* would only form galls with other Mucorales of opposite mating type thus adding credence to his belief that this was a parasitic attack that had evolved from a sexual origin. Satina and Blakeslee (1926), however, showed that gall formation was not always sex-linked and that these structures could be formed, for example in interactions with similar mating strains of *Absidia glauca*. The situation is further complicated by the fact that in some interactions the galls are smaller and less well differentiated, and there is no fusion of protoplasts or mixing of cytoplasm, nor any development of hyphal outgrowths. This latter observation has been confirmed (Jeffries, unpublished) yet it remains probable that these interactions have a sexual basis and may represent abortive attempts at copulation rather than mycoparasitism. The interactions of *Chaetocladium* and other Mucorales would seem to represent a succeeding stage in the evolutionary process towards parasitism. *Chaetocladium* species also form galls with closely related fungi, but there does not seem to be any preference for mating type, the galls are less like abortive zygosporangia, and there is a positive growth response observed in colonies of *Chaetocladium* that have formed these structures. In a wider context, Prillinger (1987) has expressed the alternative view that mycoparasitism has played a central role in the evolution of heterothallism in the fungi. This is seen as arising from interactions of haustorial mycoparasites with their hosts, through non-haustorial interactions typified by *Parasitella* and *Chaetocladium*, to a form of sexual symbiosis resulting in heterothallism.

Host range studies within mycoparasites can sometimes provide evidence for evolutionary links between groups of fungi. For example, *Chaetocladium* has a restricted host range and only forms galls with species of other genera from the Mucorales. In this respect it resembles most species of the haustorial biotrophic mycoparasite *Piptocephalis*, which also parasitize members of this order of fungi. This emphasizes that close physiological relationships may still exist between biotrophic mycoparasites and their hosts. The genus *Piptocephalis* was formerly considered to belong to Mucorales but has recently been transferred to an associated order, the Zoopagales. Host range studies, however, are not, always a reliable indicator of close relationships. *Piptocephalis xenophila*, for example, provides an exception to the general rule and as well as attacking Mucorales it also parasitizes unrelated fungi in the Ascomycota, i.e. from well outside this taxon. It is tempting to suggest that *P. xenophila* has developed from an evolutionary line with a narrow host range, and has made an evolutionary jump in being able to recognize a wider range of potential hosts. *Dispira* is a further example of a zygomycete genus of haustorial biotrophs with a restricted host range. Of the three described species, two (*D. parvispora* and *D. xerosporica*) are known only to parasitize Mucorales, but the third, *D. simplex*, can apparently only parasitize species of *Chaetomium*, a genus in the Ascomycota. It is speculation to suggest that *D. simplex* has also made an evolutionary leap in being able to recognize less closely related hosts, but in this case it is presumed to have lost the ability to attack those more closely related to itself.

These examples provide some evidence in support of the evolution of less specialized parasitic forms from a biotrophic origin, rather than the development of mutualistic relationships from parasitic ones alluded to earlier with *Gonatobotryum* and *Graphium*. Narrow host ranges are not only found amongst the biotrophic mycoparasites; two necrotrophic fungi which uncharacteristically show a marked restriction in their respective host ranges are *Eudarluca caricis* (*Sphaerellopsis filum*; anamorph *Darluca filum*) and *Ampelomyces quisqualis* (=*Cicinobolus cesatii*). *Eudarluca caricis* is of worldwide occurrence growing within the sporulating pustules of Uredinales which cause rust diseases of the leaves and other aerial parts of higher plants. Hyphae and spores of the host are destroyed by the mycoparasite. Interestingly, *E. caricis* shows no nutritional abnormalities which might explain its restriction to such a specialized ecological niche. *Ampelomyces quisqualis* is parasitic on Erysiphales, on the other hand, and is always found in association with well-developed mildew infections of the leaves of cucumber plants. Again, there appear to be no obvious nutritional reasons why this should be so. It has been suggested that fungi such as these have probably evolved from leaf-inhabiting ancestors which have lost the ability to colonize leaf tissues and have, instead, become dependent on other fungi which do have the ability to colonize leaves. *Syncephalis* is a further example of a genus of **necrotrophic mycoparasites** with a narrow host range. *Syncephalis* (Zoopagales) is closely related to the haustorial biotroph *Piptocephalis*, and the infection process in the members of these two genera is very similar (see section 4.1.1). The main difference occurs at the post-penetration stage when the invasive hypha of

Syncephalis continues to extend and proliferate within the invaded hypha, rather than ceasing growth and differentiating into a haustorium.

Again, a close evolutionary relationship between these necrotrophic and biotrophic modes of attack can be readily envisaged although the direction of evolution is open to debate. Lewis (1974) has emphasized that the biotrophic association between plants and fungi is of great antiquity, and that most obligate biotrophs have highly developed physiological specialization and a relatively restricted host range. Adaptation to obligate biotrophy by plant-pathogenic fungi would have involved a change from high to low competitive saprotrophic ability, a diminution and eventual disappearance of the capacity to destroy cells by production of cytolytic enzymes, and a progression from indiscriminate general invasion of cells to restricted establishment as inter- or intracellular parasites. Intracellular penetration may be via haustoria, or may involve complete movement of the fungus into the parasitized cell. Similar considerations apply to mycoparasitic situations and the arguments evinced by Lewis (1974) for the origins of biotrophy from necrotrophy can also be applied to mycoparasitic relationships. Necrotrophic relationships, however, do provide scope for evolutionary speculation. Fungi that are competing for the same primary nutrient resource are usually in close contact and there will be selection pressure towards the acquisition of the ability to antagonize and kill a competing fungus. Because nutrients are derived from dead host tissues by a necrotrophic parasite, it is taken for granted that there is a strong connection between necrotrophy and saprotrophy, but it is the direction of evolutionary movement that is in doubt. Did secondary saprotrophic fungi evolve from necrotrophic ancestors, or vice versa? Secondary saprotrophic fungi are typified by the 'sugar fungi' which grow in close association with cellulose-degrading fungi which are able to utilize more complex organic substrates (Garrett, 1963). These fungi are able to exploit this ecological niche as commensals by taking up a proportion of the monomers and oligomers derived from the more complex organic polymers as they are released by the enzymes secreted by the primary colonizers. Commensalism of this type was demonstrated by Tribe (1966) for *Pythium ultimum*, *P. debaryanum* and *P. oligandrum*. These three species grow very slowly on a cellulose substrate, but when grown in association with certain cellulolytic fungi, such as *Fusarium culmorum*, all three *Pythium* species grow vigorously. *Pythium ultimum* and *P. debaryanum* are secondary saprotrophs, but unlike these two species of *Pythium*, *P. oligandrum* is mycoparasitic and could derive a proportion of its nutrients from direct parasitism of the associated mycelia.

The production of antimicrobial metabolites with fungicidal activity might also have developed as a useful component which enhances competitive saprotrophic activity. Thus the death of one of the interacting mycelia might result from toxin production relating to a strategy involving primary resource capture. Since fungal saprotrophs produce chitinolytic and proteolytic enzymes, it is possible to envisage the evolution of facultative necrotrophs from saprotrophic origins as a process involving extracellular secretion of those enzymes usually utilized exclusively in the controlled lysis of the hyphal tip during normal mycelial extension. On the other hand, it can also be argued that there could have been a direct line of evolution from facultative biotrophy to facultative necrotrophy (Cooke and Whipps, 1980). The acquisition of an ability to utilize nutrients other than those contained in diffusates from the host would allow the exploitation of dead as well as living host cells. Competitive prowess would not be of paramount importance at this stage as the fungus has already occupied the substrate. The development of an ability to degrade cell wall material of the host in the interfacial matrix, and an improvement in ability to compete saprotrophically with other fungi could then lead to the development of a truly saprotrophic existence. Mycoparasitic fungi from one phylum of the Fungi, the Zygomycota, often parasitize related fungi within the same taxon and rarely have a host range confined to members of a different subdivision. This reflects the fact that hyphal wall biochemistry is distinctly different across the subdivisions (Bartnicki-Garcia, 1968) and lends further support to this line of argument.

1.5 Host–Parasite Interfaces

One of the most interesting aspects of mycoparasitic relationships is the structure and physiology of the interface connecting the symbionts. The zone of contact is a specialized region through which, it is presumed, the transfer of nutrients is mediated. For the biotrophic parasites where growth in the absence of the host is either limited or may not occur at all, the interface is crucial to their existence. For an effective host–parasite relationship to develop, a system for the transfer of nutrients from the host to the parasite is essential.

Although it is convenient to group mycoparasitic interactions on the basis of their physiology into saprotrophic, necrotrophic and biotrophic categories, these divisions need to be further categorized on the basis of the morphology of the interface type. In some of the fungicolous and lichenicolous saprotrophs no nutrient exchange is implied and thus no complex interface between the partners is established. In necrotrophic relationships, however, there needs to be direct contact between partners so that a channel for nutrient exchange is established that is not easily accessible to competing microorganisms. In some necrotrophic relationships this interface is established solely by the close contact of adjacent hyphae. These are referred to as the **contact necrotrophs** and they are distinctive in that penetration of the mycelium of the host has not been recorded. Here again there is the problem of deciding whether or not a true parasitic relationship exists. Are nutrients taken from one fungus by the other? Is one fungus harmed by this relationship? Further investigations are needed to answer these questions.

The second type of necrotrophic relationship is easier to conceive as it involves the direct penetration of one fungus by another, accompanied almost immediately by the degeneration and death of the invaded cytoplasm. This phenomenon is referred to as **invasive necrotrophy** and it is widespread in nature.

Often, however, nutrient uptake may not occur and such examples may represent an aggressive form of antagonism that does not involve one fungus obtaining nutrients directly from the mycelium of another.

Biotrophic relationships, in sharp contrast, involve the formation of stable interfaces between partners. These range from the complex multilamellate barriers between the haustorial mycoparasites and their hosts through to the absolute fusion of the protoplasts of the partner fungi. Broadly speaking the interfaces can be categorized into three groups based on the infection structures that are formed.

First there are those in which the entire thallus of the parasite enters the hyphae of the host fungus. These are the **intracellular biotrophs** and include chytridiaceous and oomycete fungi. Many of the aquatic mycoparasitic Chytridiomycota, for example, penetrate their host and discharge the complete protoplast into the cytoplasm of the host where it remains viable and presumably absorbs nutrients directly through the host–parasite interface. The parasite grows within the mycelium of the host before eventually forming a generation of zoospores. The motile spores are released through a hole developed in the wall of the hypha of the host and swim off providing the inoculum for the next round of infections (see section 4.3).

In the other two groups most of the thallus of the parasite remains external to the host. In the haustorial biotrophs specialized hyphae form during the interaction. An appressorium develops once the parasite contacts the host, and a narrow **infection peg** develops and penetrates the hypha below. Once the peg has breached the wall it grows into a lobed haustorium which invaginates the plasmalemma of the host so establishing the mature host–parasite interface. Nutrients are presumed to be absorbed from the cytoplasm of the host by the haustorium and are then transported through the haustorial neck into the mycelium of the parasite (see section 4.1). The morphological elaborations which develop around these haustoria show close parallels with those of plant-pathogenic fungi and it is assumed that they serve the same purpose.

The third interface type found in biotrophic relationships is very unusual and involves the formation of channels of contact between closely appressed partner hyphae. Specialized hyphae or buffer cells are formed which contact the hyphae of the host but do not penetrate them. These contact elements serve as the interface for transfer of nutrients from the cytoplasm of the host to that of the parasite. Electron microscope studies of the interface have shown that the plasma membrane of the host and that of the parasite come into direct contact in the buffer zone and fuse, thus the cytoplasm of the parasite becomes contiguous with that of the host (see section 4.2). Mycoparasites with this form of symbiotic interface have been referred to as **fusion biotrophs** which emphasizes this extraordinary phenomenon. Barnett (1963) had previously termed these the **contact biotrophs**, but we would suggest that the term fusion biotroph emphasizes the intimate nature of the interface more strongly. The use of this term also avoids confusion with contact necrotrophic relationships. The channels between

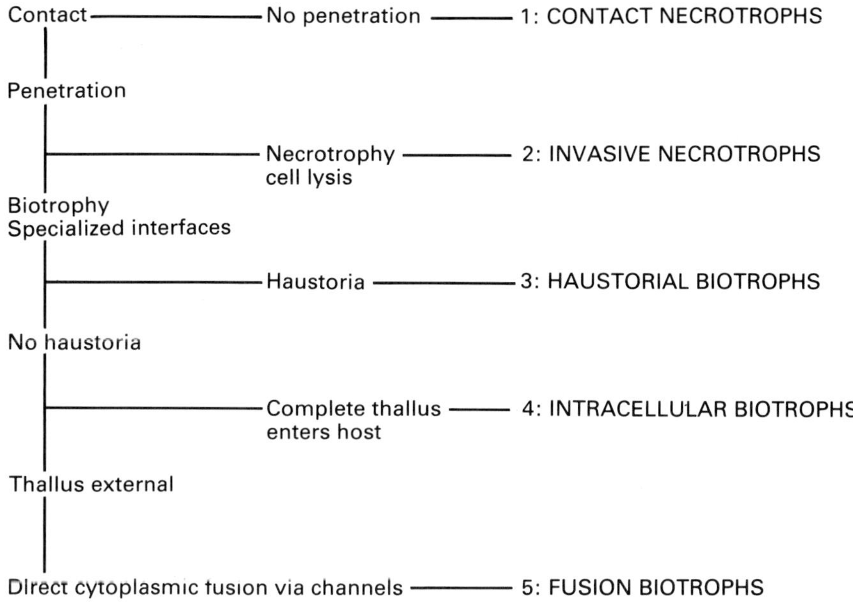

Fig. 1.3. Interface types in mycoparasitism.

partners in these relationships can be relatively narrow, and resemble the plasmodesmata formed between adjacent plant cells, or they may be broad enough to permit the migration of organelles and other cytoplasmic components. They differ, however, from the vegetative fusions of hyphae within closely related taxa as the partners are frequently distantly related, if at all. The different types of mycoparasitic interactions are summarized in Fig. 1.3 although it should be realized that the figure is constructed for convenience in interpretation and undoubtedly represents an oversimplification of the natural relationships.

The extent of the physical contact between a mycoparasite and the host may vary considerably. The only intrahyphal growth of the host by mycelial haustorial biotrophs is by means of the penetration hyphae and haustoria. This contrasts strongly with necrotrophic parasites such as *Penicillium vermiculatum*, *Schizophyllum commune* and *Trichoderma* species where the mycelium of the host is penetrated by that of the parasite and considerable intrahyphal growth occurs. The soil-inhabiting mycoparasite *Rhizoctonia solani* is itself parasitized by *P. vermiculatum* whose narrow hyphae either contact or coil round those of the host then effect entry by means of penetration pegs. The parasitic hyphae grow through those of the host, traversing each septum by passing through the central pore. Coiling of hyphae around the hyphae of another fungus is frequently reported as evidence of mycoparasitism. This is not necessarily the case as the stimulus responsible for the induction of hyphal coiling is unknown.

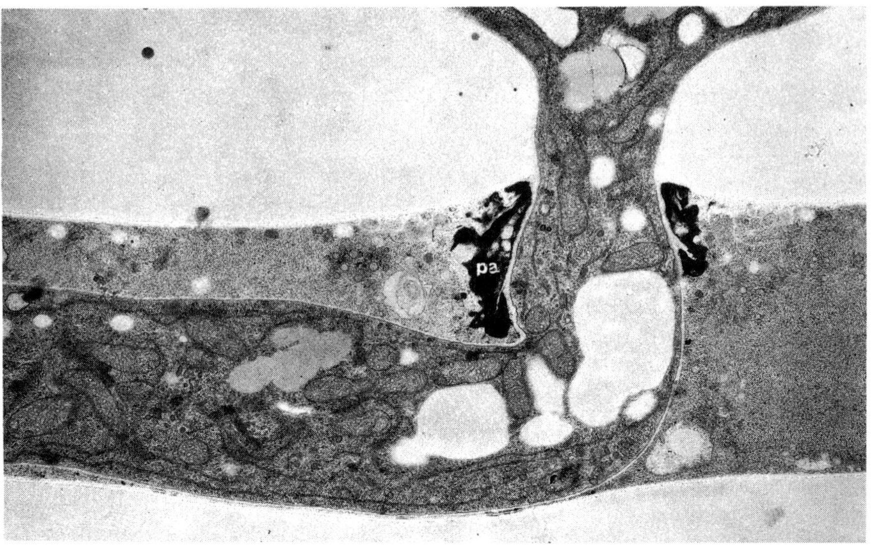

Fig. 1.4. Longitudinal section of a hypha of the basidiomycete *Corticium* invaded by *Pythium acanthicum*. A papilla (pa) is evident where the parasitic hypha passes through the hyphal wall of the host (TEM: × 14,400). (Micrograph courtesy of Dr H.C. Hoch.)

Coiling does not necessarily take place in all host–antagonist combinations as penetration may occur immediately after hyphal contact. Perhaps coiling occurs when some resistance is encountered (Deacon, 1976) thus the potentially infective hypha may be directed towards a more susceptible region. Hyphal coiling around fine glass threads or nylon fibres has been demonstrated (Inbar and Chet, 1992) which suggests that the phenomenon may be a response to contact with a surface rather than a specific response to the presence of a susceptible host hypha.

During penetration of the wall of a hypha by a mycoparasite, a reaction in the host may involve the development of localized thickenings on the inner surface of the wall at the site of penetration. These **papillae** are very common and apparently accompany every host wall penetration in certain host–parasite interactions. In addition to occurring in some of the biotrophic host–mycoparasite combinations (see chapter 4) papilla formation also occurs in many of the necrotrophic associations and seems to be a consistent feature of specific pairwise combinations. In the host–parasite interaction of *Phycomyces blakesleanus* and *Pythium acanthicum* the individual papillae may be very large, often exceeding 12 μm in diameter, but they may, nevertheless, eventually be penetrated by the invading hypha. Sometimes the papilla extends across the hypha entirely to form a pseudoseptum which apparently functions in separating the parasitized part of

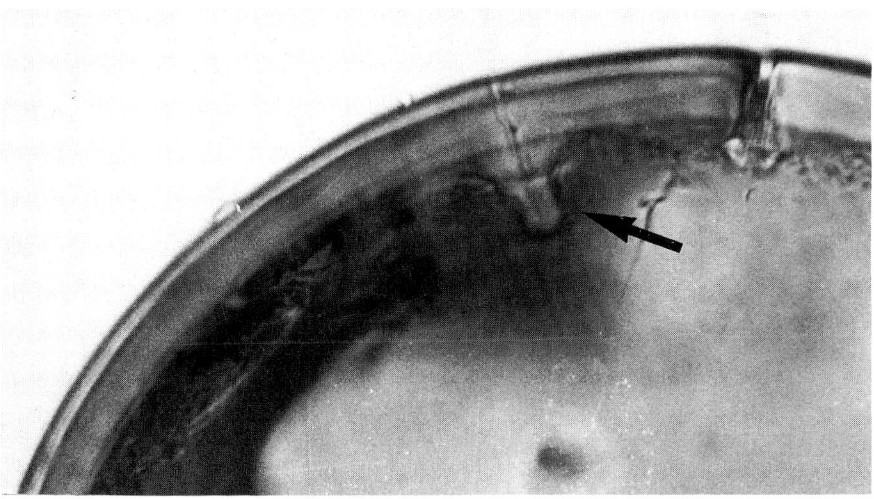

Fig. 1.5. Optical section through the wall of a chlamydospore of *Glomus* showing penetration hyphae of mycoparasites. A papilla is arrowed (LM: × 1000).

the hypha from an uninfected region, possibly limiting or reducing the spread of infection. The hyphae of *Corticum* sp. also form papillae when penetrated by those of *P. acanthicum* (Fig. 1.4). Papillae are also formed on the inner layer of the wall of the spores of **arbuscular mycorrhizal fungi** when they are penetrated by other fungi (Fig. 1.5).

2 Fungicolous and Lichenicolous Associations

2.1 Fungicolous Associations

There are numerous examples of fungi reported to grow in association with either a limited or a wide range of fungal hosts (Rudakov, 1978; Ellis and Ellis, 1985), and extensive lists and keys of species are available for some groups such as the hyphomycetes (Deighton, 1969; Deighton and Pirozynski, 1972) and the **lichenicolous** species (Hawksworth, 1983; Clauzade *et al.*, 1989). In particular there are very large numbers of interfungal relationships described from tropical environments, but the detailed biology of many of these is unknown. For this reason they will not be discussed in detail here but representative examples are provided in the Appendix. Although the fungicolous fungi are known to have a constant interfungal association, frequently assumed to be parasitic, experimental evidence in support of the assumption may be either lacking or minimal. In a survey of the fungicolous conidial fungi (Hawksworth, 1981) it has been reported that interfungal associations are known from all of the major groups of fungi on a global basis except the slime moulds (Myxomycota). The majority of the associations described, however, have been between members of the deuteromycetes and Ascomycota, including the lichenicolous fungi. Rudakov's (1978) extensive survey of physiological groups in fungicolous fungi indicated at that time that they were a widespread ecological group of about 1700 species found in the tropics, subtropics and in more northerly latitudes. Rudakov investigated 410 strains of fungicolous fungi representing 183 species from 77 genera and classified them into the five physiological groups:

1. biotrophs (e.g. *Ampelomyces quisqualis*, *Fusarium oxysporum*, *Fusidium stilbophilum*);
2. facultative necrotrophs (e.g. *Acremonium alternatum*, *Oospora ferruginea*, *Verticillium agaricinum*);

3. facultative biotrophs (e.g. *Botryoxlyon geniculatum*, *Calcarisporium antibioticum*, *Didymocladium ternatum*);
4. nectrotrophs (e.g. *Botrytis gonatobotryoides*, *Chaetomium cochliodes* (= *Chaetomium globosum*), *Geotrichum bipunctatum*);
5. semisaprophytic (= semisaprotrophic) **mycophiles** (e.g. *Actinomucor elegans*, *Cladosporium umbrinum*, *Zythia boleticola*);
6. saprophytic associates (e.g. *Acladium curvatum*, *Penicillium fellutanum*, *Stemphylium pulchrum*).

Members of groups (1)–(4) displayed some form of parasitism either through intrahyphal invasion or exogenous attachment. Most of the so-called semisaprotrophic mycophiles were antibiotic producers alone and none of the saprotrophic associates were shown to produce antibiotics or exhibit any manifestation of parasitism. A useful index has been produced by Ellis and Ellis (1985) to which the reader is referred for a survey of the wide range of fungicolous associations.

2.1.1 Associations with Myxomycota

The apparent scarcity of fungicolous associations in slime moulds is possibly related to their motility and evanescent sporangial stages, or their location in host tissue where it would appear that they are less likely to be attacked by an antagonistic fungus. On the other hand, although the myxomycetes are not normally known to be parasitized in the plasmodial phase, the stationary exposed sporangial apparatus can, at least, be colonized. Species of *Aspergillus*, *Mucor*, *Penicillium*, *Scopulariopsis* and *Trichoderma*, for example, although saprotrophic, are sometimes also found on moribund, decaying myxomycete sporangia. Colonization of the moribund fructification is reminiscent of the association of *Nyctalis* species with the decaying basidiomes of *Russula* and *Lactarius* although this is a more specific relationship. Only a few species of conidial fungi are known solely from their association with myxomycetes. For example, *Leucopenicillifera gracilis* (Hawksworth, 1981) and *Acremonium fungicola* (Ing, 1974) constantly associate with the fructification of *Fuligo septica.* The invading fungi smother the myxomycete fructifications with sporulating hyphae, which in *Verticillium catenulatum* produce conspicuous aggregations of thick-walled chlamydospores (Ing, 1974). The white hyphal mat of *Nectria candicans* (= *Nectriopsis candicans*) grows on the sporangia of the myxomycetes *Amaurochaete*, *Diachea*, *Didymium*, *Diderma*, *Fuligo*, *Physarum* and *Stemonitis* and the violet mycelium of *Nectria violaceae* (= *Nectriopsis violacea*) on the spreading fructifications of *Fuligo septica* (Dennis, 1977). Other **myxomyceticolous** conidial fungi are reported to associate with the fructifications of a wide range of myxomycetes. For example, *Aphanocladium album* has been recorded from *Arcyria cinerea*, *Comatricha nigra*, *Cribraria aurantiaca*, *C. rufa*, *C. vulgaris*, *Didymium nigripes*, *Trichia afinis* and *T. varia*; *Gliocladium album* from *Arcyria cinerea*, *Badhamia* sp., *Comatricha nigra*, *Cribraria aurantiaca*, *C. cancellata*, *C. rufa*, *Didymium* *squamulosum*, *Physarum leucophaeum*, *P. nutans*, *Stemonitis fusca*, *Trichia affinis*, and *T. varia*; *Sesquicillium microsporum* from *Craterium leucocephalum*, *Didymium melanospermum*, *Leucocarpus fragilis*, and *Physarum* spp.; *Stilbella byssiseda* from *Cribraria argillacea*, *Lindbladia tubulina*, and species of *Fuligo* and *Trichia*; and *Polycephalomyces* (= *Blistum*) *ovalisporum* (= *Stilbella ovalispora*) from *Hemitrichia serpula*, *Perichaena corticalis*, *Trichia persimilis*, and *T. varia* (Ing, 1974). Species of *Verticillium* are also myxomyceticolous (Ing, 1974). For example, *V. catenulatum* which produces conspicuous groups of thick-walled chlamydospores grows in association with *Trichia affinis*, and *V. insectorum* which, as the specific name suggests, grows on insects is also associated with *Trichia verrucosa*. *Verticillium rexianum*, the conidial **anamorph** of *Nectriopsis exigua* (= *Nectria myxomyceticola*), associates with a broader range of myxomycetes than either *V. catenulatum* or *V. rexianum* being found commonly on species of *Arcyria*, *Comatricha*, *Cribraria*, *Physarum*, *Stemonitis* and *Trichia*, in addition to *Ceratiomyxa fruticulosa*, *Fuligo septica*, and *Lycogala epidendrum*. Of the conidial fungi, members of the genus *Polycephalomyces* (*P. ovalisporum*, *P. tomentosum*) and the monotypic *Leucopenicillifera gracilis* on *Fuligo septica* are obligately myxomyceticolous whereas the majority of myxomyceticolous species are also clearly saprotrophic (Hawksworth, 1981). Members of several other saprotrophic genera which grow on myxomycetes include *Aspergillus*, *Dendryphiella*, *Harposporium*, *Mucor*, *Penicillium*, *Scopulariopsis* and *Sporotrichum* (Ing, 1974) although generally only when the sporangia have started to decay. A possibility is that the older, disintegrating myxomycete fructifications are susceptible to such fungi which, although normally strongly saprotrophic, also have a weakly pathogenic tendency and smother them with their mycelium and sporulating structures. Although the plasmodial stage in Myxomycota appears to be an unlikely host of **mycopathogenic** fungi, that of *Badhamia utricularis* seems to be a possible exception. The hemiascomycete *Dipodascus macrosporus* was consistently observed growing saprotrophically in the slime trails of the plasmodium. The plasmodium ingests the cells of *D. macrosporus* which apparently remain alive and multiply in the vacuoles. The effect of *D. macrosporus* on the plasmodium in culture depends on the virulence of the strain brought into contact with the potential host and the nature of the agar medium. On plain agar *D. macrosporus* strain F reproduces rapidly, the plasmodium becomes filled with cells and dies, although on corn meal agar the relationship is prolonged and the plasmodium continues to migrate. Strain M of *D. macrosporus*, on the other hand, is less virulent and the plasmodium continues to migrate even when plain agar is used. The suggestion is that *D. macrosporus* acts as a facultative mycoparasite. As the organisms are readily grown in culture, the system could provide a useful research tool for the investigation of the physiology of mycoparasitism (Madelin and Feest, 1982). A comprehensive review of the myxomyceticolous fungi has been recently published (Rogerson and Stephenson, 1993) and includes 11 taxa not previously reported on myxomycetes. The nomenclature of the myxomyceticolous fungi is reviewed, keys for their identification are provided and aspects of their ecology and host relationships are discussed.

2.1.2 Associations with Chytridiomycota, Oomycota and the Fungi

A wide variety of mycoparasitic relationships have been recorded from the Chytridiomycota, Oomycota and the Fungi and these form the basis of this book. In the Chytridiomycota, Oomycota and Zygomycota, fungicolous associations are not so readily observable in the field as these fungi do not produce large sporomes that act as a substrate for other species. According to Hawksworth (1981) the Chytridiomycota themselves (sensu Karling, 1977) are not known to be parasitized by many other fungi possibly because in the stationary phase of the life-cycle most are aquatic or are contained within the thallus of the host where they are unlikely to be attacked. The mycelial *Allomyces arbuscula* is, however, parasitized by the **endobiotic** *Catenaria allomycis*, both members of the Blastocladiales (Sykes and Porter, 1980). The encysted zoospore of *C. allomycis* produces a germ-tube which becomes attached to hyphae of the host by an appressorium. Ultrathin sections show that an infection peg penetrates the hyphal wall rupturing the plasmalemma allowing the thallus released from the cyst to be enclosed within the cytoplasm of *Allomyces* which subsequently degenerates as the parasite produces the zoosporangial stage. The mycelial basidiomycetous yeast *Itersonilia perplexans* is parasitized by the chytridiomycete *Rozella itersoniliae*. This example is of special interest because *Rozella* is a genus of obligate parasites previously known to parasitize only chytridiomycetes and oomycetes (Barr and Bandoni, 1979). Most of these relationships are biotrophic and are described in Chapter 4. The Oomycota, the Lagenidiales, Leptomitales and Saprolegniales (which are essentially water moulds or restricted in their growth to moist situations) do not appear to be attacked in nature by mycelial fungi. On the other hand, some hyphomycetes are able to parasitize the oospores

Table 2.1. Selected fungicolous hyphomycetes occurring on sporomes of Basidiomycota in the USA. (From Gray and Morgan-Jones, 1980; Morgan-Jones and Gray, 1979, 1980; Helfer, 1991.)

Arnoldia clavispora	*M. rosea*
Cladobotryum apiculatum	*Sepedonium ampullosporum*
C. mycophilum	*S. mycophilum*
C. varium	*Sibirina fungicola*
C. verticillatum	*S. gamsii*
Hypomyces tulasneanus	*S. lutea*
Mycogone perniciosa	*S. purpurea*
M. psilocybina	*Verticillium fungicola*

of these fungi and several examples are given in section 3.2.3. In addition, the chytrid *Phlyctochytrium planicorne* has a wide host range including saprolegniaceous hosts such as *Achlya oligacantha* and *Saprolegnia ferax* whose colonies are damaged and may be killed by the parasite (Milanez, 1967). This fungus also parasitizes several fresh water algae.

The large sporomes of some members of the Ascomycota and Basidiomycota provide ecological niches for colonization by a range of other fungi, and there is a large group of fungi that grow in association with the sporomes, or in close association with the mycelium of another fungus. Many of them are closely related, with **teleomorphic** states in *Hypomyces* (= *Apiocrea*) and *Verticillium.* For example, in an extensive study of hyphomycetes occurring on sporomes of members of the Agaricales and Aphyllophorales in the USA, over thirty different species were found, including several new species. Several of these are listed in Table 2.1.

Often the exact nature of the nutritional relationship is not known, but in some examples, especially where the commercially grown edible mushrooms are involved, the physiology of the association is relatively well understood. Although it is often assumed that mushroom compost is a selective medium for the growth of *Agaricus bisporus* it is nevertheless true that many other fungi can also grow in it. If these fungi affect the growth of the mushrooms they are known as **weed moulds** or mushroom pathogens depending on the severity of attack (see section 6.3). However, the weed moulds of mushroom beds are not necessarily parasitic but can behave as strong saprotrophic competitors. *Sepedonium niveum*, for example, has recently been shown to cause problems in the commercial cultivation of *Agaricus bisporus* in South Africa (Botha and Eicker, 1986), being responsible for reducing the crop yield. Analysis of the activity of enzymes produced in culture filtrates revealed that *S. niveum* produces β-1,4-exoglucanase and β-glucosidase, responsible for the partial breakdown of cellulose, and β-1,4-xylanase and β-xylosidase which degrade hemicellulose. *Sepedonium niveum* also utilizes cellobiose, glucose, mannose and xylose, the hydrolysis products of cellulose and hemicellulose. As *S. niveum* has been isolated from mushroom compost during spawn running an implication is that the primary mechanism of reducing the crop is through competition for the cellulose residues essential for the production of mushrooms. *Sepedonium* species are, nevertheless, known to exert necrotrophic effects on mycelium of the host (Barnett and Binder, 1973; Gray and Morgan-Jones, 1981). *Sepedonium mycophilium* antagonizes the mycelium of *Agaricus bisporus* in culture by growing over and causing it to turn brown and become fluffy. Necrosis of cytoplasm of the hyphae of the mushroom occurs, however, before contact is made and this necrotic action in advance of the mycopathogenic mycelium is presumed to be due to the secretion of diffusible **cytotoxins**.

Species of *Sepedonium*, in addition to growing with species of *Agaricus* also infect the basidiomes of Boletales in the field, an order which includes such apparently diverse genera of fungi as *Boletus*, *Suillus* and *Rhizopogon.* It is not

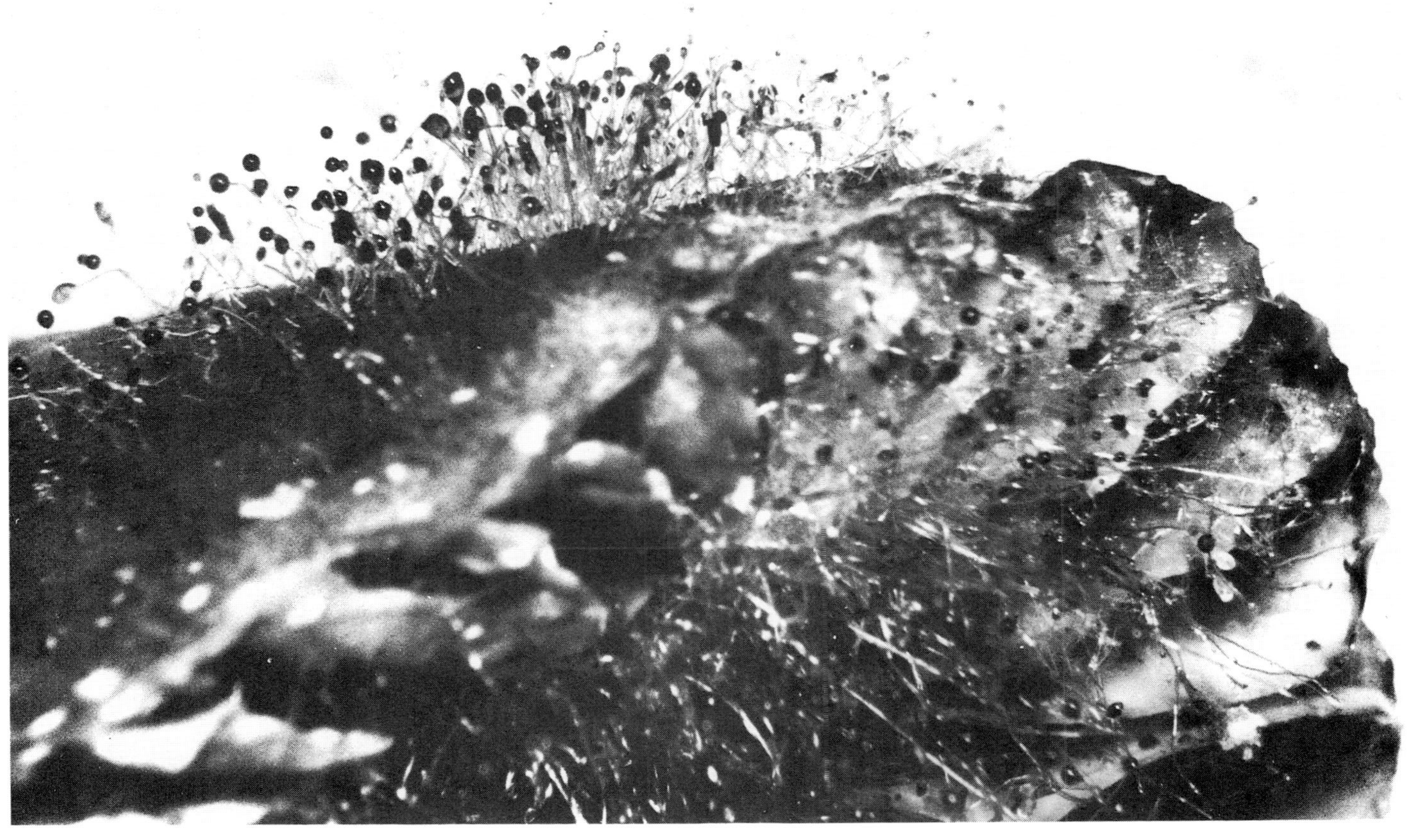

Fig. 2.1. Sporangiophores of *Spinellus* on the moribund cap of a *Russula* basidiome (× 10).

known if these basidiomes are killed as a direct result of colonization by *Sepedonium*, or whether *Sepedonium* exploits moribund toadstools as its usual ecological niche. For this reason they have been included as fungicolous saprotrophs. *Boletus* and *Suillus* species form fleshy toadstools which have a central stalk and the undersurface of the cap bears numerous tubes which open to the exterior via pores. The basidiome of *Rhizopogon* species, on the other hand, lacks a stalk and is reduced to a subglobose structure, 1–6 cm in diameter, being epigeous or hypogeous in the surface layer of the soil, often attached to fine rhizomorphs and always associated with the roots of trees. Boletales are physiologically important because many are ectomycorrhizal fungi of coniferous and deciduous trees. The mycelium is responsible for the improved uptake of mineral elements, especially phosphorus in soils where the supply is poor, and the relationship may be necessary to establish the tree in the soil and to obtain improved growth. *Sepedonium chrysospermum* can be found growing on the caps of moribund boleti and agarics in the field which appear to be covered with the powdery golden yellow conidia. The entire toadstools are finally reduced to a powdery bright yellow mass. *Sepedonium* species also infect *Rhizopogon* when the sporome becomes permeated by the mycelium which produces a mass of golden yellow spores both in the glebal chambers and over the surface of the gasterocarp which eventually collapses. Possibly, an effect of the *Sepedonium* is to reduce the quantity of inoculum of Boletales available for establishment of the mycorrhizal associations. Many *Sepedonium* species are anamorphs of the genus *Hypomyces* and will be discussed later in this section. *Spinellus* and *Dicranophora* are examples of other genera associated with the decaying caps of members of the Basidiomycota. Species of the mucoralean fungi *Syzygites* and *Spinellus* can often be found on foray during the autumn flush of toadstools in the United Kingdom. *Syzygites megalocarpus*, seen as a greyish mould with a dichotomously branched sporangiophore, grows on a wide range of gill-bearing and tube-bearing toadstools. *Spinellus macrocarpus* (= *Spinellus fusiger*), a pin-mould with an unbranched sporangiophore, appears to have a more restricted host range as a mass of rather stiff-looking sporangiophores is frequently found associated with the caps of *Mycena* or *Russula* species (Fig. 2.1). *Spinellus* species probably require relatively humid conditions for sporulation to occur as apparently uninfected caps placed for a few days in covered Petri dishes at room temperature may develop an infection whereas exposed caps do not. The relatively large, thin-walled sporangiospores of *Spinellus* which can be germinated on defined agar media probably also need the moist nature of the cap of the toadstool for growth to commence. The sporangiospores of *Spinellus macrocarpus* appear to bear many tiny hemispherical bumps and are covered in mucilage. Those of *Syzygites*, also relatively large, are dry and covered with minute spines (Young, 1968; Ekpo and Young, 1979). An electron microscope is required in order to see the spines clearly. Possibly the mucilage and spines enable the sporangiospores of these fungi to adhere to the surface of a potential host and may represent adaptations to the parasitic mode of life. *Dicranophora fulva* has been described from *Paxillus involutus*, *Gomphideus viscidus* and *Leccinum scaber* and grows readily in culture (Dobbs, 1938), but few studies of this fungus have been made. The overgrowth of a fungicolous fungus can be mistaken for an alternative spore form of the 'host'. This has been the case for the yeast-like hyphomycete *Zakatoshia* (Gams, 1986). Two species of this genus have been described, both found in nature only on decaying larger fungi. The conidiophores of *Z. erikssonii* were found covering a basidiome of *Sistotrema brinkmannii*, whereas *Z. hirshiopori* was described from very old, barely recognizable basidiomes of *Trichaptum abietum* (= *Hirschioporus abietum*) (Gams, 1986).

Relatively few homobasidiomycetes are recorded as fungicolous. The association of some hymenomycetes which themselves grow on the macroscopic basidiomes of other hymenomycetes are examples that can be readily observed in the field. For example, *Nyctalis parasitica* (anamorph = *Asterophora parasitica*) and *N. asterophora* (anamorph = *A. lycoperdoides*) are found growing on the caps and stalks of moribund basidiomes of species of *Russula* and *Lactarius* on which they are often said to be parasitic (Fig. 2.2). *Nyctalis asterophora* is more common than *N. parasitica* in the United States of America where it has been found growing on species of *Armillaria*, *Cantharellus* and *Clitocybe* as well as *Lactarius* and *Russula*. The fleshy caps of *N. asterophora* produce white basidiospores on the hymenium of the gills and sporulation is also effected through stellate conidia which develop on the surface of the cap. On germination, the conidia produce the basidial form (Ramsbottom, 1965) and it would appear that the conidia are therefore responsible for colonizing the agaric basidiomes. *Nyctalis parasitica* differs as the chlamydospores are smooth and ovoid being produced on the gills where they completely replace the basidiospores which further emphasizes the physiological importance of the chlamydospore in the establishment of the symbiotic state in this genus. *Nyctalis parasitica* grows in clusters on old *Russula* basidiomes, especially *R. nigricans*. *Russula nigricans* is readily recognizable in the field as the cap, widely spaced gills and stem blacken with age. It would appear that this association is highly specialized and it is likely that these relationships are mycoparasitic although it is worth noting that the host toadstools often look moribund when found in association with the supposed parasite. Whether the moribund appearance is due to the effects of mycoparasitism or whether the invading fungi are able to do so because the fructifications are either dying or dead, remains to be determined. *Nyctalis asterophora*, however, has been demonstrated to grow on the young, developing basidiomes of *Russula densifolia* (McMeekin, 1991) and it is clear that further physiological research is required to determine the symbiotic nature of these fungi. Both species can be readily isolated from agarics and grown in culture on malt agar and various other nutrient media (Thompson, 1936). Mature basidiomes develop in culture after incubation for seven days and the characteristic chlamydospores can be found in the rudimentary lamellae of *N. parasitica*, on the surface of the pileus of *N. asterophora*, and on the aerial and submerged mycelium. *Nyctalis asterophora* grows well in culture on a variety of conven-

Fig. 2.2. Basidiomes of *Nyctalis asterophora* growing from the decaying cap of a *Russula* species (× 1.5). (Courtesy of R. Phillips.)

tional agar media, and can be induced to form basidiomes in Petri dishes in response to light and chlamydospore formation is promoted when the concentration of glucose in the growth medium exceeds 2% (McMeekin, 1991). It is also **dimorphic** and the yeast-form can be generated from chlamydospores (Koller and

Jahrmann, 1985). Associations between fungicolous fungi and their associates are often difficult to analyse, and parasitism must not be taken for granted. Although the basidiomes of *Xerocomus parasiticus* are always found growing on those of earthballs such as *Scleroderma citrinum* (= *S. aurantium*) it seems that the latter fungus may not be as parasitic as the specific name might suggest (Rayner *et al.*, 1985). The apparent parasitism of *X. parasiticus* is noteworthy owing to the mycorrhizal activity of close relatives of this species in Boletales (Beaton *et al.*, 1985). Thus the symbiotic state appears to be the rule rather than the exception in Boletales. *Volvariella surrecta* also appears to be unusual for members of the Pluteaceae which are usually either lignicolous or terrestrial. This toadstool is found growing on old basidiomes of fleshy members of the Tricholomataceae, especially *Clitocybe nebularis* or other species of *Clitocybe* and *Tricholoma* (Orton, 1986). The association of fungi such as *Nyctalis* species and *V. surrecta* with moribund basidiomes implies that if they are parasitic at all, they are only weakly so.

Other fungicolous hymenomycetes are found less frequently. For example, *Claudopus parasiticus* (= *C. subdepluens*) is itself a toadstool-producing species, and infects the brackets of *Polyporus perennis* (= *Coltricia perennis*) on which it produces its basidiomes, although the basidiomes of the host do not appear to be unduly affected by the presence of the symbiont. On the other hand, *Stropharia epimyces* which is said to parasitize the inkcap toadstools *Coprinus atramentarius* and *C. comatus* reduces the growth of the basidiomes and their basidiospore-producing ability (Buller, 1924). Fifteen species of wood-rotting Basidiomycota, including *Flammulina velutipes*, *Pleurotus ostreatus* and *Polyporus sensu lato*, were reported as antagonistic towards conidia and mycelia of *Ceratocystis fimbriata* and *C. fagacearum* (Griffith and Barnett, 1967). No penetration of living cells occurred, but penetration of killed cells was sometimes observed. Another example is *Episphaeria fraxinicola* which produces groups of small sessile, cup-shaped basidiomes, 0.2–0.7 mm diam., on the surface of the **ascomata** of stromatic pyrenomycetes such as *Diatrype stigma* which may be covered with the ochraceous to brown spores of the crepidotaceous symbiont.

A particularly unusual interfungal association occurs in *Entoloma abortivum* (Watling, 1974). This basidiomycete is known to produce two forms of basidiome: a normal **agaricoid** form typical of the genus and a second carpophoroid form which is a subglobose mass of tissue. The carpophoroid form is apparently composed of hyphae from both *E. abortivum* and the agaric *Armillaria mellea*; the latter is lacking from the agaricoid basidiome. There does not seem to be any obvious nutritional relationship between the two fungi, but they are nevertheless found growing together in this way, with the *Armillaria* disorganizing the normal mechanisms which control pileus expansion of *Entoloma*. It is evidently not unique to find situations in which two basidiomycete fungi grow together in the exploitation of a particular ecological niche (Watling, 1974) but it is unusual to find the two mycelia forming a basidiome. In *Coprinus* species, however, the situation can be more complex and a single basidiome can apparently be formed from hyphae of two different species. The hyphae of one species forms basidiospores but the other apparently does not. Three other hymenomycete fungi recorded as fungicolous are *Dacrymyces palmatus* and *Dacryospinax spathularia* (Dacrymycetales) and the aphyllophoraceous *Gloeophyllum saepiarium* reported from China to parasitize the mucoralean *Cunninghamella lanceolata* (Bo and Bau, 1980).

In contrast to the situation with the homobasidiomycetes, there are many heterobasidiomycete fungi reported to be fungicolous. An interesting but little-mentioned group are the Tulasnellales which grow as fine, sometimes velvety, films on the surface of moist, rotting plant debris. They are difficult to see without the aid of a hand lens and, even then may be readily overlooked. They produce structures customarily regarded as basidia on which are developed spore tetrads, regarded as basidiospores. Most are saprotrophic and recently (Roberts, 1992) the spiral-spored British species have been described. Two species, however, have been reported as naturally fungicolous, *Tulasnella inclusa* on Corticiaceae *sensu lato* including *Athelia*, *Phlebia*, *Repetobasidium* and *Sistostrema* and *T. danica* on Tremellaceae and *Myxosporium nucleatum* although both are also recorded as occurring on the bark of dead wood and mosses (Julich, 1984). Other orders of related fungi are the Tremellales and Auriculariales. Many species from these orders have been described as mycoparasitic on a range of hymenomycetes, but the true relationships are unknown (Oberwinkler and Bandoni, 1982, 1983; Bandoni, 1984, 1987; Oberwinkler *et al.*, 1984; Metzler *et al.*, 1989; Bauer and Oberwinkler, 1990a,b, 1991). Where detailed studies have been carried out, very interesting host–parasite interfaces have been described (see sections 4.1.2 and 4.2). In other cases, the interactions are only known from fungicolous associations recorded on field material. Because these may all represent interactions which involve a haustorium-like interface they are discussed in more detail in Chapter 4.

Camptobasidium hydrophilum, the teleomorph of *Crucella subtilis*, a litter and stream-dwelling inhabitant of decaying leaves, is a heterobasidiomycete placed in the Auriculariales sensu lato (Marvanova and Suberkropp, 1990). Conidia of the anamorph have a tetraradiate form typical of many so-called aquatic hyphomycetes associated with the decay of a range of leaves found in freshwater streams. In pure dual culture on 1% malt extract agar the hyphae of *C. hydrophilum* contact and coil around those of the aquatic hyphomycetes *Anguillospora filiformis*, *Articulospora tetracladia*, *Lemonniera aquatica*, *Lunulospora curvula*, *Taeniospora gracilis* var. *gracilis*, *Tricladium chaetocladium*, *Tetracladium marchalianum* and *Triscelophorus septatus*. Although no hyphal penetration was detected, it was suggested that *C. hydrophilum* might behave primarily as a biotrophic mycoparasite on the mycoflora in the stream environment, but no evidence was presented to support this hypothesis. In contrast, *Naiadella fluitans*, also an aquatic hyphomycete with tetraradiate (and triradiate) conidia produces tremelloid haustorial branches in culture and is presumed from a range of morphological characters to be the anamorph of a heterobasidiomy-

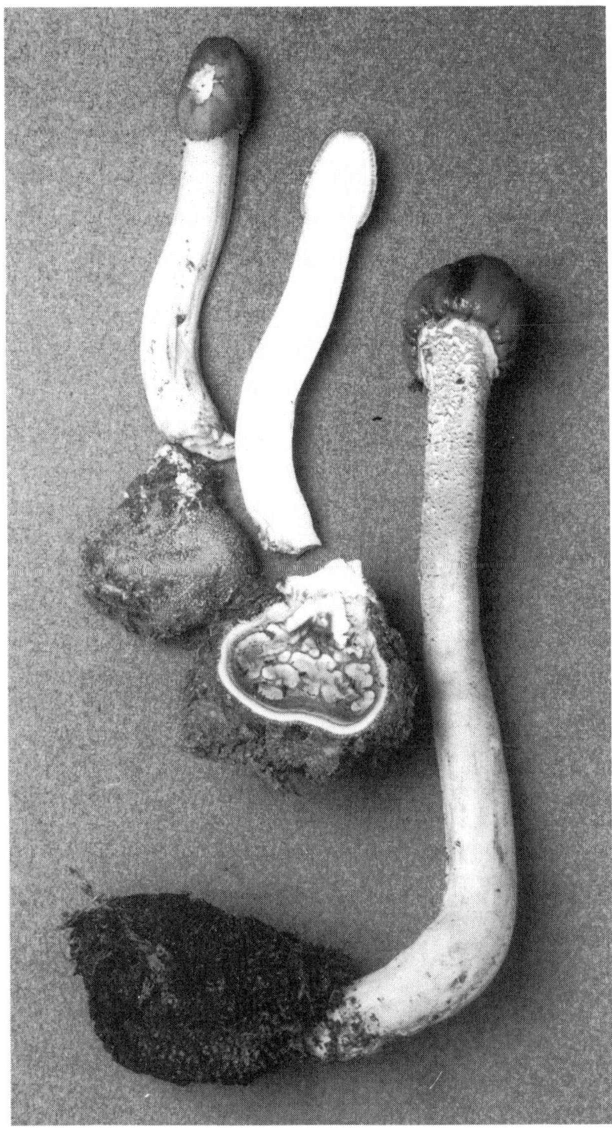

Fig. 2.3. Ascomata of *Cordyceps canadensis* growing from the truffles of *Elaphomyces granulatus* (× 1.25). (Courtesy of R. Phillips.)

cete although the teleomorph is unknown (Marvanova and Bandoni, 1987). The occurrence of these haustorial branches in culture is evidence for a biotrophic relationship in nature and is dealt with in more detail in section 4.1.2. The occurrence of these interfungal relationships within the aquatic hyphomycetes indicates that this could be a potentially fruitful area of research into mycoparasitism.

The relatively large ascomata of discomycetous Ascomycota, analogous to the macroscopic basidiomes of hymenomycetes are also colonized by other fungi which are usually regarded as parasitic. The stalked apothecia of the helvellaceous *Gyromitra esculenta* are said to be parasitized by the perithecial *Sphaeronaemella helvellae.* Basidiomes of the fairy clubs *Clavulinopsis cinerea* and *C. cristata* are colonized by a web of dark-brown hyphae of the perithecial *Helminthosphaeria clavariarum* and *H. corticiorum* occurs on the resupinate basidiomes of *Phanerochaete* (= *Peniophora*) spp. In *Battarrina inclusa* (Hypocreales, Nectriaceae) the perithecia are produced within the fleshy part of the truffle *Tuber puberulum.* Many members of the Ascomycota have been reported to grow in association with other fungi, especially Basidiomycota and other Ascomycota. In the yeast-like genera, a number of species have been described growing among the hyphae making up the sporomes of higher fungi. For example, *Endomyces* contains at least three species found growing on agaric basidiomes (Redhead and Malloch, 1977; Von Arx, 1977); *Helicogonium jacksonii* has been found on *Ceraceomyces sublaevis* (= *Corticium microsporum*) (White, 1942); and *Dipodascus polyporicola* was described from brackets of *Piptoporus soloniensis* (Schumacher and Ryvarden, 1981). None of these fungi appear to cause adverse effects on their hosts. In the mycelial Ascomycota, a specific association is that of *Cordyceps* species which have brightly coloured aerial ascomata, the subterranean mycelium associating with the hypogeous ascomata of truffles. In order to find truffles it is necessary to grub them up from the surface layer of soil around the roots of trees with which they associate as mycorrhizal symbionts. *Cordyceps canadensis* (Fig. 2.3) and *C. capitata* parasitize *Elaphomyces granulatus* and *E. variegatus* whereas *C. ophioglossoides* parasitizes in addition the ascomata of *E. muricatus* (Ramsbottom, 1965). From the infected truffle, however, the stalked ascoma of *C. capitata* grows up through the soil and the bright yellow coloured stalk of the sporulating stroma, which may be several centimetres long, provides the clue to the presence of the truffle below ground. These fungi have to be collected very carefully in order not to break the stalk of the fructification of *C. capitata* from the truffle. *Cordyceps* species may be found in British woodlands during September and October, *C. ophioglossoides* being more common than *C. capitata* (Dennis, 1977). The sclerotia of *Claviceps* species, fungi which are themselves parasites or endophytic mutualists of plants such as the Gramineae, are also parasitized by *Cordyceps* species, including *C. clavicipitis*, the stromata of which have been noted within the stromata of *Claviceps purpurea* and suppressing them (Kobayashi, 1941). *Barya aurantiaca*, a very rare species in Britain, has been reported to parasitize the germinating sclerotia of *Claviceps purpurea* (Dennis, 1977). *Claviceps*, *Cordyceps* and *Barya* are closely related, all being members of the Clavicipitales in which the perithecia are brightly coloured or light in colour and parasitic on plants, insects, arachnids or other fungi.

Several other Ascomycota reported to be of constant interfungal association, described and beautifully illustrated as British ascomycetes by Dennis (1977), are outlined here and in section 2.2. *Letendraea helminthicola* produces inconspicuous flesh-coloured or light-yellow perithecia among the densely crowded dark-brown to blackish conidiophores of the saprotroph *Helminthosporium velutinum.* A fungus also associated with ascomycetes is *Berlesiella nigerrima* which is found growing on old stromata of *Eutypa* species. It produces numerous partially immersed pseudothecia in its minute stromata giving the impression of tiny blackberries. In *Tubeufia helicomyces*, which grows saprotrophically on the debris of marsh plants and grasses which grow in damp areas, the pseudothecia which develop on a brown mycelial mat occur in association with the pinkish-grey mould *Helicosporium phragmitis*, whereas *T. cerea* develops on old stromata of *Diatrype stigma* a saprotroph of the wood of deciduous trees, especially *Crataegus.* The small scattered shallow cup-shaped apothecia of *Pithyella ilicicola* develop on various fungi, including the stromata of *Myriangium*, which itself parasitizes insects. In a similar fashion, the small ovoid ascomata of *Catulus aquilonius* grow on the stromata of *Seuratia millardetii* (Malloch and Rogerson, 1978). These fungicolous relationships can become more complex when more than two fungi are involved. For example, *Keissleriella bavarica* has been described growing on the ascomatal stroma of *Ascodichaena rugosa*, a parasite of bark of *Fagus* and *Quercus* (Butin, 1981). Various other fungi have been found in the presence of *K. bavarica.* Some appeared to colonize the *A. rugosa*, whereas others colonized the perithecia of *Keissleriella.* Thus spore masses of *Tripospermum myrti* were frequently established in the ascomata of *A. rugosa*, which became sterile and showed signs of disintegration. Others, such as *Eriomyces* sp., grew in the perithecia of *K. bavarica* causing premature disintegration (Butin, 1981).

There are many interfungal relationships known only from little more than the brief descriptions contained in the taxonomic reports of the genus. More detailed studies are required. For example, the genus *Titaea* contains five species which overgrow other fungi, including Uredinales and especially ascomycetes (Sutton, 1981). They are described mainly on the basis of their conidia and mode of conidiogenesis, which itself lacks detailed information (Sutton, 1981). *Titaea callispora* for example, was described by Saccardo (1877) from material growing on the *Sarcinella* anamorph of *Dimerosporium pulchrum* (= *Schiffnerula pulchra*). *Scopinella gallicola* is another example of an ascomycete fungus described from rust-infected material collected in the field (Tsuneda and Hiratsuka, 1981). It occurs on *Pinus contorta* galls caused by the western gall rust fungus *Endocronartium harknessii.*

At least 24 genera assigned to the Hypocreales, mainly neotropical, are known to have fungicolous, lichenicolous and myxomyceticolous species (Samuels, 1988). Some are known only from the type collections and it seems likely that many more species have yet to be found. Of the fungicolous Hypocreales, some are associated with substrates colonized by other fungi and it is assumed that these Hypocreales could parasitize the substrate mycelium. The white to yellow perithecia do not turn red in potassium hydroxide and these species are said not to be closely related to *Nectria episphaeria* and its relatives which have red perithecia in potassium hydroxide. *Trichonectria* species are characterized by light-coloured perithecia which arise from thick-walled setae and two-celled ascospores which fragment at the septum (phragmospores). Fungicolous species of *Trichonectria* are *T. albidopilosa* (= *Nectria albidopilosa*) (on the hymenium of inoperculate discomycetes), *T. horrida* (on perithecia and associated mycelium of a pyrenomycete, possibly *Chaetosphaeria* sp.), and *T. rectipila* (on stromata of *Diatrype stigma*). *Peristomialis* species are frequently found on decaying herbaceous debris although it is suspected that the true substrate may be other fungi. *Peristomialis leucocarpa*, for example, known only from the type material from Jamaica was found on old perithecia of *Nectria jungneri* (anamorph = *Cylindrocarpon victoriae*) and *P. parapilis* is known only from black mycelium on a bamboo culm.

Species of *Hypocrea*, as well as growing on dead wood and other plant debris are also found growing on moribund basidiomes (Dennis, 1977). The lignicolous *Hypocrea rufa* also grows on the old basidiomes of various species, whilst *H. pulvinata* is common on the moribund basidiomes of the common birch polypore *Piptoporus betulinus*. *Peckiella viridis*, which produces masses of nearly black ovoid perithecia on a thin hyphal network, and *P. lateritia* are found on decaying agarics especially *Lactarius*. Species of the closely related *Hypomyces* (= *Apiocrea*) which also produce perithecia on a hyphal web or subiculum also occur on old hymenomycete basidiomes. The densely crowded golden-yellow to orange–red perithecia of *Hypomyces aurantius* occur on the old brackets of polypores throughout the year in Britain, and *H. broomeanus* is found on the brackets of *Heterobasidion annosum* (= *Fomes annosus*) which is responsible for producing a serious root rot in coniferous trees. *Hypomyces aurantius* is also reported as a destructive contact mycoparasite of aphyllophoraceous fungi (Kellock and Dix, 1984a). *Hypomyces rosellus* (anamorph = *Cladabotryum dendroides*) is often found on the basidiomes of mushrooms and toadstools and on the decaying basidiomes of *Stereum hirsutum*, a very common colonizer of the bark of moribund trees. It can be a nuisance in the mushroom growing industry causing cobweb disease (see section 6.6.3). Infected fruit bodies become engulfed by the rapid advance of the grey–white cobweb-like mycelium, the gills may fail to develop and the basidiomes appear distorted. The mycelium becomes pink on ageing and the parasitized fructifications turn brown as they develop a wet rot. The cobweb fungus produces perithecia which appear as dark-coloured specks over the surface of the parasitized basidiome. In an interesting tripartite relationship, the perithecia of a related species, *Hypomyces asterophora*, develop on the mycoparasitic *Nyctalis*. The hyphal mat of *Hypomyces ochraeceus* which grows on the soil where infected species of *Russula* and *Lactarius* have disintegrated is infected by the mycelium of *Sphaeroderma episphaeria* (= *Sphaerodes episphaeria*) which develops clusters of perithecia which dry dark brown to black on the hyphal mat. *Melanospora fusispora* (*Sphaeroderma fusisporum*) is found on *Isaria farinosa* and *Apiocrea* species develop on Boletales, for example species of *Paxillus*, *Rhizopogon* and *Xerocomus*. The orange–yellow perithecia of *Hypomyces chrysospermus* (= *Apiocrea chrysosperma*) develop on a yellow hyphal mat on the decaying basidiome. This is the teleomorph of *Sepedonium chrysospermum*, which produces masses of powdery yellow-coloured warted chlamydospores on boleti although the asexual and sexual states are rarely found together (Dennis, 1977). The anamorph and teleomorph of *Hypomyces tulasneanus* (= *Apiocrea tulasneana*) are also associated with the basidiomes of boleti, and five new boleticolous species of *Hypomyces*, *H. badius*, *H. boletiphagus*, *H. chlorinigenus*, *H. melanochlorus* and *H. microspermus* have been described (Rogerson and Samuels, 1989). In the boleticolous *Hypomyces* the subiculum, initially white and lax becomes cottony then coloured. The colour of the subiculum which may cover the stipe, pileus and tube surfaces of the basidiome is characteristic ranging at maturity from white to buff (*H. microspermus*), golden yellow, remaining white or becoming pink, vinaceous to ferruginous (*H. chrysospermus* = *Apiocrea chrysosperma*), yellow–green becoming brownish (*H. tulasneanus*), pale brown (*H. boletiphagus*, *H. chlorinigenus*), pale brown to dark olivaceous brown (*H. melanochlorus*), ochraceous to brick red (*H. transformans*), to black (*H. completus*). From experiments involving the growth of cultures from isolated ascospores, it has been shown that five species of *Hypomyces* definitely have a *Sepedonium* anamorph and that others are associated with an anamorph referable to *Sepedonium* (Rogerson and Samuels, 1989). *Hypomyces aurantius*, on the other hand, has the anamorph *Cladobotryum varium* (Gams and Hoozemans, 1970). Heim (1948) considered that as *Hypomyces chrysospermus* grows on certain species of *Rhizopogon*, *Boletus* and *Paxillus*, this demonstrated that these three genera were related. Subsequently the discovery (Hawker, 1955) that *H. chrysospermus* also grows on *Melanogaster* (Basidiomycota, Melanogastrales) threw doubt on this reasoning although the current view is that the three genera are all members of the Boletales (Beaton *et al.*, 1985). Although several species of Hypocreales are known to develop on Basidiomycota and other Ascomycota, few have been recorded on discomycetes (Rogerson and Samuels, 1989). The mycoparasites are widespread geographically and an illustrated key to the species of *Hypomyces* and *Nectria* occurring on discomycetes has been provided by Rogerson and Samuels (1989). Examples from the former genus include *Hypomyces leotiicola* on ascomata of *Leotia lubrica* from Switzerland, eastern USA and New Zealand; *H. mycogones*, on Geoglossaceae, from Ecuador, Bermuda and USA; *H. papulosporae* var. *americanus*, on geoglossaceous ascomycetes from the USA and China; *H. papulosporae* var. *papulosporae* on geoglossaceous ascomycetes, known only from New Zealand; *H. sepulchralis*, on a discomycete, from Guadeloupe and the Dominican Republic and *Hypomyces stephanomatis*, on apothecia of *Humaria hemisphaerica*, from eastern USA. Species of *Nectria*, which are saprotrophs of, and also parasitic on, a wide range of higher plants, are characterized by perithecia which occur in groups on a stroma which breaks through the surface layers of fallen and living branches and which are often associated with a conidial stage (Dennis, 1977). Some species of *Nectria* similarly exhibit mycoparasitic relationships. Some have wide host ranges but others with an apparently more restricted host range include *N. magnusiana* on the stromata of *Diatrypella*, *N. purtonii* on perithecia of *Valsa*, *N. leptosphaeriae* on and among the ascomata of *Leptosphaeria acuta*, *N. aureola* on the mycelium of *Meliola niessleana*, and *N. lecanodes* on the thalli of foliose lichens especially *Lobaria* and *Peltigera* species. *Nectria polyporina* produces light-red coloured perithecia on old basidiomes of *Heterobasidion annosum* (= *Fomes annosus*) (Dennis, 1977). *Nectria brunneostriata* has been described from ascomycete perithecia, *N. peziza* from various substrates including *Polyporus* sp., and *N. suffulta* from *Xylaria* sp. (Samuels, 1988). Examples of other fungicolous species of *Nectria* are *Trichonectria albidopilosa* (*N. albidopilosa*) on a discomycete from Venezuela, *N. discicola* on the hymenium of *Bioscypha* sp. from Ecuador, *Nectriopsis discophila* (= *Nectria discophila*) on the apothecia of *Lachnum* sp. from Indonesia and Venezuela, *Nectriopsis ostiolorum* (= *Nectria ostiolorum*) on perithecia of *Nectria* and *Xylaria*, *Nectriopsis perpusilla* (= *Nectria perpusilla*) on perithecia of *Hypoxylon*, *Nectria* and *Xylaria*, and *Geopora arenicola* (= *Sepultaria arenicola*) on the hymenium of the discomycete *Nectriopsis sepultariae* (= *Nectria sepultariae*) from Germany (Samuels, 1988). Some species of the related *Nectriopsis*, *Nectriella*, and *Paranectria* also have mycoparasitic associations. For example, the golden-yellow perithecia of *Sphaerostilbella aureonitens* (= *Nectriopsis aureonitens*; anamorph = *Gliocladium penicillioides*) develop on old basidiomes of *Stereum* spp., whereas *Paranectria affinis* occurs on that of the lichen *Ephebe lanata* (Dennis, 1977). *Nectriopsis mindoensis* may be both fungicolous on *Nectria peziza* or lichenicolous on *Dimerella* sp. Other fungicolous species of *Nectriopsis* are *N. bactridioides*, *N. byssotecta*, *N. cephalosporii*, *N. cordiae*, *N. cupulata*, *N. epichloe*, *N. epimyces*, *N. epimycota*, *N. epinectria*, *N. guarapiensis*, *N. hyperbiota*, *N. hypocrellicola*, *N. lasioderma*, *N. lasiodermopsis*, *N. leucorrhodina*, *N. lilliputia*, *N. macroepichloe*, *N. oidiodes*, *N. oxyspora*, *N. pipericola*, *N. puiggarii*, *N. sensitiva*, *N. septofusidiae*, *N. sibicola* and *N. soroicola* (Samuels, 1988). Although some species of *Nectriopsis* may not be fungicolous, e.g. *N. albofulta* and *N. flavella*, it would appear that the majority of species in this genus are fungicolous. *Hypocreopsis episphaeria* (= *Nectria episphaeria*) is found on the effete perithecia and stromata of other ascomycetes especially Diatrypaceae and Sphaeriaceae, and *H. lichenoides* is associated with species of *Hymenochaete* and *H. xylariicola* with the old perithecia of *Xylaria* sp.

The Centrostomataceae (= Melanosporaceae), with perithecia which grow superficially on the substratum and contain masses of olive-brown spores at maturity, contains mycoparasitic species. *Syspastospora parasitica* parasitizes *Isaria farinosa* and *Beauvaria bassiana* which are themselves entomopathogenic. *Melanospora lagenaria* occurs on the decaying basidiomes of the bracket fungi *Polyporus*, *Polystictus* and *Stereum* and *M. brevirostris* parasitizes the apothe-

Table 2.2. Range of distribution of fungicolous hyphomycetes (after Deighton, 1969; Deighton and Pirozynski, 1972).

Mycoparasite	Known geographical distribution
Acremoniula sarcinellae	Central Africa, Cuba, Guinea, India, Panama, Senegal, Sierra Leone, Tanzania, Uganda, Zambia
Acrostaurus turneri	Sarawak, Sierra Leone
Annellodochium ramulisporum	Sierra Leone
Atractilina parasitica	Congo, Ghana, Guinea, India, Mauritius, Nigeria, Philippines, Puerto Rico, Sierra Leone, South Africa, West Africa, Taiwan, Tanzania, Uganda, USA, Venezuela
Calcarisporium setiphilum	Tanzania
Chionomyces meliolicola	Africa (East and West), Dominican Republic
Cladosporiella asterinae	Sierra Leone
Cladosporiella uredinis	India, Malaysia, Trinidad, Philippines
Cylindrocarpon macrosporum	Australia, Brazil, Paraguay, Sierra Leone, Tanzania, Uganda
Domingoella asterinum	Australia, Dominican Republic, Ghana, Indonesia, Nigeria, Sierra Leone, Solomon Islands, South Africa, Tanzania, Trinidad, Uganda
Elletevera parasitica	USA, Zambia
Eriocercospora balladynae	Costa Rica, Ghana, Malaysia, Panama, Sierra Leone, Uganda
Eriomycopsis bonplandii	Argentina, Dominican Republic, Ghana, Sierra Leone, South Africa, Uganda
Irpicomyces schiffnerulae	Malaysia
Monosporiella meliolicola	Argentina
Paratrichonis chinensis	China
Pseudofusidium hansfordii	Uganda
Redbia pucciniicola	Tanzania
Scolecobasidium pusillum	Sierra Leone
Spermatoloncha maticola	Uganda
Spermosporella aggregata	Sierra Leone
Stenospora uredinicola	Sierra Leone
Sympodiophora stereicola	Russia
Trichoconis caudata	Cameroon, Sierra Leone
Triposporina uredinicola	Indonesia
Tuberculispora jamaicensis	Jamaica, USA
Vermispora grandispora	Sierra Leone

cium of *Sepultaria*. The uredinia and telia of the plant-pathogenic rust *Phragmidium mucronatum* are themselves parasitized by *Micropodia oedema*, the minute pale-brown apothecia arising from an erumpment stroma in the sori of the rust (Ellis and Ellis, 1985).

Mycoparasitism by members of the deuteromycetes has been reviewed (e.g. Madelin, 1968; Hawksworth, 1979). On a global basis, many of the these fungi have been reported to grow in association with other fungi. They have been included here when the nature of the relationship is unknown. They occur often on other deuteromycetes although fungi from several other groups may be affected. Numerous hyphomycetes are said to parasitize other fungi, especially the tropical foliicolous ascomycetes in the Meliolales (Deighton, 1969; Deighton and Pirozynski, 1972) and these, with their hosts are outlined in the Appendix. Table 2.2 shows the wide range of geographical distribution of selected genera of the hyphomycetes described as mycoparasitic (or hyperparasitic) by Deighton (1969) and Deighton and Pirozynski (1972). There are also many reports of fungicolous associations involving the genera *Amblyosporium* and *Cladobotryum* but there are few details of the actual relationships.

Several species of *Cercospora* have been described as mycoparasitic, and *C. uromycestri* is an example, being found growing among and penetrating the aeciospores of *Uromyces cestri* (Pollack, 1971). A related fungus, *Cladosporiella cercosporicola* has itself been found to parasitize the spores of *Cercospora* (Pollack, 1971). Species of *Spiropes* are dematiaceous tropical hyphomycetes which overgrow and are apparently parasitic on Meliolinae and other leaf-inhabiting Ascomycota (Ellis, 1971). *Acremonium sordidulum* f.sp. *colletotrichum-dematii* described from India parasitizes the acervuli of *Colletotrichum dematium* f. sp. *truncatum* both *in vivo* on the pods of *Albizzia lebbek* and *in vitro* in experimentally produced necrotic lesions (Singh *et al.*, 1978). Some of the tropical fungicolous hyphomycetes are known only from single collections, for example, *Spermosporella aggregata* and *Scolecobasidium pusillum*, and their true distribution could be wider than would appear from Table 2.2. On the other hand, the mycoparasite may be rarely found although geographically widespread. In the main, the tropical mycoparasitic hyphomycetes are found overgrowing and in association with other superficially growing ascomycetes although fungi from other groups may also be parasitized. Many more examples would probably be discovered, if detailed examinations, such as that of Deighton (1969), were made of the interfungal associations occurring on the leaves of tropical plants.

Several fungicolous hyphomycetes are constantly associated with the sori of rust fungi. *Stenospora uredinicola* has been described from the sori of *Puccinia kraussiana*, *Cladosporiella uredinicola* from the uredia of *Puccinia eucomi* and the uredinia and telia of *Ravenelia albigiae* var. *zygiae* and *C. uredinis* from the sori of *Uredo* sp. and the uredinia of *Puccinia polygoni-amphibii* and *P. thaliae* and the telia of *P. solmsii* (Deighton, 1969). *Redbia pucciniicola* overgrows the telia of *Puccinia holosericea* and *Triposporina uredinicola* is associated with *Puccinia periodica*. *Trichoconis caudata* is found on the wilting and diseased pods of

Theobroma cacao where it overgrows *Lasiodiplodia theobromae* (coelomycetes) and *Eriomycopsis minuta* is known from the hyphomycete *Melanographium citri. Dissoconium dekkeri* is possibly a mycoparasite of Erysiphaceae, mainly occurring on the leaves of conifers and other evergreen shrubs (de Hoog *et al.*, 1991). Other examples are discussed in section 6.4. *Trichoconis hibernica* is notable as it has only been found overgrowing *Appendiculella calostroma* on stems of *Rubus fruticosus* in the Republic of Ireland and appears to be the first record of *Trichoconis* which is not either tropical or subtropical. Similarly, *Sympodiophora stereicola* (= *Pseudohansfordia stereicola*) is non-tropical, being found on the basidiome of *Stereum hirsutum* in the Leningrad (St Petersburg) region of Russia (Deighton and Pirozynski, 1972). *Sympodiophora mycophila* (= *Pseudohansfordia mycophila* = *Dactylaria mycophila*) has been found to grow on the ascomycete *Bulgaria* sp. and the basidiomycetes *Hirneola* and *Marasmius. Sympodiophora meliolae* on the other hand, is a tropical species found overgrowing the foliicolous *Meliola paulliniae*, thus it appears that both the host range and geographical distribution are wide in this genus.

A range of British fungicolous fungi, many of them Ascomycota, has recently been reviewed from the point of view of their identification (Ellis and Ellis, 1988). In the discomycetes, nine genera recorded having fungicolous species are *Ascophanus, Bisporella, Cistella, Hyalopeziza, Micropodia, Phaeohelotium, Pithyella, Polydesmia* and *Pyrenopeziza. Ascophanus consociatus* is found on old *Chaetosphaeria cupulifera, Bisporella sulfurina* is very common on stromatic pyrenomycetes such as *Diatrype stigma* and *Diatrypella favacea*, being found from September to February in the United Kingdom. *Cistella incrustata* occurs on *Hypoxylon rubiginosum*, whilst the closely related *C. stereicola* grows on the basidiomes of *Stereum* species. Also basidiomyceticolous are *Micropodia oedema*, occurring on the uredinia and telia of *Phragmidium mucronatum*, and *Pyrenopeziza kolaensis* on Corticiaceae and *Stereum* species. *Phaeohelotium extumescens* occurs mainly on *Diatrype stigma* and *Eutypa* species, and the apothecia of *Polydesmia pruinosa* are found on stromatic pyrenomycetes such as *Diatrype disciformis, D. stigma, Diatrypella favacea, D. quercina, Hypoxylon multiforme* and *H. rubiginosum.* The apothecia of *Hyalopeziza ilicincola* develop in clusters on *Myriangium* spp. and other fungi.

Many other Ascomycota are described as fungicolous (Ellis and Ellis, 1988). Several have been described elsewhere in this book and only an outline of the species with their associated symbionts is provided in Table 2.3 based on the information from Ellis and Ellis (1988).

2.2 Lichenicolous Associations

Although the lichenicolous fungi have until recently received little attention (Hawksworth, 1977a,b, 1982a,b, 1983; Clauzade *et al.*, 1989) they are biologically significant. They are considered here because of the importance of the

Table 2.3. A range of fungicolous Ascomycota and their symbionts in the UK, excluding fungicolous discomycetes (after Ellis and Ellis, 1988).

Fungicole	Symbiont
Barya aurantiaca	*Claviceps purpurea*
Berlesiella nigerrima	*Diatrype stigma, Eutypa acharii, Hypoxylon multiforme*
Calosphaeria parasitica	*Quaternaria dissepta*
Chaetosphaerella fusca	*Diatrype stigma, Eutypa flavovirens*
Cordyceps canadensis	*Elaphomyces* sp.
C. capitata	*Elaphomyces granulatus, E. muricatus,*
C. ophioglossoides	*E. variegatus*
Didymosphaeria conoidea	*Leptosphaeria doliolum*
D. futilis	*Diatrypella favacea*
D. winteri	*Phyllachora*
Dimerium meliolicola	*Appendiculella calostroma*
Eudarluca caricis	*Puccinia* spp.
Helminthosphaeria clavariarum	*Clavulinopsis cinerea, C. cristata*
Hypocrea pulvinata	*Piptoporus betulinus* and other decaying polypores
Hypocreopsis episphaeria	*Diatrype stigma, Diatrypella* spp., *Eutypa* spp., *Hypoxylon fragiforme, Lopadostoma turgidum, Melanconis alni, Melanomma pulvis-pyrius, Melogramma spiniferum, Quaternaria quaternata*
Hypomyces aurantius	*Armillaria mellea* and polypores, e.g. *Bjerkandera adusta, Coriolus versicolor, Ischnoderma benzoinum, Phaeolus schweinitzii, Piptoporus betulinus, Polyporus badius, P. squamosus, Pseudotrametes gibbosa*
H. broomeanus	*Heterobasidion annosum*
H. ochraceus	*Lactarius* and *Russula* spp.
H. rosellus	*Coriolus versicolor, Physisporinus sanguinolentus, Piptoporus betulinus, Polyporus squamosus, Stereum hirsutum*
Letendraea helminthicola	*Helminthosporium velutinum*
Litschaueria corticiorum	*Phanerochaete sordida, Stereum hirsutum*
Melanospora brevirostris	*Geopora arenosa* and other discomycetes
M. caprina	*Armillaria mellea* (old rhizomorphs), *Stereum hirsutum, Tomentella ferruginea*
M. fusispora	*Beauveria bassiana, Paecilomyces farinosus*
M. lagenaria	*Bjerkandera adusta, Coriolus versicolor* and other polypores
Nectria aureola	*Meliola niessleana*
N. berkeleyana	*Corticium* spp., *Polyporus* spp., *Stereum hirsutum*

Table 2.3. continued.

Fungicole	Symbiont
N. leptosphaeriae	*Leptosphaeria acuta, L. doliolum*
N. magnusiana	*Diatrypella favacea, D. quercina*
N. peziza	*Coriolus versicolor, Polyporus squamosus*
N. polyporina	*Heterobasidion annosum*
N. purtonii	*Melanconis stilbostoma, Valsa* spp.
N. wegeliniana	*Hapalocystis bicaudata, Pseudovalsa modonia*
Nitschkia collapsa	*Diatrype stigma*
N. confertula	*Hypoxylon rubiginosum*
N. cupularis	*Botryosphaeria stevensii*
N. grevillii	*Peroneutypa heteracantha*
N. parasitans	*Nectria cinnabarina*
Peckiella lateritia	*Lactarius deliciosus, L. torminosus* and other agarics
P. viridis	
Protocrea farinosa	*Junghunia nitida, Perenniporia medulla-panis*
Pyxidiophora asterophora	*Nyctalis asterophora*
Scotiosphaeria endoxylinae	*Endoxylina pini*
Sepedonium chrysospermum	*Boletus* spp., *Paxillus involutus, Scleroderma citrinum*
Sepedonium tulasneanum	*Boletus* spp.
Sphaeroderma episphaeria	*Hypomyces ochraceus*
Sphaerodes fimicola	*Sclerotinia sclerotiorum*, old polypores
Sphaerostilbella aureonitens	*Stereum hirsutum*
Syspastospora parasitica	*Beauveria bassiana, Paecilomyces farinosus, Sarcopodium circinatum*
Trichothyrina parasitica	*Chaetosphaerella fusca, Diatrype stigma, Eutypa acharii*
Tubeufia cerea	*Diatrype stigma, Lasiosphaeria hirsuta, Polydesmia pruinosa*

fungal partner in the structure of a lichen. Although a lichen is a mutualistic symbiotic association between an alga or cyanobacterium (the **photobiont**) and a fungus (the **mycobiont**) the form of the lichen and its reproductive capacity is determined by the mycobiont. About 24 Chlorophyta (green algae), including species of *Chlorella*, *Trebouxia* and *Trentepohlia*, and 13 cyanobacteria (blue-green bacteria) including species of *Anabaena*, *Anacystis*, *Nostoc* and *Stigonema*, are known to act as the photobionts whereas well over 13,500 species of fungi are involved, most of them belonging to the Ascomycota (Hawksworth and Hill, 1984). In fact, members of five orders of the Ascomycota comprise entirely lichenized species, the Graphidales, Gyalectales, Peltigerales, Pertusariales and Teloschistales (Hawksworth and Hill, 1984). Basidiolichens are formed by species in various orders of the Basidiomycota, e.g. *Omphalina* (Agaricales), *Multiclavula* and *Dictyonema* (Aphyllophorales). Over 55 species of lichen-forming deuteromycetes, *Geosiphon* (probably Oomycota) and several **myxomyceticolous lichens** have been reported (Hawksworth and Hill, 1984).

There are many examples of lichenicolous fungi. For example, *Pronectria robergei* parasitizes the thallus of *Peltigera canina* and *Paranectria affinis* occurs on that of the lichen *Ephebe lanata* (Dennis, 1977). According to Hawksworth *et al.* (1980) 183 lichenicolous fungi were recorded from the British Isles, and over 300 genera and 1000 species of obligately lichenicolous fungi probably occur on a global basis (Hawksworth and Hill, 1984). A distinction has been made between saprotrophic lichenicolous fungi which develop on decaying lichen thalli, e.g. *Niesslia cladoniicola*, commensals or secondary colonizers which develop a stable relationship with the lichen but do not appear to cause harm, e.g. *Arthonia punctella* (= *Arthonia varians*) found in the apothecia of *Lecanora rupicola*, and parasites which adversely affect the growth and development of the lichen, e.g. *Athelia arachnoidea* which grows over and kills the thallus of *Lecanora conizaeoides*. Some lichenicolous fungi, e.g. *Arthrorhaphis citrinella* on *Baeomyces rufus*, *Blarneya hibernica* on *Enterographa* and *Lecanactis*, *Chaenothecopsis consociata* on *Chaenotheca chrysocephala*, *Diploschistes caesioplumbeus* on *Lecanora gangaleoides* and *D. muscorum* on *Cladonia* species are also able to displace the original mycobiont in an established lichen thus forming a new lichen in combination with the original algal symbiont. Whatever the precise nature of the lichenicolous association, however, the mycobiont is presumed to derive its energy-containing carbon compounds required for metabolism from the photosynthetic activity of the photobiont. It is also worth mentioning that some lichens are themselves lichenicolous thus constituting four-membered symbiotic associations (Hawksworth and Hill, 1984). Whether mycoparasitism is involved in such associations, however, remains to be determined.

It appears that the recognized fungal symbiont(s) in specific lichen thalli may not necessarily be the only members in the association as it has recently been shown that the thalli of a range of fruticose lichens harbour an extensive range of filamentous fungi, normally predominantly associated with soil and litter (Petrini *et al.*, 1990). The role of such fungi in the life of these lichens is unknown although it seems that at least the dominant or most regularly encountered species have developed some form of association with the 'host' lichen and this is an area which requires further research (Petrini *et al.*, 1990). As an example, from 17 samples of *Cladonia* and *Stereocaulon*, 506 fungal strains were isolated, 166 were isolated more than once and, of these, 62 sporulated and were identified to genus or species (Petrini *et al.*, 1990).

Taking the genus *Peltigera* as an example it can be seen from the key to obligately lichenicolous fungi occurring on *Peltigera* (Hawksworth, 1980) that members of this genus can be colonized by basidiomycetous, ascomycetous and conidiogenous symbionts. *Hobsonia santessonii*, a conidial hyphomycete, was recorded as the only species of *Hobsonia* known to occur on a species of *Peltigera* (*P. scabrosa*; Lowen *et al.*, 1986) and it seems probable that some species specificity may occur in the lichenicolous fungi.

Most lichenicolous fungi are probably saprotrophic although they may be confined for either the whole or part of the early phases of their life cycle to the lichen thallus. The lichen association becomes three-membered when invasion of the lichen thallus by another fungus occurs. The secondary invader may enter into a stable symbiotic relationship with the pre-existing symbionts of the lichen. Such fungi, previously referred to as 'parasymbionts' (Hawksworth, 1988), are presumed not to damage the other symbionts in any way, but owing to the difficulties in demonstrating the true relationship, this term becomes redundant. The physiological balance in this relationship alters, however, when the invader parasitizes the original fungal symbiont, eventually replacing it and establishing a new lichen thallus (Hawksworth, 1983). Where some or all of the nutrients of the invading symbiont are derived from the lichen symbionts then the invader is, by definition, a lichen parasite. In extreme cases complex relationships between several fungi can occur where, for example, one is parasitic on a second which itself obtains nutrients parasitically from a third fungus, the original lichen symbiont.

The parasitic lichenicolous fungi often but not always visibly affect the thallus of the host lichen. The symptoms are expressed in various ways such as the development of malformations which are frequently gall-like, or by forming holes, or discolouring the thallus, or through various degrees of necrosis of the cytoplasm, sometimes culminating in death of the thallus (Hawksworth, 1977a). Of ten species of *Lichenoconium* all, with the exception of *L. boreale* are exclusively lichenicolous, occurring on at least 58 host lichens (Hawksworth, 1977a). Up to three species of *Lichenoconium* are known to parasitize a single host although when this occurs, the symptoms induced may be different. On *Parmelia saxatilis*, for example, *L. erodens* and *L. lecanorae* cause the production of necrotic spots with a black margin on the thallus. In *L. erodens*, however, there are several pycnia per spot, the centre of which erodes leaving a hole whereas in *L. lecanorae* there is only one pycnium per spot and the centre of the spot does not erode. *Lichenoconium usneae* on *P. saxatilis* produces pycnia in apothecia which become discoloured, usually blackened. *Lichenoconium usneae* also attacks *Parmelia olivacea* causing discoloration of pycnia in apothecia, *Cladonia arbuscula* causing decolourization of the podetia and *Hypogymnia physodes* where the infection spreads, not being limited to spots. *Lichenoconium erodens* also attacks *Parmelia galbina* causing pycnia in apothecia to become discoloured, *Pertusaria pertusa* causing infected parts of the thallus to discolour and die, *Evernia prunastri* causing thallus discoloration, and *Hypogymnia physodes* causing necrotic spots with black margins on the thallus with several pycnia per spot, the centre eroding leaving a hole. *Lichenoconium lecanorae* also attacks moribund thalli of *Evernia prunastri* although no particular symptoms result from the infection. On *Parmelia galbina*, *L. lecanorae* causes discoloration of apothecia. *Lichenoconium erodens* and *L. echinosporum* are pathogenic being able to infect healthy, actively growing thalli where extensive necrotic lesions are formed and the thalli turn brown or whitish-brown. *Lichenoconium lecanorae*, *L. pyxidatae* and *L. usneae* appear less strongly pathogenic and often produce localized infections on thalli which are older and unhealthy or damaged, and *L. xanthoriae* also causes little damage unless the infected thallus is already unhealthy (the related *Laeriomyces pertusariicola*, on the other hand, appears to be commensalistic as it seems to cause little damage to the host lichen). *Lichenoconium* species growing on unhealthy tissue are frequently associated with other lichenicolous fungi, e.g. *Abrothallus parmeliarum* with *L. usneae*, *Arthonia clemens* with *L. lecanorae*, and *Phoma* species with *L. erodens* (Hawksworth, 1977a).

Other symptoms can be found associated with lichenicolous pathogenic fungi (Hawksworth, 1983). Galls are formed, for example, by *Endococcus parietinus* on the thallus of *Xanthoria parietina*; *Polycoccum peltigerae* on the thalli of *Peltigera* species; *Refractohilum galligenum* on *Nephroma laevigatum* forms ochre-yellow convex galls, and *R. peltigerae* causes reddish-brown pustulate galls on *Peltigera* species; and *Skyttea lettaui* forms galls on *Evernia prunastri*. In *Polycoccum trypethelioides* on *Stereocaulon* species the galls are 2.5 mm diameter; in *P. galligenum* on *Physcia caesia*, wart-like; in *Thamnogalla crombei* on *Thamnolia vermicularis*, bullate; tuberculate-convex in *Epicladonia sandstedei* on *Cladonia conoidea*; convex in *Bachmanniomyces uncialicola* on *Cladonia uncialis* and in *Ramularia peltigericola* on *Peltigera* species. *Clypeococcum cladonema* which parasitizes *Cetrelia olivetorum* and *Parmelia pulla* causes neat round holes to develop in the thallus. Damage to the thallus may be extensive as in *Nectria lecanodes* on *Lobaria* and *Peltigera* species; bleaching of the thallus is caused by *Arthonia punctella* (= *A. varians*) on *Diplotomma alboatrum* and *D. chlorophaeum*; *Pronectria santessonii* on *Anaptychia fusca* (= *A. runcinata*); *Nesolechia oxyspora* on *Parmelia saxatilis* and *P. sulcata*; *Lauderlindsaya borreri* on *Normandina pulchella*; apothecia of *Corticifraga* (*Phragmonaevia*) *peltigerae* form in circular groups in bleached white patches on thalli of *Peltigera* species. Several species cause discoloration of the thallus of the host. *Pezizella epithallina* causes discoloration of the thallus of *Peltigera* species to bluish-green; *Refractohilum achromaticum* causes the lobes of the thallus of *Parmelia* species to discolour and become slightly pustulate but not galled; podetia of *Cladonia* species are discoloured brownish by *Lichenoconium pyxidatae*; apothecia may be decolourized from yellow-green to pale fawn in *Lecanora strobilina* by *Diplolaeviopsis ranula* (Giralt and Hawksworth, 1991) or are turned black in *Xanthoria polycarpa* by *Lichenoconium xanthoriae*, jet black in *Lecanora conizaeoides* by *L. lecanorae*, and sooty black in *Xanthoria parietina* by *Xanthoriicola physicae*. In *Lecanora* species, especially *L. chlarotera*, infected apothecia are decolourized by *Vouauxiella lichenicola* giving the thallus a piebald appearance; pale brownish necrotic patches are produced on the thalli of *Evernia prunastri* by *Everniicola flexispora*, and fawn-coloured necrotic patches on thalli of *Thelotrema subtile* by *Opegrapha brevis* and on *T. lepadinum* by *O. thelotrematis* (Coppins, 1987). Apothecia of the lichenicolous Lecanorales are frequently erumpent, e.g. *Nesolechia oxyspora* on *Parmelia* species, and *Dactylospora* (= *Leciographa*) parellaria on thalli of *Ochrolechia parella*. The ascomata of *Rhagadostoma lichenicola* on the thallus of the arctic-alpine lichen *Solorina crocea* are also erumpment.

The extent of the infection appears characteristic being limited to neat circular grey-black patches by *Arthopyrenia microspilum* on *Graphis scripta*. *Cornutispora lichenicola* causes the development of black-margined decolourized necrotic patches on a wide range of foliose lichens, e.g. *Lobaria pulmonaria*, *Platismatia glauca*, *Parmelia borreri* and *P. sulcata*. Raised pustular black patches due to '*Vouauxiella*' *uniseptata* on the thallus of *Parmelia laevigata* finally become converted to holes whereas almost the entire hymenium of *Opegrapha herbarum* is eventually colonized by *Laeviomyces opegraphae*. The thallus is eventually killed in *Xanthoria parietina* by *Xanthoriicola physicae*, and in *Peltigera hymenina* by *Omphalina cupulatoides*. Isidium production (vegetative reproductive papillae) in *Pertusaria coccodes* is inhibited by *Cyphelium sessile*, and in *P. corallina* and *P. pseudocorallina* by *Sclerococcum sphaerale*.

Clearly, fungicolous and lichenicolous associations are widespread on a global basis, of common occurrence in the terrestrial habitat and likely to be advantageous to at least one of the symbionts in most of the associations. The examples highlighted in this chapter have been selected as representative of mycoparasitic relationships being taken from a few of the numerous studies in this area. The current total of fungi described from all taxonomic groups is approximately 69,000 which probably represents about 5% of the fungi estimated to exist worldwide (Hawksworth, 1991). The probability is, therefore, that many further fungicolous and lichenicolous associations remain to be discovered and this will undoubtedly prove to be a fruitful area of research in mycology.

3 NECROTROPHIC ASSOCIATIONS

3.1 Contact Necrotrophs

3.1.1 Characteristics of contact necrotrophs

Members of this category of fungi are found in close contact with other fungi but differ from fungicolous associations in that the host fungus distinctly appears to suffer directly from the association. In some associations, such as those involving *Sepedonium* discussed in section 2.1.2, it is not easy to make the distinction. In most examples, however, the different relationships are clear, although in contact necrotrophic interactions no penetration of the attacked hyphae has been observed and the degree of exchange of metabolites is not known. 'Parasitic' hyphae are frequently much finer than those of the host (this is also true of mycoparasites that do penetrate the host) and can usually be clearly distinguished from those of the 'host'. Antagonism of the hyphae of *Sphaerotheca fuliginea*, the causal agent of powdery mildew of cucumber, by a species of *Tilletiopsis*, a common phylloplane fungus, provides a typical example. The hyphae of the antagonist are about 1–1.8 μm in diameter, approximately one-tenth of those of the host (Hoch and Provvidenti, 1979), and similarly in *Gliocladium virens* attacking *Rhizoctonia solani* the host hyphae are approximately three times the diameter of the parasitic ones (Tu and Vaartaja, 1981). This is not always so, however, and *Fusidium parasiticum*, which grows within the perithecia of *Xylaria oxycanthae* has hyphae which are generally of slightly greater diameter than those of the host (Backus and Stowell, 1953).

The situation is further complicated in that some fungi can penetrate the hyphae of their hosts in some interactions but apparently do not in others. *Pythium oligandrum* is an invasive necrotroph of several other fungi, but when *Mycocentrospora acerina* (the causal agent of cavity spot disease of carrots) acts as the host its hyphae are apparently not penetrated (Lutchmeah and Cooke, 1984) and hyphal interference results (see section 3.1.2). A similar phenomenon is exhibited by *Gliocladium virens* which can parasitize *Rhizoctonia solani* by coiling round and penetrating the hyphae. *Gliocladium virens* does not antagonize *Pythium ultimum* in this way but strongly inhibits it through antibiosis (Howell, 1982). Necrotrophic interactions can vary considerably and various types of attack and susceptibility have been observed even between the members of a single host-parasite combination. The outcome of these interactions depends on the particular isolates of the fungi, environmental conditions, age of mycelium and time of observation.

3.1.2 Hyphal interference

Hyphal interference, described by Ikediugwu and Webster (1970a), occurs when the mycelium of a fungus growing either in very close proximity to ($< 50\ \mu m$) or in contact with that of another species of fungus affects the latter by reducing its rate of growth and causing cytoplasmic disruption. Most of the observations of hyphal interference relate to fungi grown in Petri dishes on nutrient agar. Hyphal inteference reactions can be determined on microscope slides coated with a layer of nutrient agar sufficiently thick to allow the fungi to grow but thin enough to enable microscopic observations to be made. Many observations of the fine structure of interacting hyphae have also been made on material prepared in this way.

The principal effect seems to be a marked alteration in the permeability of the plasmalemma of the antagonized fungus. Loss of turgor occurs, accompanied by granulation, vacuolation and death of the cytoplasm in the zone of the interaction. The factor(s) that induces the interference effect can diffuse through a cellophane membrane separating the two fungi. This suggests that it is a low molecular weight metabolite and not an enzyme (Ikediugwu and Webster, 1970a). In those cases where septate hyphae are affected, the interference effect is limited to the hyphal compartment in contact with the interfering hypha and the effect does not spread into the mycelium. This is consistent with the fact that the interference factor does not diffuse any appreciable distance in agar, but exerts its effect directly at the region of contact with the affected hypha. The rate of interference action, as measured by the time between hyphal contact and cessation of linear growth differs from interaction to interaction. A typical example occurs between *Ascobolus crenulatus* and *Coprinus heptemerus*, in which growth of the former fungus ceases around 21–24 min after contact. Younger hyphae, especially hyphal tips, are more susceptible than older regions of the mycelium. Hyphal lysis does not seem to be a common characteristic of classic hyphal interference. Although hyphal contact is not necessarily a prerequisite for the reduction in growth of the antagonized fungus, appressed growth of the hyphae of the antagonist along the surface of the hypha of the recipient, hyphal coiling and penetration usually occur. Hyphal interference constitutes the basis for one of the most successful examples of biocontrol of a plant-pathogenic fungus to be used in commercial practice (see Chapter 7). Butt-rot of pine trees, caused by the bracket fungus *Heterobasidion annosum*, is prevented if newly exposed surfaces of the stumps are treated with a spore suspension of *Phanerochaete gigantea. Phanerochaete gigantea* is a wood-inhabiting fungus which quickly colonizes the stumps and prevents the subsequent establishment of *H. annosum* by a combination of hyphal interference and direct competition. The interference reaction involves a reduction in the number of mitochondria in the cytoplasm of *H. annosum* in the contact zone (Ikediugwu, 1976a) which suggests that the process of respiration is adversely affected.

Hyphal interference was first proposed as an explanation why the basidiomycete *Coprinus heptemerus* was able to inhibit the development of sporomes of *Ascobolus crenulatus* and *Pilobolus crystallinus* growing on pellets of rabbit dung. Laboratory studies showed that hyphal interference by *C. heptemerus* prevents growth of the competing fungi and the antagonism probably also occurs under natural conditions (Ikediugwu and Webster, 1970a). Loss of permeability was demonstrated by the absorption of the vital dye neutral red by the antagonized hyphae of *Pilobolus* and *Ascobolus*, the interference effect being most marked at the point of contact of the sensitive hypha with the antagonist. Further evidence that *C. heptemerus* destroys the selective permeability of cytoplasmic membranes is derived from the effect of plasmolysing solutions of glucose and glycerol on the antagonized hyphae of *A. crenulatus*. The cytoplasm of unaffected hyphae rapidly plasmolyses whereas that of affected hyphae does not plasmolyse and takes up neutral red incorporated in the plasmolysing solution. Nutrients are presumed to leak to the environment and may be taken up by the antagonistic fungus. There is no direct evidence, however, that the antagonist derives any nutrients from the affected cells, thus it is difficult to say if this phenomenon is a truly parasitic relationship. What advantage the antagonist might derive from the association is obscure in the absence of experimental evidence. Possibly, however, vitamins may be absorbed by the antagonist from materials leaked from the cytoplasm as *C. heptemerus* is deficient for thiamine. Thus fungi sensitive to interference may provide a source of essential growth substances to the antagonist. *Coprinus heptemerus* is itself antagonized by the coprophilous agarics *Coprinus stercorarius* and *Bolbitius vitellinus*. The phenomenon of hyphal interference has been described for many fungi and it seems that although many coprophilous fungi are sensitive, non-coprophilous fungi are also affected (Ikediugwu and Webster, 1970b).

Detailed descriptions of the cytological changes occurring during hyphal interference have now been described for several fungal interactions (Ikediugwu, 1976a,b; Traquair and McKeen, 1978). When *Mycocentrospora acerina* is antagonized by *Pythium oligandrum*, for example, tip to tip contact of leading hyphae of those of the host and the parasite occurs (Lutchmeah and Cooke, 1984). This results in a rapid loss of opacity in the apical region of the hyphae of the host, the process taking less than 5 min after contact. The apices of the transformed hyphae may also disintegrate, cytoplasm flowing freely from them. A

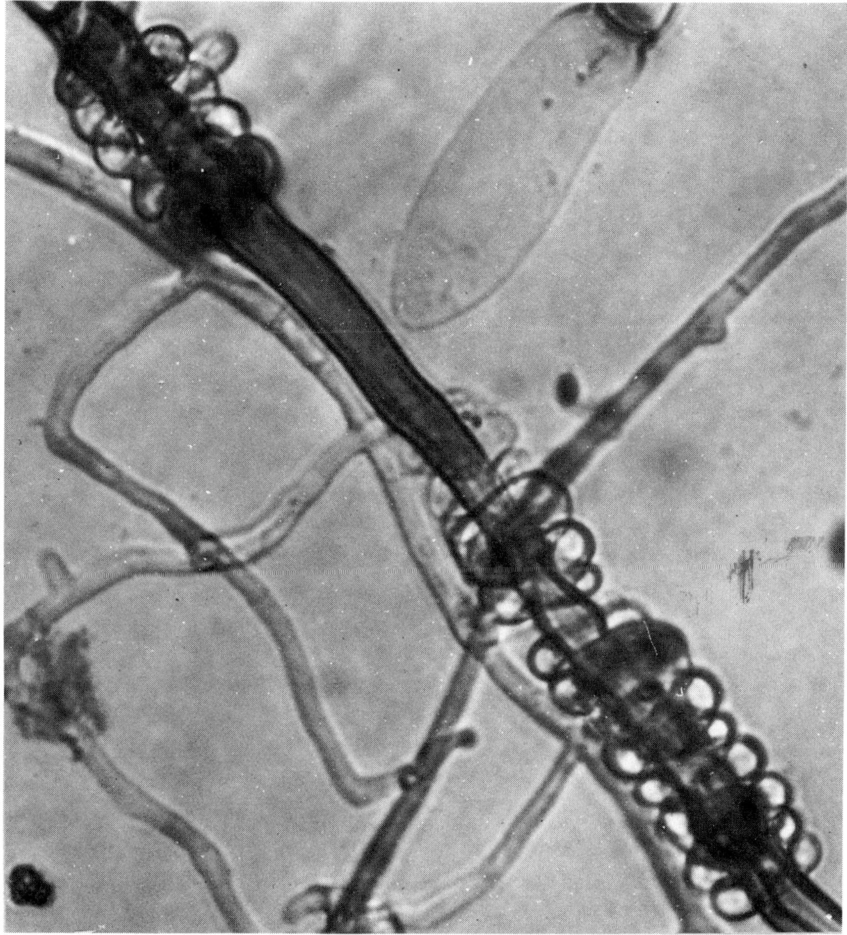

Fig. 3.1. (see also Fig. 3.2) Interactions of hyphal coils of *Arthrobotrys oligospora* and *Rhizoctonia solani*: hyphal coils (LM: × 350). (Micrograph courtesy of Dr Y. Persson.)

physiological change, concomitant with the morphological alterations in the hyphae of the host, is a reduction in cellulolytic ability of the host, further indicating the harmful effect of the parasite on the host.

The phenomenon of hyphal interference by basidiomycetes growing within woody substrates might be extremely common. Griffith and Barnett (1967) reported that 15 species of wood-rotting basidiomycetes antagonized spores or hyphae of *Ceratocystis* species in this manner and suggested that this action may represent a stage in the normal succession of fungi on freshly cut logs and

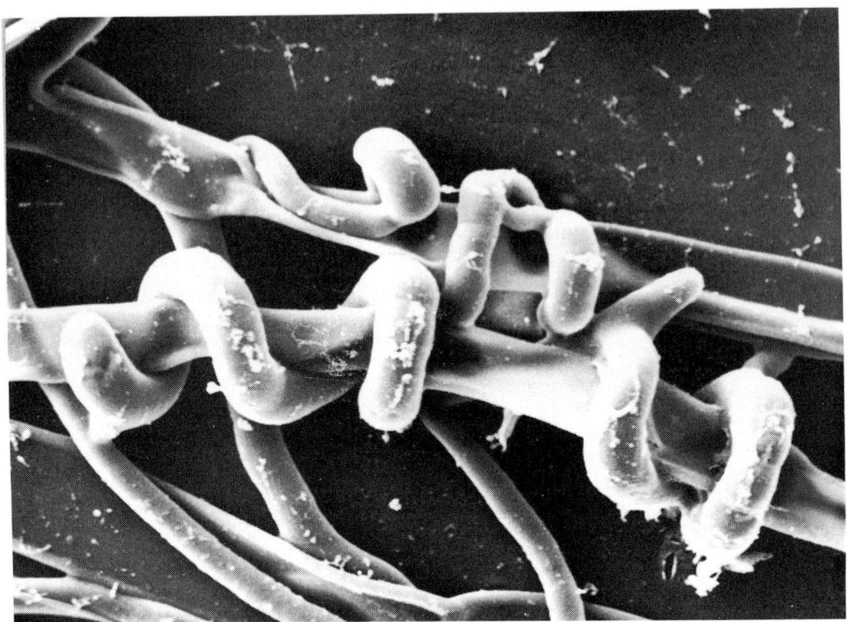

Fig. 3.2. (see also Fig. 3.1) Interactions of hyphal coils of *Arthrobotrys oligospora* and *Rhizoctonia solani*: detail of hyphal coils (SEM: × 600). (Micrograph courtesy of Dr Y. Persson.)

stumps. In the interaction of *Hirschioporus pargamenus* and *Ceratocystis fimbriata*, the necrotrophic reaction in hyphae of *C. fimbriata* is rapid when they are either in close proximity to or in contact with the opposing hyphae of *H. pargamenus* (Traquair and McKeen, 1978). Invagination of the plasmalemma of the host and the deposition of electron-dense extraplasmalemmal deposits occurs within 30 min of hyphal contact. Mitochondria and cytoplasmic vesicles in the contact region enlarge and become disrupted and, after 1–2 h, disruption of the cytoplasm of the membrane system and nuclei of the host is extensive. After 3 h, necrosis of the cytoplasm is complete and osmiophilic aggregates of cytoplasm and fragments of membrane are scattered within the hyphae. As the cytoplasm degenerates, the infected hyphae tend to collapse. These fungi are readily distinguished on the basis of septum structure and hyphal width. *Hirschioporus pargamenus* is a bracket polypore and the fine hyphae possess dolipore septa whereas those of *C. fimbriata* are wider, with septa typical of the Ascomycota. Penetration of the hyphae of the host does not normally occur which suggests that degeneration of the cytoplasm is due to a toxic principle produced by the mycopathogen. Intrahyphal growth, if it occurs at all, is seen mainly after 72 h when the cytoplasm has undergone complete necrosis. Once the necrotrophic reaction is underway it seems likely that autolytic enzymes within the cytoplasm of the host itself contribute to cytoplasmic degradation. In this particular fungus–fungus interaction it appears that hyphal contact is not always necessary, and the cytological events described above can be mediated at a distance by an exotoxin secreted by *H. pargamenus. Hirschioporus pargamenus* is a tree pathogen causing a white rot in the sapwood of living and dead trees. It is also subject to hyphal interference itself when antagonized by *Trametes hispida* (= *Inonotus hispidus*) (Traquair and McKeen, 1977).

An extensive *in vitro* study of hyphal interference between the leaf-infecting pathogen *Septoria nodorum* and many cereal **phylloplane** fungi indicated that the phenomenon is also likely to occur on the leaf surface (Skidmore and Dickinson, 1976). Hyphal interference was also suggested to be the mechanism of antagonism of *Erysiphe graminis* f.sp. *hordei* by the yeast-like **epiphyte** *Tilletiopsis pallescens* (Klecan *et al.*, 1990), and may thus also account for the inhibition of *Sphaerotheca fuliginea* by *Tilletiopsis* sp. (Hoch and Provvidenti, 1979). Both *Erysiphe* and *Sphaerotheca* are obligate parasites of higher plants and cause powdery mildew diseases. Hyphal interference by leaf surface yeasts and yeast-like fungi may thus be important in natural biocontrol of these diseases. The phenomenon of hyphal interference is probably significant in many ecosystems where fungi are competing for nutrients and space. A potential application, however, is in the biocontrol of plant-pathogenic fungi where the effect may either prevent the pathogen from colonizing a potential host or mitigate the effects of the pathogen to the extent that the results of its pathogenicity are negligible.

Hyphal interference may also underlie the antagonistic interactions of several soil fungi. For example, species of *Arthrobotrys* are soil-inhabiting fungi capable of trapping nematodes by means of specialized hyphal networks to which they adhere strongly. These fungi are also capable of parasitizing the hyphae of *Rhizoctonia solani* and certain other fungi by coiling tightly around them (Figs 3.1 and 3.2) and inducing death of the cytoplasm (Tzean and Estey, 1978a; Persson *et al.*, 1985). Of 34 fungi tested as potential hosts, Tzean and Estey (1978a) found only three, *Geotrichum* sp., *Matruchotia varians*, and *R. solani*, were susceptible which suggests specificity in the response of the mycopathogen to the host. Persson *et al.* (1985), however, showed that *Arthrobotrys* spp. had a wider host range and six out of 13 fungi from a broad taxonomic spectrum were coiled around. A reciprocal reaction also occurs with *Geotrichum* sp. as, on agar plate culture, the hyphae of *Arthrobotrys* spp. are partially lysed although *Geotrichum* does not form hyphal coils. The parasitized fungi are responsible for the induction of hyphal coiling in the nematophagous mycopathogens which may form between one and ten coils at a time (Tzean and Estey, 1978a). Ultrathin sections through the hyphal coils show that although the wall of the host is not penetrated it is compressed and may become lysed whereas that of the coil remains intact. Hyphal constriction is succeeded by alterations in the cytoplasm of the host which also undergoes lysis and eventually disappears. With *R. solani* as the potential host growing on half-strength corn meal agar, it may take up to 40 days to destroy 50% of the host colony, thus the process is relatively slow. Further, with other media such as malt extract, the effect of the mycopathogen is negligible. This type of hyphal interaction with the nematode-trapping species of *Arthrobotrys* appears to be dependent on mutual hyphal contact and on the nutritional status of the medium on which the interacting fungi grow. On corn meal agar, for example, increasing concentrations of the substrate resulted in an increased frequency of coiling (Persson and Baath, 1992). However, there is no mutual contact of the cytoplasm of the antagonists, unlike the fusion biotrophs nor has hyphal penetration been recorded. Although there is no directed growth of the *Arthrobotrys* towards the hyphae that it subsequently coils around, it appears that growth of *Trichoderma harzianum* parasitizing *Rhizoctonia solani* is directed (Elad *et al.*, 1987). It is not unreasonable, therefore, to consider this interaction as a case of hyphal interference.

In order to observe antagonism *in situ* between *Arthrobotrys oligospora* and *Rhizoctonia solani* in more detail, a piece of dialysis membrane was placed on the surface of agar and inoculated at each end with one of the two fungi (Persson *et al.*, 1985). Both fungi grew over the surface of the membrane which was lifted off the agar and immersed in the vital stain methylene blue in buffer solution to observe the living mycelia. A similar technique was used to examine the zone of interaction by means of fluorescence microscopy. Mycelia on the membrane were stained with fluorescein diacetate which fluoresces under ultraviolet illumination. Fluorescence indicates an area of intense metabolic activity. In this particular interaction, coiling occurred a few hours after hyphal contact, irrespective of the concentration of nutrient in the medium. The hyphal coils of *A. oligospora* fluoresced whereas the hyphae of *R. solani* both in the region of contact and beyond the coils did not, indicating the cessation of enzyme activity and cytoplasmic death. As hyphal contact is a prerequisite to cytoplasmic death in this example of hyphal interference it seems that diffusible antifungal antibiotics are not produced prior to contact. Observation of the fine structure of the parasitized hyphae in the region of contact showed the development of papillae similar to those found in certain biotrophic associations involving the mycoparasitic genus *Piptocephalis* (see section 4.1.1). Furthermore, cytoplasm of the coil contained a concentration of membranous organelles, indicating intense metabolic activity. Variation in the production of hyphal coils in relation to the nature of the substratum on which the antagonists are growing implies that mycoparasitism is not an obligatory feature in the life of *Arthrobotrys* spp. and may, indeed, be transitory. In a comparison of hyphal coils formed around other fungi with the hyphal traps that this fungus forms when attacking nematodes, Persson (1991) found that the coils and traps differ both in structure and in their physiology. For example, coils do not become coated in the mucilaginous material that is important in the interaction of traps and the surface of nematodes. Neither are they induced by addition of peptides to axenic cultures of the fungus. Coils have only been observed in cultures that have come into contact with an appropriate 'host' fungus (Persson, 1991).

3.1.3 Involvement of extracellular diffusates in antagonism

Hyphal interference is mediated through a diffusible metabolite(s), but this material does not appear to be released until interacting hyphae are in close proximity and is not found in culture filtrates of the antagonistic fungus. In contrast, several necrotrophic interactions involve freely diffusible material which can affect hyphae some distance away. For example, *Hypomyces aurantius* produces a non-specific toxin which induces rapid and irreversible permeability changes in the plasmalemma of its basidiomycete hosts (Kellock and Dix, 1984a). Culture filtrates from *Hypomyces aurantius* contain this material and can also be used to inhibit mycelial extension of these fungi in agar culture. The toxin is not an enzyme and is produced equally well either in the presence or absence of a potential host. During interactions with *Stereum hirsutum*, this substance damages membranes, and causes cytoplasmic vacuolation and bursting of host hyphae, changes associated in some instances with the deposition of brown pigments in mycelial interaction zones. Brown discoloration is also observed when *Heterobasidion annosum* and *Coniophora puteana* are antagonized by *Scytalidium* spp. and the production of pigments in interaction zones is associated with the secretion of polyphenol oxidases (Klingström and Johansson, 1973). One isolate of *S. lignorum* apparently produces a diffusible inhibitory compound, but other species of *Scytalidium* do not, indicating an alternative mechanism of antagonism. The fact that the effects of a toxin occur some distance from the interaction zone has also been noted in interactions involving *Trichoderma* spp., when vacuolation and coagulation in host hyphae were observed when mycelia were up to 3–5 mm away (Dennis and Webster, 1971c). This differs from hyphal interference in this respect, even though the cytological effects that result appear similar. The cytoplasm of hyphae of *Stereum hirsutum* antagonized at a distance by *Hypomyces aurantius* is severely disrupted (Kellock and Dix, 1984b). Lipid bodies accumulate, the cisternae of mitochondria and the endoplasmic reticulum swell and the plasmalemma withdraws from the hyphal wall. It is not known whether these destructive effects of *Hypomyces* occur in some of the other interactions of members of this genus with other fungi (see section 2.1.2).

Where extracellular diffusates only act over relatively short distances the phenomenon may be equivalent to hyphal interference. Cell lysis is a common feature of such interactions which may be mediated through the action of cell wall degrading enzymes. A typical example occurs with *Gliocladium roseum*, a necrotroph capable of parasitizing a wide range of hosts. When attacking *Botrytis allii*, the hyphae of *G. roseum* do not normally either coil around or penetrate those of the host although both of these phenomena may occur. Hyphal contact or close contact seems to be necessary for the necrotrophic response which includes coagulation of the cytoplasm followed by disintegration of the hyphal wall. Coagulation of the cytoplasm which occurs without wall penetration implies the production of toxic material. Diffusible toxic substances of low molecular weight which are effective over short distances only, probably owing to their production in low concentration, have been detected in parasitized cultures. Wall breakdown, however, is probably due to β-1,3-glucanase and chitinase activity of *G. roseum* as it has been demonstrated that the production of these hydrolytic enzymes increases significantly in dual cultures of *G. roseum* and *B. allii* in comparison with that detected in cultures of either fungus grown alone. It has been suggested that, under natural conditions, the hydrolytic enzymes are inactive and the walls of living hyphae escape hydrolysis. However, when the cytoplasm of the host becomes moribund through the action of a toxin, the hydrolytic enzymes released, probably by the mycoparasite, degrade the hyphal walls. Thus a general feature of necrotrophic mycoparasitism may be initial hyphal interference through the action of a metabolic poison such as a toxin or antibiotic which precedes death of the cytoplasm (Pachenari and Dix, 1980). *Gliocladium roseum* is also the most common mycoparasite found in Dutch potato fields baited with *Rhizoctonia solani* mycelium (Jager *et al.*, 1979). Its hyphae coil around those of *R. solani* but only rarely penetrate. Water-soluble diffusates are produced which inhibit the growth of *R. solani*. Similarly *Hormiactis fimicola* grows through a colony of *R. solani*, without attacking the hyphae, and the diffusate accumulates in the sclerotia.

The necrotrophic mycoparasite *Trichoderma hamatum* which attacks species of *Pythium*, *Rhizoctonia* and *Sclerotium*, causes vacuolation, shrinkage, collapse and disintegration of cytoplasm of the host. The lytic enzymes β-1,3-glucanase, chitinase and cellulase are possibly responsible for the induction of cell death through the digestion of important wall polymers. The antibiotics known to be produced by strains of *T. hamatum* are not, apparently, involved in these hyphal interactions. Other research, however, has shown that antibiotic production can be important in the process of mycoparasitism. In one study, for example, of 80 isolates of *Trichoderma* investigated 70 were capable of coiling around the hyphae of various potential host fungi. Bursting of the tips of the hyphae of the hosts *Heterobasidion annosum* and *Rhizoctonia solani* was noted prior to contact with the hyphae of *Trichoderma*. It was suggested that cyclic peptide antibiotics such as alamethicine, known to be produced by *Trichoderma* isolates, could induce leakage of cell contents, and that the action of extracellular enzymes might complement antibiotic activity (Brückner and Przybylski, 1984). Recent investigation of the parasitic relation between two strains of *Trichoderma harzianum* and the plant-pathogenic potential hosts *Fusarium oxysporum* f.sp. *vasinfectum*, *F. oxysporum* f.sp. *melonis*, *Rhizoctonia solani* and *Pythium aphanidermatum* shows that *T. harzianum* is unable to parasitize the mycelium of *Fusarium* (Sivan and Chet, 1989). The hydrolytic enzymes β-1,3-glucanase and chitinase appear to be ineffective against the wall of the living hypha of *Fusarium*. If, however, the hyphal wall of *Fusarium* is treated with proteolytic enzymes prior to treatment with hydrolytic enzymes, breakdown of the chitin occurs. The hyphal wall of *F. oxysporum* contains a high protein content (7–28%) in comparison with other mycelial fungi (usually around 5%). An explanation advanced to explain the results is that proteins associated with the outer layer of the hyphal wall of these strains of *Fusarium* protect the fungus against the hydrolytic activity of the enzymes from *T. harzianum* and hence from mycoparasitic attack. Nevertheless, one of the strains of *T. harzianum* (T-35) has been used to achieve significant biocontrol of *F. oxysporum* under field conditions which suggests that other physiological activities such as antibiosis and the competition for nutrients may be involved. A similar result was obtained for mutants of *Gliocladium virens* which were no longer mycoparasitic in culture, yet still caused a reduction in viable sclerotia of *Rhizoctonia solani* sclerotia in soil (Howell, 1987). This also suggested that mycoparasitism was not a primary mechanism in biocontrol, but that it may function in concert with antibiosis in a natural ecosystem. The development of a transformation system for *Trichoderma* (Sivan *et al.*, 1992) should allow more precise determination of the mode of action of strains used for biocontrol.

The capacity to produce wall-degrading enzymes is not unusual to parasitic fungi and is common among purely saprotrophic species. For example, more than 90% of 160 fungi examined by Chesters and Bull (1963) were capable of producing β-1,3-glucanase. *Zygorhynchus moelleri*, a common soil fungus, is reported to antagonize a range of other soil fungi through production of β-1,3-exoglucanases and endoglucanases (Brown, 1987). This phenomenon seems likely to be important in competitive situations, but has no role in mycoparasitism unless some nutritional benefit is derived by the antagonistic fungus. Enzyme production can be induced *in vitro* by the respective substrate, and *Aphanocladium album*, for example, produces two endochitinases and seven exochitinases when grown on a colloidal chitin preparation (Studer *et al.*, 1992). The primary role of hyphal wall-degrading enzymes in contact mycoparasitism remains in doubt. They are certainly released in many fungus–fungus interactions, but whether this is at a later stage and is merely a saprotrophic activity associated with the breakdown of a dead organic polymer is difficult to determine. It has been shown that the production of these enzymes increases in the presence of the host (Pachenari and Dix, 1980; Elad *et al.*, 1982), but whether enzyme activity is the primary cause of death of the host is not known. The interaction of the rose powdery mildew fungus *Sphaerotheca pannosa* and the antagonist *Stephanoascus flocculosus* (anamorph=*Sporothrix flocculosa*) has been studied using chitin-specific lectins (Hajlaoui *et al.*, 1992). Chitinase activity was thought less important in antagonism than antibiosis, which resulted in cytoplasmic degeneration prior to hyphal contact. In the invasive necrotrophs, on the other hand, hyphal wall-degrading enzymes have a primary role in mycoparasitism as they facilitate the penetration of host hyphae by the mycoparasite (see section 3.2.2).

Considerable evidence has been accumulated for the role of non-enzymic antifungal metabolites in the antagonistic interactions of fungi. That the toxic effects are important in mycoparasitism is an important consideration, and is paralleled in those situations where phytotoxins have been shown to be a primary determinant of disease in some plant-pathogenic associations. The ecological significance of antibiotic production, and its relevance to competitive interactions of fungi remains obscure. Direct evidence that fungi such as *Gliocladium* and *Trichoderma* inactivate or destroy the hyphae of other fungi by antibiosis is difficult to obtain. Indirect evidence from *in vitro* studies, however, suggests that these fungi produce metabolites that affect membrane permeability. Antibiosis was shown to be the dominant factor in the antagonistic activities of *Gliocladium virens* against *Rhizoctonia solani* (Aluko and Hering, 1970). Culture filtrates of the antagonist contained two inhibitory compounds, gliotoxin and viridin, and the role of direct parasitism of hyphae of *R. solani* appeared insignificant in the antagonistic process. This finding is also in agreement with the view of Howell (1987) described earlier, where mycoparasitism was not considered of prime importance in biocontrol of *R. solani* by *G. virens*. Dennis and Webster (1971a,b,c) showed that several antibiotic metabolites inhibit the mycelial growth of a variety of fungi. Some of the antibiotics are volatile and others are not. The isolates that were most effective were associated with a strong 'coconut' aroma and, in the case of the two strains of *T. harzianum*, the major volatile metabolite has been identified as 6-n-pentyl-2H-pyran-2-one or 6PP (Claydon *et al.*, 1987). This alkyl pyrone has potent inhibitory properties to a wide range of fungi. In a review of the antifungal antibiotics produced by *Trichoderma* species, Ghisalberti and Sivasithamparam (1991) have noted that at least four separate strains of *T. harzianum* can be recognized on the basis of the antifungal metabolites that they produce. Two of these produce 6PP, but they are clearly distinguished by the nature of their other metabolites, such as a pyridone compound active against *Botrytis cinerea* and *Rhizoctonia solani* (Dickinson *et al.*, 1989). The other two strains produce, respectively, anthraquinones and butenolides (Ghisalberti and Sivasithamparam, 1991). *Trichoderma harzianum* is known to coil around and penetrate the hyphae of other fungi, and the production of volatile antibiotics may be an important first stage in the process of mycoparasitism. *Trichoderma* species have also been reported to produce unidentified water-soluble antibiotics specifically effective against *Neolentinus lepideus* (= *Lentinus lepideus*) (Bruce *et al.*, 1984; see also section 7.4), but it is not known whether volatile metabolites are also produced and are alone likely to be responsible for the observed inhibition and lysis of the mycelium. Another compound produced by *T. harzianum* has recently been identified as 3-(2-hydroxypropyl)-4-(2-hexadienyl)-2(5H)-furanone (Ordentlich *et al.*, 1992). This compound was the one produced in largest amounts of three compounds secreted into the growth medium of liquid cultures of *T. harzianum* which showed inhibitory activity against *Fusarium oxysporum*. Reports of the production of antifungal metabolites by *Trichoderma* have been confused by taxonomic difficulties in relation to *Gliocladium*. For example, *G. virens* is morphologically similar to *Trichoderma* and has been wrongly identified in the past as *G. fimbriatum* (= *Myrothecium verrucaria*) and *T. viride* (= *T. lignorum*). Many of these isolates produce secondary metabolites that have biological activity *in vitro* against plant pathogens, and that are likely to be involved to some degree in the mechanism of action of antogonism *in vivo*. Additionally, a mode of action involving hyphal penetration has also been suggested. However, it is often difficult to determine whether the antifungal metabolite is an important component in the process of mycoparasitism.

The induction of cytoplasmic leakage from an antagonized host could be a crucial event in the acquisition of nutrients by a mycoparasite. Cell membrane damage caused by an extracellular diffusate was originally implicated in the biocontrol of plant pathogens by Weindling (1934). He suggested that a diffusible 'toxic principle', later identified as gliotoxin (Webster and Lomas, 1964), produced in culture filtrates of *Trichoderma viride* causes vacuolation and coagulation of cytoplasm in *Rhizoctonia* and *Pythium* hyphae with only slight damage to the hyphal wall. More recently, the diketopiperazine antibiotic gliovirin, produced by *Gliocladium virens*, was shown to cause disintegration and coagulation of the cytoplasm of *Pythium ultimum* (Howell and Stipanovic, 1983).

Gliotoxin and gliovirin are non-enzymic metabolites of low molecular weight (<1 kDa). Gliotoxin, also produced by *Gliocladium virens*, prevents nutrient uptake in *R. solani* and *P. ultimum* by selective binding to the thiol groups of the plasmalemma. It appears to be responsible for causing cytoplasmic leakage in *R. solani* resulting in the loss of soluble protein, amino acids, carbohydrates and salts from the hyphae (Lewis *et al.*, 1991). Water extracts of actively growing hyphae of the biocontrol agent *G. virens* G-21 from wheat bran cultures contain gliotoxin. In culture there is a direct linear relationship between an increase in gliotoxin concentration and increased cytoplasmic leakage by *R. solani*. Gliotoxin levels in soil have been correlated with disease suppression towards *R. solani* and *Pythium ultimum* in non-sterile growing media (Lumsden *et al.*, 1992). In addition to gliotoxin, *Gliocladium* species also produce 1.5–2.1 kDa antibiotic-like polypeptides such as alamethicin, paracelsin and trichotoxin (Brückner and Przybylski, 1984) but these apparently do not cause cytoplasmic leakage (Lewis *et al.*, 1991). Nevertheless, a second unidentified factor produced by *G. virens* may also be involved in inducing leakage. It is of higher molecular weight (>10 kDa) and may be an enzyme. Isolates of *G. virens* have been shown to produce the hydrolytic enzymes β-1,3-glucanase, β-1,4-glucanase, chitinase and protease (Erbeznik *et al.*, 1986; Roberts and Lumsden, 1990) all of which could act on cell wall integrity and induce cytoplasmic leakage from *R. solani*.

Deoxyphomenone is a fungistatic sesquiterpene produced by the hyphomycete *Dicyma pulvinata* (= *Hansfordia pulvinata*), a destructive mycoparasite of *Fulvia fulva* (= *Cladosporium fulvum*), which is itself a leaf parasite of tomato plants (Tirilly *et al.*, 1991). Maximum production of this antibiotic occurs in the young parts of hyphae in zones where contact between the mycoparasite and its host occur. Experiments using cellophane sheets were used by Gandy (1979) to demonstrate the involvement of an extracellular diffusate, possibly a cephalosporin, in the inhibition of growth of the mycelium of *Mycogone perniciosa* by *Acremonium strictum*.

3.2 Invasive Necrotrophs

3.2.1 Characteristics of invasive necrotrophs

Invasive necrotrophs always penetrate the hyphae of their host fungi. There is often extensive lysis of the hyphal wall of the host and the cytoplasm is disrupted and killed as the penetration process occurs. Some invasive necrotrophs invade any of the host structures that they attack, but most are specialized and colonize either vegetative hyphae, sclerotia or spores of their respective hosts. Examples of each are given in Table 3.1.

As with the contact necrotrophs, there are several examples of invasive necrotrophs with a wide host range. For example, *Schizophyllum commune*, a bracket fungus commonly found growing on the stumps and felled trunks of hardwood trees, parasitizes the nematode-destroying fungus *Arthrobotrys oligospora*, the plant-pathogen *Rhizoctonia solani*, and saprotrophic species of the Mucorales commonly found in soil such as *Cunninghamella*, *Rhizopus* and *Zygorhynchus*. Hyphae of *S. commune* coil around the hyphae and sporulation apparatus and penetrate them either by means of unspecialized hyphae or from penetration pegs developed from terminal, adhesive swellings termed appressoria. The parasitized structures tend to collapse through lysis of the cytoplasm (Tzean and Estey, 1978b) and there is no evidence for the production of extracellular diffusates which act in advance of the hyphal front.

Invasion of the host cell appears to involve a recognition event and the differentiation of an appressorium at an early stage of the infection process. Hyphal

Table 3.1. Examples of invasive necrotrophs.

Structure attacked	Parasite/host
Hyphae	*Nectria inventa*/*Alternaria brassicae* *Pythium acanthicum*/*Phycomyces blakesleeanus* *Rhizoctonia solani*/Mucorales *Syncephalis californica*/*Rhizopus oryzae*
Sclerotia	*Coniothyrium minitans*/*Sclerotinia sclerotiorum* *Talaromyces flavus*/*Sclerotinia sclerotiorum*
Spores	*Anguillospora* sp./*Glomus deserticola* (chlamydospores) *Cladosporium uredinicola*/*Puccinia violae* (uredospores) *Eudarluca caricis*/*Puccinia graminis* (uredospores) *Fusarium merismoides*/*Pythium ultimum* (oospores) *Humicola fuscoatra*/*Phytophthora megasperma* (oospores) *Mycogone perniciosa*/*Rhopalomyces elegans* (conidia) *Nectria inventa*/*Alternaria brassicae* (conidia)

contact of the host and parasite seems usually to be a random event, but sometimes directed growth, or the stimulation of hyphal branching of the mycoparasite is induced by the presence of the host fungus as in the interaction of *Trichoderma hamatum* and *Rhizoctonia solani* (Chet and Baker, 1981). There are several accounts of the fine structure of the processes associated with the penetration of host hyphae by mycoparasites and the details of penetration and cytoplasmic reaction are broadly similar in each case (e.g. *Gliocladium virens*/*Rhizoctonia solani*, Tu and Vaartaja, 1981; *Trichoderma* spp./*Corticium rolfsii*/*Rhizoctonia solani*, Elad *et al.*, 1983a,b; *Corticium rolfsii*/*Trichoderma harzianum*, Ferrata and D'Ambra, 1985). The wall of the invaded hypha appears to be breached by a combination of enzymatic hydrolysis and mechanical pressure, and may show evidence of papilla formation during the initial stages of attack. Once the invasive hypha breaches the wall, the cytoplasm of the host becomes disorganized and finally appears necrotic.

A typical sequence of events is illustrated when hyphae of *Gliocladium virens* contact those of *Rhizoctonia solani* (Tu and Vaartaja, 1981). Hyphal tips of the mycoparasite become swollen, forming small appressoria; this occurs either directly on contact, or sometimes only after the hyphae have grown for some distance in a zig-zag pattern along the surface of the host hypha. The attacked hyphae lose turgidity and collapse. Intrahyphal penetration was demonstrated by ultrastructural studies of parasitized sclerotia where mycelia of *G. virens* were found in the sclerotial cells of *R. solani*.

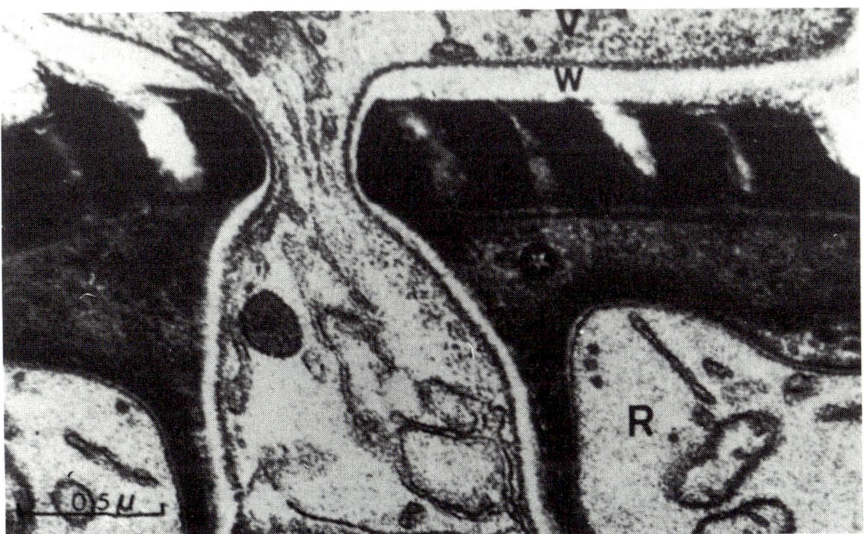

Fig. 3.3. (see also Fig. 3.4) Hypha of *Rhizoctonia solani* infested by *Verticillium biguttatum*. Penetration point where a hypha of *V. biguttatum* (W) has entered a hypha (R) of *R. solani* (TEM: × 3000). (Micrograph courtesy of Dr P.H.J.F. van den Boogert.)

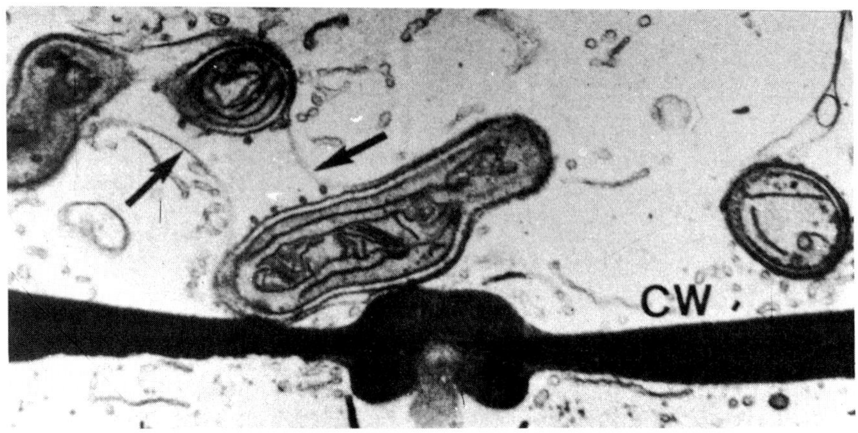

Fig. 3.4. (see also Fig. 3.3) Hypha of *Rhizoctonia solani* infected by *Verticillium biguttatum*. Intracellular growth of hyphae (arrows) of *V. biguttatum* inside a hypha of *R. solani*. A septum of the host fungus is visible (CW) (TEM: × 15,000). (Micrograph courtesy of Dr. P.H.J.F. van den Boogert.)

Attachment of the antagonistic hypha to the surface of the host may involve lectin binding. *Corticium rolfsii*, for example, produces a lectin (Barak *et al.*, 1985) as a component of the extracellular polysaccharide on the hypha (Barak and Chet, 1990). The ability of different mycoparasitic *Trichoderma* isolates to attack *C. rolfsii* is correlated with the capacity of *C. rolfsii* to agglutinate conidia of *Trichoderma*. The lectin binds with β-1,3-glucan in the wall of *C. rolfsii*. The purified lectin subjected to isozyme analysis using gel electrophoresis comprises two protein bands with molecular weights of 55 and 60 kDa (Barak and Chet, 1990). It has been suggested that the lectin might be involved in the recognition step which leads to the further successive events in mycoparasitism. Lectins also stimulate coiling of hyphae of *T. harzianum* around inert surfaces. The incidence of coiling around nylon fibres treated with concanavalin A or a lectin from *C. rolfsii* was significantly higher than that around untreated fibres (Inbar and Chet, 1992) further indicating the role of lectins in initiation of fungal interactions.

Some mycoparasitic interactions do not involve appressorium formation. In a range of mycoparasites which infect hyphae of *Rhizoctonia solani*, for example, the penetration phase is achieved by other means (Turhan, 1990). In *Verticillium biguttatum*, penetration of the host occurs directly via a fine infection tube which is believed to breach the wall through the action of hydrolytic enzymes (Boogert *et al.*, 1989). The evidence for enzymic degradation of the wall of the hypha of the host is based upon transmission electron micrographs which show an apparent zone of lysis around the fine penetration peg. The perforation in the wall due to the entrance of the parasite also appears circular and lacks the ragged edge which might be expected if the penetration were due to mechanical pressure (Fig. 3.3). Electron-dense papilla material develops between the plasmalemma and the wall of the host in the region of penetration. After penetration, the hypha expands, depresses the plasmalemma, and may grow and branch to form the so-called trophic hyphae. Later, the plasmalemma is disrupted and the cytoplasm degenerates and dies (Fig. 3.4). Investigations using light and scanning electron microscopy demonstrate the penetration of hyphae of *Rhizoctonia solani* by a wide range of mycoparasites (Turhan, 1990). The infection process of *Botryotrichum piluliferum*, *Coniothyrium sporulosum*, *Dicyma olivacea* (= *Ascotricha erinacea*), *Gliocladium catenulatum*, *Stachybotrys chartarum*, *S. elegans*, *Stachylidium bicolor*, *Trichothecium roseum*, *Verticillium chlamydosporium* and *V. tenerum* (= *Nectria inventa*) is similar in each species. Loose to profuse coiling of the parasitic hyphae around those of the host is generally observed. Cytoplasm of the host hyphae becomes granulated and disintegrates as penetration occurs. The walls of the parasitized hyphae appear to persist even when most of the contents have been utilized (Turhan, 1990).

Some destructive mycoparasites attack different structures in a range of hosts. *Geotrichopsis mycoparasitica*, the hyphomycete anamorph of a basidiomycete, is a wide spectrum necrotrophic mycoparasite capable of attacking a range of fungi, although of 50 potential host fungi tested no members of the Ascomycota (*Byssochlamys fulva*, *Emericella variecolor*, *Sordaria fimicola*) or the Basidiomycota (*Heterobasidium annosum*, *Ischnoderma resinosum* (= *Polyporus resinosus*)) were infected (Tzean and Estey, 1991, 1992). Tzean and Estey (1992) have made a light and electron microscope study of this fungus. Hyphae of *G. mycoparasitica* coil around those of the host and although penetration may occur it has not been determined whether biotrophy, in addition to necrotrophy, is involved in the mycoparasitic reaction. Potential host susceptibility varied from complete immunity to highly susceptible resulting in overgrowth and death of the host colony. Any part of a susceptible host including the mycelium and sporulation structures could apparently be parasitized. Highly susceptible hosts were the nematode-trapping fungi in cultures of which the mycoparasite was first observed (Tzean and Estey, 1991) and they included *Arthrobotrys dactyloides*, *A. oligospora*, *A. pyriformis* and *A. superba*. Other heavily parasitized hosts included *Cunninghamella elegans*, *Fusarium oxysporum* f.sp. *lycopersici*, *Mortierella hygrophila*, *Papulospora dodgei*, *Pythium dissotocum*, *P. monospermum*, *Rhizopus stolonifer* and *Thielaviopsis paradoxa*, fungi from a range of taxonomic groups. Species of *Curvularia* isolated from surface-disinfected *Sorghum* grains have been shown to parasitize and sporulate within the mycelia and sporulating structures of the zygomycetes *Absidia glauca* (both + and − strains), *Rhizopus arrhizus*, *R. sexualis*, *R. stolonifer* and *Syncephalastrum racemosum* (El Shafie and Webster, 1979). All species of *Curvularia* tested, i.e. *C. andropogonis*, *C. cymbopogonis*, *C. intermedia*, *C. lunata*, *C. leonensis*, *C. pallescens*, and in addition, *Alternaria tenuis* (= *Alternaria alternata*) and *Drechslera spicifera* were very parasitic on *R. arrhizus*. On the other hand, *C. pallescens*, a seed-borne plant pathogen, did not infect *Mucor hiemalis*, *Micromucor* (= *Mortierella ramanniana*) *ramannianus* or *Zygorhynchus moelleri*. The mechanism of parasitism was the same irrespective of the host. Vegetative, rhizoidal hyphae, aerial sporangiophores and sporangia of the hosts were penetrated and hyphal coiling often occurred. Conidiophores sporulated on the surface of the sporangia. After utilization of most of the cytoplasm of the host, chlamydospores were formed within the host elements. Although the host cultures were damaged they were not killed as it was possible to transfer spores and hyphae from parasitized hosts in order to establish fresh cultures.

It is not always clear whether a particular fungal interaction is invasive or merely involves contact. Coiling of antagonistic hyphae around those of the host can obscure cytological details. For example, investigation of the interaction of various antagonists of the root-infecting pathogen *Rhizoctonia solani* shows that hyphae of a species of *Chaetomium* frequently coil around and penetrate those of *R. solani* and may grow within them. Granulation of the cytoplasm is recorded in about 50% of the observations but the hyphae are never lysed. With *Trichoderma harzianum*, on the other hand, coiling around the hyphae of *R. solani* occurs in about 20% of the interactions and penetration in 10%, whilst intrahyphal growth rarely happens. Granulation and vacuolation of the cytoplasm always occur, however, and about 25% of the hyphae of *R. solani* are lysed (Chand and Logan, 1984). The hyphae of *Chaetomium* and *T. harzianum* are seen to coil tightly round those of *R. solani* prior to penetration. Coiling is a response both to the physical contact between hyphae and directed growth through the influence of chemicals. Wall penetration is again thought probably to be due to a combination of physical pressure and the action of hydrolytic enzymes. Alterations in the state of the cytoplasm are likely to be brought about by the action of antibiotics, enzymes and other metabolites produced by the aggressive fungus.

Study of the interaction of the mycelium of *Verticillium* species with *Rhizoctonia solani* shows that appressed growth, coiling or penetration of the hyphae of *R. solani* occur (Kuter, 1984; Boogert *et al.*, 1989). Appression of the hyphae of *Verticillium* occurs more frequently than either coiling or penetration and may be seen on host hyphae either containing or devoid of cytoplasm. Close coiling is characteristic of interacting aerial hyphae whereas lax coiling occurs around submerged hyphae. The reaction of the hyphae of different species of *Verticillium* to those of *R. solani* varies, showing some species specificity. For example, hyphae of *V. tenerum* can penetrate those of *R. solani* directly without first coiling around them whereas those *V. lecanii* typically show appressed growth and loose coiling. Hyphal penetration is considered to be evidence that the antagonist behaves as a mycoparasite but it is not unequivocal, however, as it is also necessary to demonstrate the uptake of nutrients from the cytoplasm of the host. Nutrient transfer from the host to the parasite is difficult to demonstrate and requires, for example, the use of radioactive tracers in carefully controlled conditions. Hyphal coiling may be a sign of prolonged encounter and resistance on the

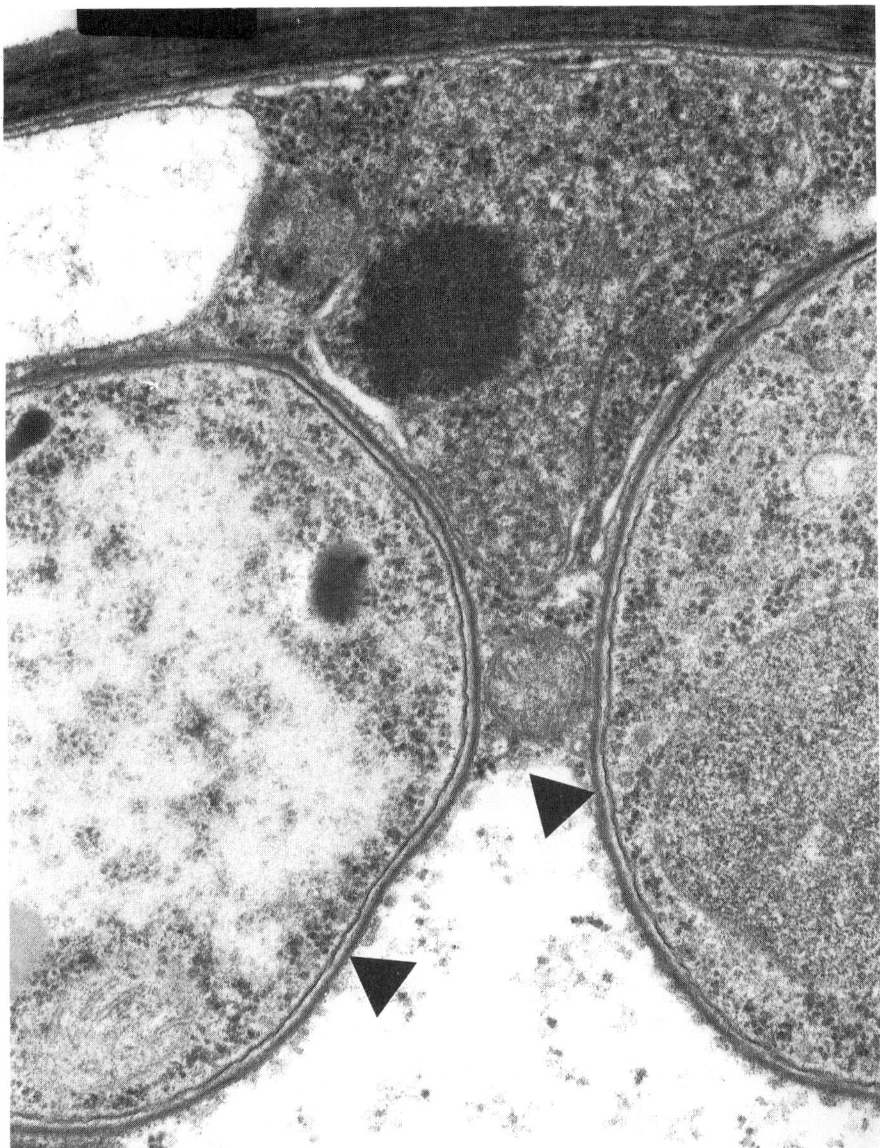

Fig. 3.5. (see also Fig. 3.6) Hyphae of *Syncephalis nodosa* growing inside a hypha of *Cokeromyces recurvatus*. An early stage of infection showing hyphae of *S. nodosa* in direct contact with the invaginated plasmalemma (arrow) of the host (TEM: × 7500).

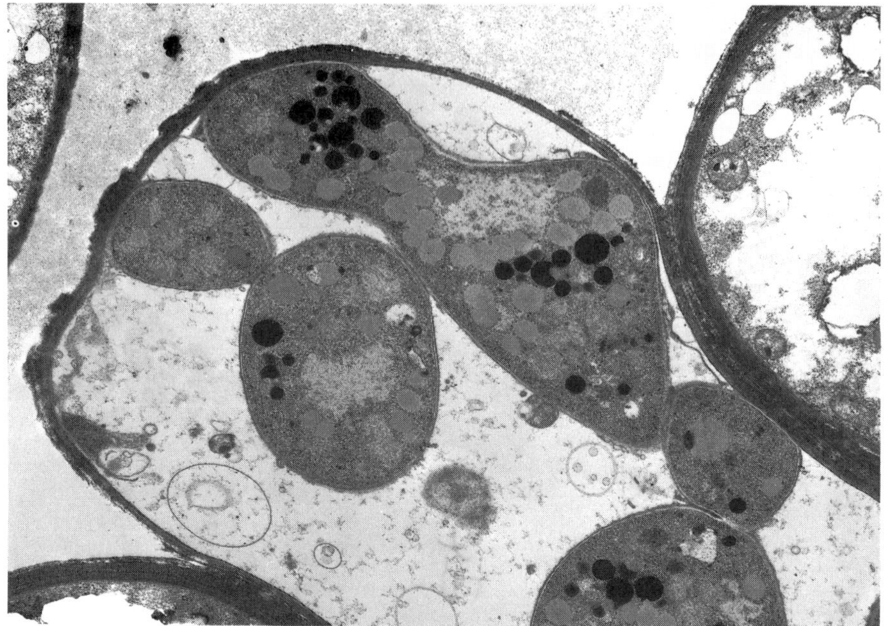

Fig. 3.6. (see also Fig. 3.5) Hyphae of *Syncephalis nodosa* growing inside a hypha of *Cokeromyces recurvatus*. A late stage of infection showing lipid-rich hyphae inside a degenerate hypha of the host (TEM: × 30,000).

part of the potential host (Deacon, 1976; Pachenari and Dix, 1980) and could be a response to chemicals produced by the host. With *V. biguttatum* (Boogert *et al.*, 1989) the growth of *R. solani* ceases after contact and the mycelium of *V. biguttatum* overgrows that of *R. solani.* If discs of agar are excised from the zone of overgrowth and plated out, only *Verticillium* is recovered which indicates that *V. biguttatum* is a potent antagonist.

3.2.2 Parasitism of vegetative hyphae

There are many examples of necrotrophic interactions of the mycelia of fungi from various taxonomic groups. Certain Zygomycota, for example species of *Syncephalis*, display invasive necrotrophic characteristics. Although *Syncephalis* species are classified in Piptocephalidaceae with *Piptocephalis*, a genus of haustorial biotrophic mycoparasites, the mode of infection differs. The initial stages in the process of infection include the penetration of the hypha of the host from an appressorium, similar to *Piptocephalis.* However, once inside the cytoplasm of the host, the penetration peg does not differentiate into a lobed haustorium but the hypha expands and apical growth is resumed. The developing internal hyphae (Figs 3.5 and 3.6) frequently branch in colonization of the mycelium of the host. No extrahaustorial matrix develops (see section 4.1) and the hyphae appear to be free within the cytoplasm of the host which rapidly degenerates in their vicinity. Material may thus be transferred between partners through the intact hyphal wall and plasmalemma of the parasite (Fig. 3.6). It is rare in dual cultures in the laboratory for the mycoparasite to kill the mycelium of the host completely even though severe inhibition of the growth and sporulation of the host is evident (Benjamin, 1959; Jeffries, 1984). Nevertheless, it is probable that the basic nutritional requirements of *Syncephalis* are provided by absorption from the necrotic cytoplasm of the host. Sometimes, infection due to *Syncephalis* is responsible for growth abnormalities in the hyphae of the host, and similar morphological changes have also been recorded in field samples. Hyphal swellings in *Rhizopus oryzae* have been used in this way to infer mycoparasitic activity of *S. californica* in both naturally infested and artificially infested agricultural soils (Hunter and Butler, 1975; Hunter *et al.*, 1977). The host range of *Syncephalis sphaerica* has been studied by Baker *et al.* (1977) who found that 20 genera of Mucorales supported growth and sporulation of the parasite. An ascomycetaceous yeast and two deuteromycetes also supported growth. *Syncephalis* species are common in soils and can sometimes be recognized in isolation plates by the clasp-like hyphal enlargements of the base of the erect sporangiophore which surround and firmly adhere to the hyphae of the host.

One of the most studied mycoparasitic interactions is that between *Pythium* species and a wide range of host fungi, including many of the non-mycoparasitic plant-pathogenic species of the same genus. Mycoparasitic species include *P. acanthicum*, *P. oligandrum*, *P. nunn*, *P. periplocum*, and *P. mycoparasiticum* (Deacon *et al.*, 1991). *Pythium oligandrum*, for example, has been widely studied as it is commonly isolated from agricultural soils and is known to be an aggressive parasite of other fungi, including important plant-pathogenic fungi (Deacon, 1976; Deacon and Henry, 1978). Hosts differ in susceptibility: *Pythium graminicola* and *P. vexans* are highly resistant, *Trichoderma aureoviride* and *Fusarium oxysporum* are susceptible, and *Botrytis cinerea*, *Fusarium culmorum*, *Rhizoctonia solani* and *Botryotrichum piluliferum* show intermediate responses (Laing and Deacon, 1990). Each host fungus, itself, shows varying degrees of susceptibility, young hyphae being more susceptible than older hyphae. The hyphae of *P. oligandrum* may penetrate those of the host and grow in an intrahyphal fashion but sometimes penetration does not occur. The hyphal interaction of *Pythium oligandrum* with a range of fungal hosts studied on cellulose film overlaying 2% distilled water agar shows that the reaction occurs at markedly different rates in different combinations of species pairs and may involve lysis of the hyphae or granulation of the cytoplasm (Lewis *et al.*, 1989). Of 14 fungi tested, the hyphae of seven, including the important plant pathogens *Botrytis cinerea* and *Sclerotinia sclerotiorum*, were rapidly lysed. Fast lysis involves total degeneration of the cytoplasm of the host and liberation of the contents of the parasitized hypha within seconds of penetration. In contrast, rapid granulation but not rapid lysis of the cytoplasm occurs with *Rhizoctonia solani* some minutes after hyphal contact. The cytoplasm loses opacity and the antagonized hypha ceases to grow, yet lysis rarely occurs. In a third type of hyphal interaction, found to occur only when three species of *Pythium* were the hosts, the reaction is slow as the cytoplasm loses opacity and degenerates from 1–8 hours after contact with the hyphae of *P. oligandrum*. The hyphae of *P. oligandrum* were found subsequently to grow throughout the mycelium of the host, irrespective of the type of hyphal interaction involved. Videomicrography has recently been used to record in detail the interactions of three mycoparasitic *Pythium* species, *P. acanthicum*, *P. nunn* and *P. oligandrum* and a range of hosts grown on water agar (Laing and Deacon, 1991). The mycoparasites have an identical mode of parasitism. There was no precontact inhibition of the host hyphae, nor was directed growth of mycoparasitic hyphae observed. Host hyphae were contacted at random, and many 'near-misses' were recorded. Susceptible hyphae stopped extending soon after contact (253 s on average after contact between host tips and subapical regions of parasitic hyphae; 439 s for contacts between parasitic tips and subapical regions of the host). Cessation of growth was followed by host lysis at the point of contact (mean 324 s postcontact) or by coagulation and vacuolation of the hyphal contents (mean 568 s postcontact). Penetration of the host hypha (mean 594 s postcontact) occurred in 47% of the interactions from a branch formed behind the apex of the hypha which made the initial contact. *Pythium nunn* was much less aggressive than the other two species as it antagonized fewer hosts, the antagonism was less consistent and less rapid, and growth within host hyphae was slower. Lifshitz *et al.* (1984a) also reported that mycoparasitism by *P. nunn* could be divided into two types, a 'quick reaction' and a 'slow reaction'. In the former, host hyphae were extensively coiled and subsequently lysed, but in the latter case *P. nunn* penetrated the hyphae and slowly grew within them. These two reactions, though, are different from those reported by Laing and Deacon (1991), as even in the 'quick reaction' it took several hours for the cytoplasm to disappear or for hyphae to lyse. Thus it is evident that a range of interactions occurs between mycoparasitic species of *Pythium* and their hosts and the relative timing of events is important in comparing species. Deacon and Laing (1991) have argued against the involvement of parasite-derived wall-degrading enzymes in causing hyphal lysis. Their evidence was based on three points. First, *P. nunn* was much less aggressive than *P. oligandrum*, yet is reported (Elad *et al.*, 1985) to produce more inducible wall-degrading enzymes in culture. Second, mucilage was involved in determining susceptibility or resistance. Third, as lysis occurs as quickly as 55 s after contact it seems unlikely that the release and activity of inducible enzymes could occur in so short a time. In view of the rapidity with which hyphal branching or swelling can be induced in actively growing mycelia however, a process which may simply require the release of preformed lytic enzymes from inactive zymogenic forms constitutively present within the wall matrix, the latter point may require re-evaluation. It was suggested that the findings might be more compatible with the involvement of host-derived wall-lytic enzymes in the early stages of mycoparasitism. This view has also been expressed by Adams and Ayers (1983) to explain how *Sporidesmium sclerotiovorum* induces breakdown of the sclerotia of its hosts, and also to account for the widespread general phenomenon of **mycolysis** in which fungal mycelia are known to lyse when placed in unsterile soil (Lloyd and Lockwood, 1966).

The role of wall-degrading enzymes in other instances of invasive mycoparasitism seems clearer. For example, enzyme involvement has been shown to be important during mycoparasitism by *Trichoderma viride.* It seems that the process of infection of *Rhizoctonia solani* by *T. viride* occurs in several stages, probably commencing with the action of diffusible and probably volatile antibiotics in advance of the mycelium, followed by physical contact of and chemical attraction by the potential host. Enzymic breaching of the wall of the host may involve the secretion of β-1,3-glucanase, chitinase and protease as these and possibly other extracellular enzymes are induced in a mycoparasitic strain of *T. viride* when grown on the isolated hyphal walls of *R. solani* (Ridout *et al.*, 1988). These enzymes remain localized in their digestive action which results in the development of circular holes in the wall of the host where hyphae of the antagonist penetrate. Infrared photography of the interacting regions of the hyphae of *T. harzianum* and *R. solani* show bright regions of infrared irradiation which could indicate intense parasitic activity due to localized extracellular enzyme action (Elad *et al.*, 1983c). Low levels of chitinase activity occur in cultures of either fungus alone (Elad *et al.*, 1983b) indicating that hyphal interaction stimulates the production of this hydrolytic enzyme. Chitinase activity has been implicated in the colonization of the wheat rust *Puccinia graminis* f.sp. *tritici* by the mycoparasite *Aphanocladium album* (Srivastava *et al.*, 1985a; Studer *et al.*, 1992). In these examples, the mycoparasite is stimulated to produce chitinase when grown axenically on a medium containing chitin and a chitinase enzyme has recently been purified and characterized (Kunz *et al.*, 1992). Other unidentified enzymes which apparently enhance the activity of chitinase have also been shown to be produced by *A. album.* The wall of the germ-tube of the rust attacked by the mycoparasite is lysed. Although the lysis could be due to the release of enzymes by degenerating cytoplasm of the host which autolyse the wall, it has been suggested that chitinase production by the mycoparasite is an important factor, but not the only one, in wall lysis. Purified wall material has also been used to implicate enzymes in interactions of *Pythium*, and when *P. nunn* is grown in the presence of wall material of the potential host *Rhizoctonia solani*, the production of the wall-degrading enzymes chitinase and β-1,3-glucanase is stimulated and this phenomenon appears to be an adaptation to mycoparasitism because fungi which are not capable of being parasitized by *P. nunn* do not stimulate the secretion of wall-digesting enzymes (Elad *et al.*, 1985).

3.2.3 Parasitism of spores

A range of spore types, both sexual (basidiospores, oospores and zygospores), and asexual (sporangiospores and conidia) can also be parasitized by specific mycoparasites. For example, the arthroconidia of the basidiomycetous *Dacrymyces stillatus* are parasitized by an **epibiotic chytrid**, possibly referable to *Rhizophlyctis* (Canter and Ingold, 1984). The *Verticillium* anamorph of *Nectria inventa* is a necrotrophic parasite of *Alternaria brassicae*, the causal agent of leaf spot of cabbage and cauliflower plants (Tsuneda and Skoropad, 1977, 1978a). When spores of the host fungus are attacked, the invading hyphae develop appressoria and fibrous material, which is probably adhesive, is produced in the region of contact. Fibrous material is also deposited at the region of contact in several other parasite–host combinations including *Ampelomyces quisqualis*/*Sphaerotheca fuliginea* (Sundheim and Krekling, 1982) and *Eudarluca caricis*/*Puccinia graminis* (Carling *et al.*, 1976). The mature, septate conidia of *A. brassicae* tend to be penetrated where septa occur (Fig. 3.7; Tsuneda and Skoropad, 1977) whereas young conidia are usually penetrated at the basal pore (Fig. 3.8). Conidia of *A. brassicae* immersed in distilled water tend to leak nutrients primarily from the septal and basal pore regions and this has been suggested as a possible explanation of the preferential adhesion of the hyphae of *N. inventa* to these zones as a result of directed growth towards the source of leakage (Tsuneda and Skoropad, 1978b). Short branches of the hyphae of the parasite growing close (within 15 μm) to the elongate conidia turn towards the conidia and penetration occurs primarily at either the septa or the basal pore of the conidium. *Nectria inventa* also parasitizes a wide range of other fungi (Tsuneda and Skoropad, 1980), attacking the vegetative hyphae in some species as well as the spores. The degree of susceptibility of host fungi varies with differences in genera, species, the kind and age of fungal structures, and the culture medium used. Profuse directed growth appears to be a prerequisite for vigorous parasitism but sometimes the hyphae are not penetrated although the cytoplasm coagulates and becomes vacuolate in a reaction more closely resembling hyphal interference (see section 3.2.1).

The spores of rust fungi are also infected by a variety of mycoparasites. For example, an invasive necrotrophic reaction occurs in parasitism of the uredospores of *Puccinia violae*, a pathogen of *Viola odorata*, by *Cladosporium uredinicola* (Traquair *et al.*, 1984) resulting in death of the spores soon after the reaction begins. In this interaction, the hypha of *Cladosporium* penetrates the wall of the uredospore although necrotic action in advance of the hypha occurs because the cytoplasm coagulates and disintegrates before physical contact. Necrosis is not restricted to the spores in this fungus as the mycelium is also sometimes infected. Several fungi have been reported as secondary colonizers of rust pustules or galls. A typical example is *Cladosporium gallicola* which parasitizes the spores of *Endocronartium harknessii* (western gall rust) and the aeciospores of *Cronartium comandrae* (Tsuneda and Hiratsuka, 1979). Parasitic hyphae

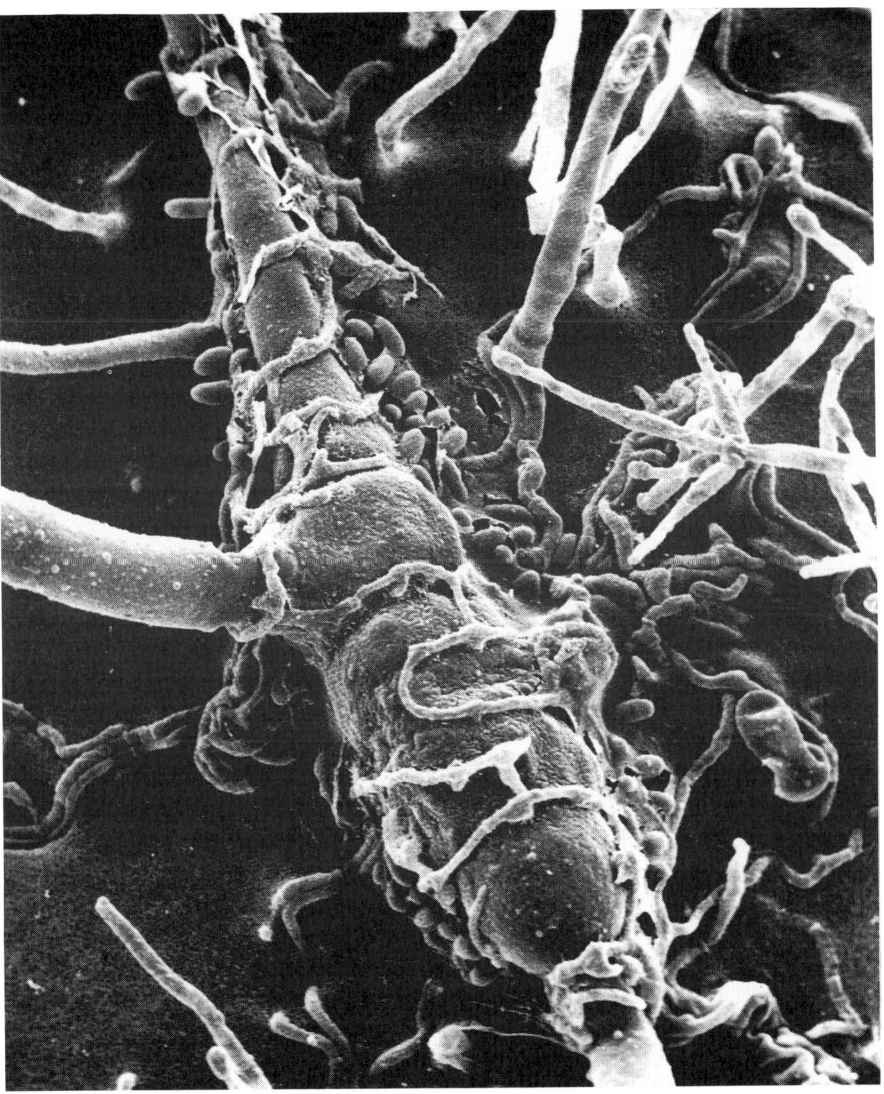

Fig. 3.7. (see also Fig. 3.8) Conidium of *Alternaria brassicae* attacked by hyphae of *Nectria inventa*. Conidium on leaf surface surrounded by hyphae of the mycoparasite (SEM: × 2000). (Micrograph courtesy of Dr A. Tsuneda.)

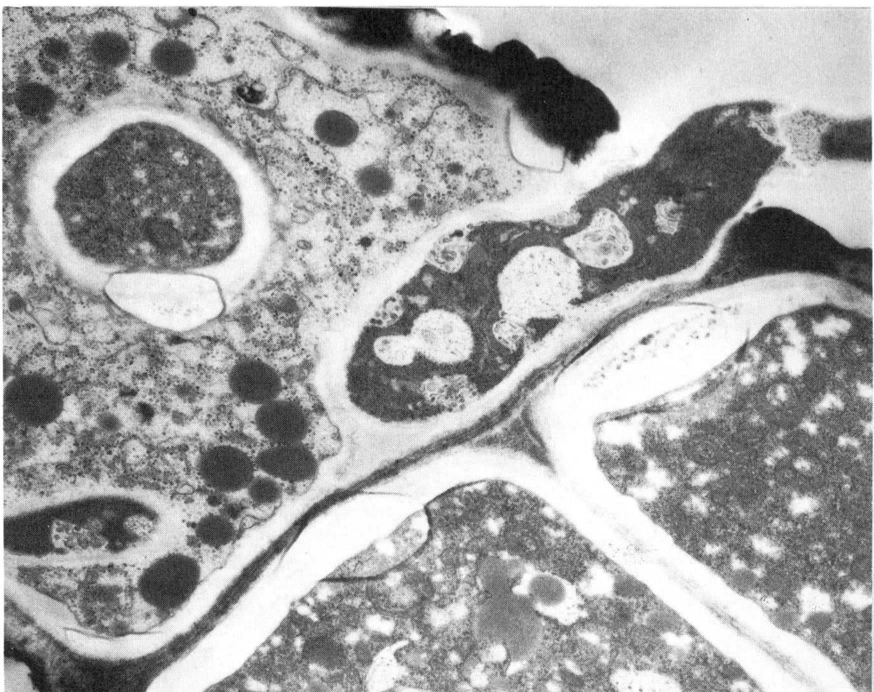

Fig. 3.8. (see also Fig. 3.7) Conidium of *Alternaria brassicae* attacked by hyphae of *Nectria inventa*. Entry point of parasitic hypha into the conidium close to the basal region (TEM: × 7500). (Micrograph courtesy of Dr A. Tsuneda.)

grow towards rust spores and parasitism occurs either by means of contact without penetration or, more usually, by penetration of the rust spores from an appressorium. The warty surface layer of the fungal host often disintegrates after being contacted by the parasitic hyphae and host cytoplasm coagulates and eventually disappears. In the late stages of parasitism, the hyphae of the parasite usually form conidiophores and conidia directly on the remnants of the host spores (Tsuneda and Hiratsuka, 1979). Similar observations have been made for the mycoparasitism of uredospores of *Melampsora larici-populina* (poplar leaf rust) by *Cladosporium tenuissimum* (Sharma and Heather, 1978), which, however, also exhibits antibiosis toward germinating uredospores. Antifungal metabolites have also been implicated in the parasitism of rust spores by *Monocillium nordinii* (Tsuneda and Hiratsuka, 1980). These metabolites can also be detected in cell-free extracts of liquid cultures of *M. nordinii* and have been identified and structurally characterized as monorden, and five related compounds, the monocillins I–V (Ayer *et al.*, 1980). Monorden has also been isolated from *Nectria radicicola* (= *Cylindrocarpon destructans* var. *destructans*) cultures (Mirrington *et al.*, 1964) where it was named radicicol. Penetration of these spores usually occurs after the host cells have been killed.

Eudarluca caricis, the teleomorph of *Darluca* (=*Sphaerellopsis*) *filum*, is a cosmopolitan mycoparasite associated with many hundreds of species of rust fungi (Carling *et al.*, 1976; Stahle and Kranz, 1984). It is most commonly observed as the anamorph, producing clumps of black, shiny, spherical pycnia situated among the spores of uredial sori, where it is presumed to derive nutrients from direct hyphal penetration of uredospores. *Eurdarluca caricis* may also penetrate pycnial, aecial and telial spore stages (Kuhlman *et al.*, 1976). This fungus is important from the point of view of its potential as a biocontrol agent. *Verticillium lecanii* and *Aphanocladium album* also occur frequently as parasites of rusts and colonize sori of *Puccinia* and *Uromyces*. In a comparative study, *V. lecanii* and *A. album* infected 90–95% of the teliospores of *Puccinia horiana*, the major pathogen of *Chrysanthemum* within 5 days of being sprayed onto infected plants. Two other parasites of rusts, *Cladosporium sphaerospermum* and *C. uredinicola* were less effective than *V. lecanii* and *A. album* in infecting the rust species tested. *Verticillium lecanii* is of interest as it also attacks powdery mildews (Erysiphales), aphids and the cysts of certain plant-parasitic nematodes (Hall, 1981). Infection hyphae penetrate teliospores of rusts through the germ-pore and rapidly utilize the spore contents (Srivastava *et al.*, 1985). *Aphanocladium album*, on the other hand, penetrates the intact uredospore wall of *Puccinia graminis* as well as entering the spore via the germ-pore (Koc *et al.*, 1981). On plants inoculated with both rust spores and *Aphanocladium*, the development of uredia is apparently reduced although the formation of telia is stimulated. Immature teliospores have also been reported to be formed within 12–14 days after *A. album* was added to axenic cultures of *P. graminis* but the mechanism of this induction is unknown (Yaniv *et al.*, 1979). Several other examples of mycoparasites which attack rust spores, or grow on rust pustules, are described in section 6.4.

In the Oomycota and Zygomycota the sexual spores are often thick-walled, resting structures, full of storage materials and which act as a reservoir of the fungus during conditions adverse to mycelial growth. In plant-pathogenic fungi the resting spores act as the source of inoculum following the period between successive susceptible crops. Being nutrient-rich, such spores are a valuable substrate for any fungus capable of attacking them and penetrating the thick walls. The oospores of plant-pathogenic *Pythium* species have been particularly studied in this respect. Plant-pathogenic pythia are capable of growing vigorously in soils in the saprotrophic phase of the life cycle. In the parasitic phase, the root systems and hypocotyls of seedlings of the host plants are infected. When the infection is so severe that the shoot system of the host plant is prevented from developing above ground, the seedlings are said to succumb to a pre-emergence blight. The stems above ground of infected seedlings tend to collapse at the junction of the stem with the soil, especially when the seedlings are grown in wet soil. *Pythium* species produce their thick-walled oospores as a result of a sexual process within the cells of the host in the parasitic phase of growth. Oospores of some species of *Pythium*, such as *P. myriotylum* and *P. ultimum*, can be invaded by *Hyphochytrium catenoides*, a mycoparasite which is also found in agricultural soils (Ayers and Lumsden, 1977). The oospores of *P. myriotylum* are more highly susceptible to infection by the mycoparasite than those of *P. ultimum.* Infection of the oospore is brought about by the uniflagellate zoospore of the mycoparasite which is capable of movement in water films which surround soil particles. Zoospores, which are released in masses from the sporangia of the mycoparasite, aggregate around the oospores where they encyst and apparently become attached to the wall. A peg which penetrates the thick wall of the oospore develops from the cyst and the contents of the cyst pass into the oospore. Within the parasitized oospore, from two to six thin-walled zoosporangia of *Hyphochytrium* develop, from which exit tubes emerge when the infected oospores are immersed in water. Zoospores released via the exit tubes may infect further oospores, thus repeating the infection cycle. Owing to the widespread occurrence of the mycoparasite in soils and its ability to reproduce rapidly in infected oospores, it has been suggested that *H. catenoides* may be responsible for reducing the natural population of susceptible oospore-producing fungi in nature. It could, therefore, be an important biocontrol agent of *Pythium* species which are themselves serious root-infecting pathogens of a wide variety of crop plants. Mycelial fungi have also been reported as mycoparasites of oospores of *Pythium.* Drechsler (1938, 1943, 1952, 1961, 1963) described the invasion of oospores of *Pythium* by a variety of hyphomycete fungi including *Dactylella* spp., *Trinacrium subtile* and *Trichothecium* spp., and Hoch and Abawi (1979a) described the mycoparasitism of oospores of *P. ultimum* by *Fusarium merismoides.* Other *Fusarium* species have also been reported as oospore mycoparasites. *Fusarium semitectum* (= *F. pallidoroseum*) parasitizes the oospores of *Sclerospora graminicola* that develop within the leafy proliferations in the earhead of pearl millet infected with this plant pathogen (Raghavendra Rao and Pavgi, 1976), and *Fusarium oxysporum* has frequently been isolated from parasitized oospores of *Phytophthora sojae* (Sneh *et al.*, 1980).

Phytophthora is another genus of important plant pathogens, and soil-borne oospores are an important residue of inoculum during the intercrop period. An interesting experimental study in which the oospores of *Phytophthora cinnamomi* and *P. nicotianae* were used as bait in experimental soils has shown that in susceptible oospores a proportion varying from about 5% to over 70% were invaded and colonized by zoospores of the chytridiomycete *Catenaria anguillulae* (Daft and Tsao, 1984). A number of chytridiomycete and hyphomycete fungi described as mycoparasitic on oospores of *Phytophthora* are summarized in Table 3.2. Several of these fungi, such as *Hyphochytrium catenoides* and *Humicola fuscoatra*, have been reported as parasitic in a variety of soils over a wide geographic area. They appear to be well adapted as mycoparasites of oospores, although it is sometimes difficult to prove that the oospores were alive when attacked and thus that true parasitism has occurred (Wynn and Epton,

Table 3.2. Mycoparasites of oospores of *Phytophthora*.

Species of *Phytophthora*	Mycoparasite	Reference
P. cinnamomi	*Anguillospora pseudolongissma* *Catenaria anguillulae* *Humicola fuscoatra* *Hyphochytrium catenoides*	Daft and Tsao (1983, 1984)
P. erythroseptica	*Alternaria tenuis* *Fusarium oxysporum* *Gliocladium roseum* *Humicola fuscoatra* *Hyphochytrium catenoides* *Rhizidiomycopsis japonicus*	Wynn and Epton (1979)
P. megasperma f.sp. *glycinea*	*Actinoplanes missouriensis* *Humicola fuscoatra*	Sutherland *et al.* (1984)
P. nicotianae	*Catenaria anguillulae*	Daft and Tsao (1984)
P. sojae	*Acremonium* sp. *Alternaria tenuis* *Canteriomyces stigeoclonii* (= *Anisolpidium stigeoclonii*) *Dactyella spermatophaga* *Diheterospora chlamydosporia* *Fusarium oxysporum* *Humicola fuscoatra* *Hyphochytrium catenoides* *Leptolegnia* sp. *Pythium* spp. *Pythium monospermum* *Rhizidiomycopsis japonicus*	Sneh *et al.* (1977); Humble and Lockwood (1981)
P. syringae	*Microdochium fusarioides*	Harris (1985a)

1979; Daft and Tsao, 1983, 1984). Nevertheless, it seems clear that soils do contain a large number and diversity of oospore parasites, which may have the potential to reduce the inoculum potential of plant-pathogenic Oomycota in soil. Their presence may be a factor in the suppressiveness of some soils towards these pathogens, although it has been suggested that the frequency of mycoparasitism is not correlated with the disease potential of soils for *Phytophthora* root-rot of soybean as determined in seedling tests on flooded soil samples (Humble and Lockwood, 1981).

Other examples of invasive necrotrophs of the spores of plant-pathogenic fungi include a sterile hyaline basidiomycete which colonizes the chlamydospores of *Thielaviopsis basicola* (= *Chalara elegans*) the cause of blackhull disease of groundnuts (Baard, 1988). Several other examples, including *Ampelomyces quisqualis*, a destructive mycoparasite of several powdery mildew fungi will be discussed in more detail in relation to their use in biocontrol (see Chapter 7). The spores of many endomycorrhizal fungi, currently included in Zygomycota, are also prone to mycoparasitic attack and a more detailed account of this phenomenon is given in section 4.2.

Many hymenomycetes have been found specifically to attack and kill the basidiospores of other hymenomycetes. Such necrotrophic mycoparasitic fungi have been termed '**sporophagous**' (Fries and Swedjemark, 1985). The fusiform cylindrical basidiospores of *Leccinum aurantiacum* are among the largest in the Boletales, measuring 12–20 × 4–5.5 μm, and they are attacked in culture by the hyphae of another edible toadstool, *Coprinus comatus* (the 'Lawyer's Wig'). The hyphae of *C. comatus* appear to show directed growth towards the basidiospores of *L. aurantiacum* indicating a positive chemotropic response to the presence of the potential host. The invading hypha coils around the spore, branches and then penetrates the host wall, a process which possibly involves enzymic degradation. The spore cytoplasm appears to be assimilated by the parasite. It seems that only living spores are attacked which implies that some property of the living cytoplasm is required for development of the mycopathogenic reaction. In further experiments, the large spores of *L. aurantiacum* were used to 'bait' soil to isolate the mycelium of a range of hymenomycetes, comprising ectomycorrhizal, praticolous (meadow-inhabiting), coprophilous and lignicolous species. Activity was related to the number of spores attacked and whether hyphae showed a directed growth response. Wood-decomposing Aphyllophorales bracket fungi were the most strongly parasitic whereas ectomycorrhizal forms were probably not parasitic. Further, the hosts appeared to be restricted to hymenomycetes which produce pigmented spores and dark-coloured spore prints, such as species of *Boletus* with olivaceous or brownish spores. The hyaline basidiospores of *Russula*, *Lactarius* and other genera with white to cream-coloured spore prints were not parasitized. An explanation advanced for the colour preference is that it could indicate a difference in metabolism resulting in different types of exudate from the spores, but it is equally likely that specific differences in spore wall biochemistry may protect certain spores from attack. Parasitic activity was highest on media with low levels of available nitrogen and carbon which supports the notion that the parasitic hyphae may obtain nutrients from the host spores. These interactions occur between very different fungal species and should not be confused with phenomena such as oidial-homing and the development of a lethal reaction between closely related species of hymenomycetes (for example, see Kemp, 1970). Intraspecific reactions between such fungi are related to the sexual process and are not connected with mycoparasitism.

3.2.4 Parasitism of sclerotia

Sclerotia of certain soil fungi also offer a nutrient-rich reserve of metabolites for those fungi able to attack them. Most sclerotial parasites are necrotrophs, although it has been suggested that sclerotial degradation is not necessarily always through necrotrophy alone as there may be an initial biotrophic phase followed by a necrotrophic phase. This transition from biotrophic mycoparasitism to necrotrophy is not an isolated observation applying only to sclerotial parasites, as *Dicyma pulvinata* (= *Hansfordia pulvinata*), which parasitizes the hyphae of *Fulvia fulva*, also has an initial biotrophic phase succeeded by the necrotrophic stage (Le Picard and Trique, 1987).

Coniothyrium minitans is a typical example of a mycoparasite of sclerotia and destructively parasitizes those of *Sclerotinia sclerotiorum*, a pathogen of a wide range of crop plants. Hyphal tips of the mycopathogen penetrate the hyphal walls of the sclerotium, the cytoplasm of the host disintegrates and the walls collapse (Huang and Hoes, 1976). Conidia of *C. minitans* germinate in contact with sclerotia of *S. sclerotiorum*, infect the rind hyphae, cause lysis of the cytoplasm then grow into the medulla. It seems likely that the mycoparasite secretes the hydrolytic enzymes β-1,3-glucanase and chitinase capable of degrading the hyphal walls of *S. sclerotiorum* as these enzymes have been detected when the mycoparasite has been grown in media containing the hyphal walls of the potential host (Jones *et al.*, 1974). Sporulation of the mycoparasite occurs on the surface of and inside the sclerotium, thus providing the inoculum required for further infections of sclerotia in the soil. Scanning electron microscopy of the contact zone of the hypha of *C. minitans* with that of the host *S. sclerotiorum* was reported to show that appressorium-like structures are produced (Tu, 1984) which firmly adhere to the host. Huang and Kokko (1988), however, consider that *C. minitans* invades the host by direct penetration without the development of an appressorium, and report that indentation of the host cell wall at the point of penetration is often evident. In the vegetative and sclerotial hyphae of *S. sclerotiorum* penetrated by those of *C. minitans*, the walls, in addition to the cytoplasm, are attacked by lytic enzymes from the mycoparasite (Huang and Hoes, 1976). Penetration of the rind elements of the sclerotium is said to occur by physical pressure owing to the invaginated appearance of the wall at the point of penetration and the constricted nature of the penetrant hypha. Elements of the medulla, however, are believed to be penetrated through a combination of physical pressure and enzyme lysis (Phillips and Price, 1983). Hyphae of the parasite grow both intracellularly and intercellularly within the sclerotia. Polysaccharide food reserves of the medullary hyphae disappear, presumably through utilization by the mycoparasite. *Coniothyrium minitans* also infects sclerotia of *Sclerotinia minor*, *S. trifoliorum*, *Sclerotium cepivorum*, *Botrytis cinerea* and *B. fabae*. It does not attack sclerotia of *Rhizoctonia tuliparum* or *Typhula incarnata* even though both living sclerotia of these two species, and sclerotia killed by autoclaving, stimulate conidial germination and germ-tube outgrowth of

Table 3.3. Colonization of sclerotia of *Rhizoctonia tuliparum* in soil at 15°C. (Based on Gladders and Coley-Smith, 1980.)

Colonizing fungi	Days of burial of sclerotia					
	3	7	14	28	35	56
Fusarium equiseti	+	+	+	+	+	+
Mucor hiemalis	–	+	+	+	+	+
Trichoderma viride	–	+	+	+	+	+
Gliocladium roseum	–	–	–	–	+	–
Trichoderma hamatum	–	–	–	–	–	+

C. minitans on distilled water or mineral salts agar (Whipps *et al.*, 1991). In contrast, sclerotia of a third non-host species, *Sclerotium delphinii* inhibit germination and germ-tube growth of the mycoparasite. Host sclerotia also stimulate the germination and growth of *C. minitans* and it is assumed that this is due to non-specific leakage of nutrients from them. The sclerotia of *R. tuliparum* and *T. incarnata* are also known to leak antibiotics which may be implicated in the failure of *C. minitans* to parasitize these resting structures. They are, however, parasitized by several other fungi. Sclerotia of *Typhula incarnata*, for example, are attacked by *Cylindrobasidium parasiticum*, which appears to be widely distributed throughout the UK (Woodbridge *et al.*, 1988).

Sclerotia of *Rhizoctonia tuliparum* are readily invaded by several fungi in succession (Table 3.3). The earliest colonizers include species of *Fusarium*, *Mucor* and *Trichoderma*, with *Gliocladium roseum* and *T. hamatum* appearing later in the succession. Of these, *Trichoderma* and *Gliocladium* species are known to be mycoparasitic. Temperature affects the ability of *T. viride* and *G. roseum* to colonize the sclerotia as neither were detected in a comparable experiment at 5°C. The ability of *Fusarium* species to colonize, however, was not affected by temperature. Sensitivity of the sclerotia to infection is increased if they are air-dried then rewetted as nutrients are leaked which stimulate growth of the colonizers.

Talaromyces flavus (anamorph = *Penicillium dangeardii* = *P. vermiculatum*) is another necrotrophic mycoparasite capable of destroying sclerotia of *Sclerotinia sclerotiorum* (McLaren *et al.*, 1989). In dual culture it also attacks the vegetative hyphae of this plant pathogen and coils around them before invading the tips by direct penetration in the absence of an appressorium (McLaren *et al.*, 1986). It also attacks the sclerotia and hyphae of *Verticillium dahliae* and *Rhizoctonia solani*. The mode of antagonism of *T. flavus* towards *V. albo-atrum* may involve competition and antibiosis (Dutta, 1981). As granulation of cytoplasm in invaded hyphae of *S. sclerotiorum* can extend into adjacent hyphal compartments, this is also indicative of the involvement of toxic materials. *Talaromyces flavus* is reported to produce several antifungal agents, including talaron and vermiculine, and it may be that the teleomorph *T. flavus* produces compounds destructive to hyphae of *S. sclerotiorum* (McLaren *et al.*, 1986). Hyphae of *S. sclerotiorum* eventually collapse as a result of infection by *T. flavus* but the walls remain intact even after the cell contents are completely destroyed. This suggests that wall-degrading enzymes may not play a major role in mycoparasitism in this invasion of the host mycelium. This differs from the situation during infection of sclerotia of *S. sclerotiorum* by this fungus, when etching of the walls is evident. It is also different from the situation when *Coniothyrium minitans* attacks hyphae of *S. sclerotiorum. Coniothyrium minitans*, unlike *T. flavus*, does not coil extensively around the parasitized hyphae, and the host walls are often disintegrated by the enzymes produced by the mycoparasite (Jones *et al.*, 1974). The growth of hyphae of *T. flavus* towards those of *S. sclerotiorum* has also been noted suggesting that directed growth occurs during parasitism of this fungus (McLaren *et al.*, 1986). A similar phenomenon occurs in *Gliocladium catenulatum* parasitizing *S. sclerotiorum* where growth of hyphal branches towards the host frequently occurs (Huang, 1978). Hyphae of *Gliocladium virens* and *Gliocladium roseum*, necrotrophic mycoparasites of *S. sclerotiorum* (Tu, 1980), also exhibit directed growth. Infection of developing mycelia inhibits the formation of sclerotia, but once sclerotia are formed these structures are also susceptible to mycoparasitism. Appressorial formation precedes penetration of host structures and an intrahyphal mycelium develops prior to profuse sporulation of *Gliocladium* over the surface of the affected sclerotia. Parasitized sclerotia are incapable of either myceliogenic or ascomatous germination (Tu, 1980).

Sclerotial mycoparasites can frequently be detected by the presence of their sporulating structures on the surface of the decaying host. For example, *Trichoderma harzianum*, a parasite of *Corticium rolfsii*, sporulates actively on the sclerotial surface and also within the parasitized elements whose contents have been digested during the growth phase of the parasite (Elad *et al.*, 1984). In this particular interaction, penetration of the sclerotium occurs and the medulla is lysed through autolysis and the activity of wall degrading enzymes of the mycoparasite (Henis, 1984).

Chitinase and β-1,3-glucanase are secreted by *T. harzianum* grown in media containing hyphal walls of *C. rolfsii* and *R. solani*. Parasitism of sclerotia appears to be isolate-specific, and the three isolates of *T. harzianum* (isolate 203, 250 and 110) used by Elad *et al.* (1984) were all capable of attacking hyphae of *R. solani*, but only isolate 203 could attack sclerotia of *C. rolfsii*. Several other *Trichoderma* species also attack sclerotia. Dos Santos and Dhingra (1982) tested the ability of 38 isolates of *Trichoderma* isolated from a variety of Brazilian fields. Relative pathogenicity of the isolates within each species again varied, and five out of 12 isolates of *T. koningii*, two out of five isolates of *T. harzianum* and three out of 11 isolates of *T. pseudokoningii* killed between 62 and 100% of the sclerotia that were directly inoculated. Two isolates of *T. pseudokoningii* killed 100% of the sclerotia within seven days, and one of these isolates killed 100% of sclerotia within 60 days under field conditions when soil was infested with 10^4, 10^6 or 10^8 conidia per gram of soil.

One noteworthy host response to mycoparasitism is the production of microconidia. Sporulation by the host under poor conditions of nutrient supply may be stimulated in some fungi and it has been suggested that the effect of the necrotrophic reaction in *S. sclerotiorum* is to deprive the host mycelium of nutrients, resulting in the production of microconidia (Tu, 1984).

4 BIOTROPHIC ASSOCIATIONS

4.1 Haustorial Biotrophs

Biotrophic mycoparasites that form haustoria have been described attacking either the sclerotia or the mycelia of other fungi (Table 4.1). Most of the available information relates to a group of mycoparasites within the Zygomycota that parasitize the vegetative mycelium or sporangiophores of their hosts which themselves are often, but not invariably, related. Extensive studies have been carried out on their morphology and physiology and will be described here or in subsequent chapters. Less information is available for other types of haustorial mycoparasites and this is summarized in section 4.1.2.

4.1.1 Haustorial biotrophs from the Zygomycota

Most of the haustorial biotrophic mycoparasites that attack the mycelium of their hosts belong to a discrete group of fungi in five genera from the Piptocephalidaceae (Zoopagales) and Dimargaritaceae (Dimargaritales), two families in the Zygomycota. The Piptocephalidaceae contains three mycoparasitic genera: *Piptocephalis* and *Kuzuhaea* (Benjamin, 1985) form haustoria, whereas *Syncephalis* forms intrahyphal infection structures and has been discussed earlier (section 3.2.2). In their mode of parasitism, vegetative development, and sporulating structures Piptocephalidaceae correspond in many ways to Zoopagaceae, which led Kreisel (1969) to combine these families in the Zoopagales, a move supported by Benjamin (1979). The Dimargaritaceae also contains three mycoparasitic genera, *Dimargaris*, *Dispira* and *Tieghemiomyces*, all of which form morphologically similar haustoria. This family was placed in the Mucorales until transferred to a separate order, the Dimargaritales (Benjamin, 1979).

Table 4.1. Biotrophic mycoparasites reported to produce haustoria.

Mycoparasite	Reference
Caulochytrium protostelioides	Powell (1981)
Dimargaris spp.	Barnett and Binder (1973)
Dispira spp.	Barnett and Binder (1973)
D. simplex	Brunk and Barnett (1966)
Filobasidium floriforme	Oberwinkler and Bandoni (1982)
Filobasidiella neoformans	Oberwinkler and Bandoni (1982)
Kuzuhaea moniliformis	Benjamin (1985)
Piptocephalis spp.	Jeffries (1985)
P. xenophila	Dobbs and English (1954)
Rhynchogastrema coronata	Metzler *et al.* (1989)
Sporidesmium sclerotiovorum	Bullock *et al.* (1986)
Syzygospora alba	Oberwinkler and Bandoni (1982)
Tieghemiomyces spp.	Brain *et al.* (1982)
Tremella spp.	Oberwinkler and Bandoni (1982)
Trimorphomyces papilionaceus	Oberwinkler and Bandoni (1983)

Reports of mycoparasitism by these organisms made an early entry into the mycological literature. For example, *Piptocephalis freseniana* was first observed by Fresenius in 1864 and was subsequently described and illustrated very briefly the following year (see Benjamin, 1959). Brefeld (1872) made a thorough study of this fungus and also investigated the interaction of *Chaetocladium brefeldii* (which he called *C. jonesii*) and *Mucor mucedo*. Appressorium formation and haustorium formation were noted and illustrated for *Piptocephalis* and reappeared in de Bary's monograph of 1887 (Fig. 4.1). Subsequently several other examples of mycoparasitic fungi have been found within the Zygomycota and are discussed further here. Some of these examples, along with the phenomenon of mycoparasitic attack on members of the Zygomycota, have been reviewed elsewhere (Jeffries, 1985).

All the mycoparasites in this group, with the exception of *Syncephalis*, form appressoria and haustoria and the infection cycle is broadly similar. Differences exist, however, in the ultrastructure of the mature host-parasite interface. Prepenetration responses in these fungi involve directed growth of the germ-tubes towards young, actively growing regions of the hyphae of potential host fungi although sometimes directed growth does not occur (Manocha, 1988). On an agar medium, this phenomenon may be noted up to a distance of several hundred micrometres, even against a background of a complex medium such as malt extract agar (Fig. 4.2). Thus the parasite must be able to recognize a readily distinguishable diffusate from host hyphae and adjust its growth pattern accordingly. The recognition of fine filaments (fimbriae) on the surface of the hyphae

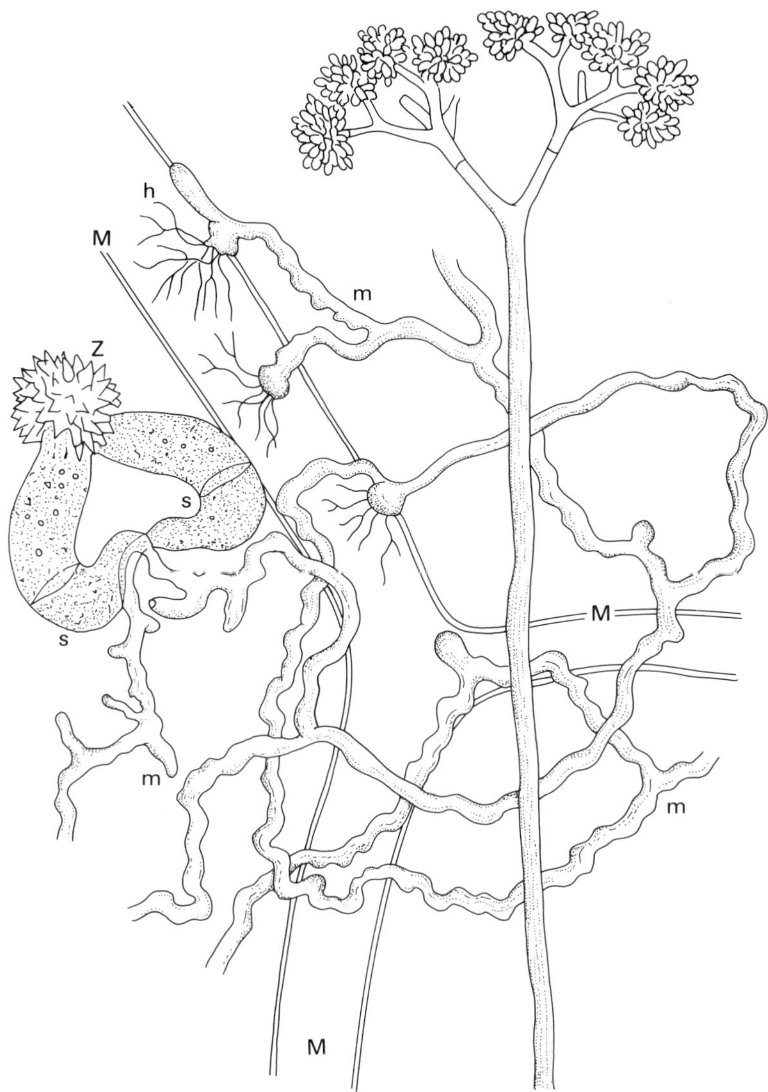

Fig. 4.1. *Piptocephalis freseniana* (from de Bary, 1887). h, haustorium; m, mycelium of the mycoparasite; M, mycelium of the host; s, suspensor; Z, zygosporangium.

of the host by the mycoparasite may also play a role in the initiation of directed growth (Rghei *et al.*, 1992). The strength of the tropic response varies from host to host, and between the various mycoparasites. Nevertheless, this response is of paramount importance in initiation of the parasitic attack as directed growth has

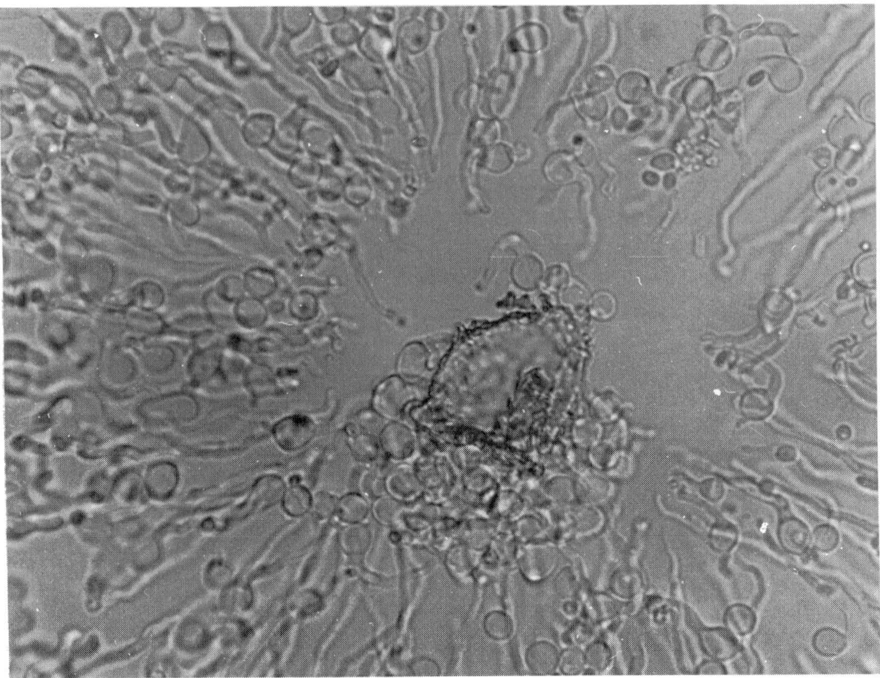

Fig. 4.2. Directed growth of germ-tubes of *Dimargaris cristalligena* towards a hyphal fragment of *Cokeromyces recurvatus* (LM: × 500).

never been observed towards a fungus that does not subsequently elicit the formation of a primary haustorium. The converse situation is not true, however, and there are certain fungi, for example *Circinella mucoroides*, which do not induce positive directed growth of the parasitic hyphae but are nevertheless hosts of mycoparasitic zygomycetes (Evans and Cooke, 1982). The substance which induces directed growth of the hyphae of *Piptocephalis fimbriata* is a non-volatile, heat-labile diffusate of high molecular weight. A detailed biochemical analysis has not been carried out although it was suggested that the compound is proteinaceous or possibly bound to protein (Evans and Cooke, 1982). In a natural situation, where background nutrients are at a low level, the diffusate may also have a role in stimulating spore germination. In some other examples the presence of the host stimulates growth of the mycoparasitic hyphae over and above any directed response. For instance, the extent of germ-tube development of *Dispira cornuta* is greatly increased in the presence of a susceptible fungus prior to contact with the host (Jeffries, 1985). In most species the host range of the haustorial mycoparasites is confined to members of the Mucorales, and there is often further specificity within this order, depending on the species of mycoparasite. *Piptocephalis unispora*, for example, parasitizes many mucoralean fungi

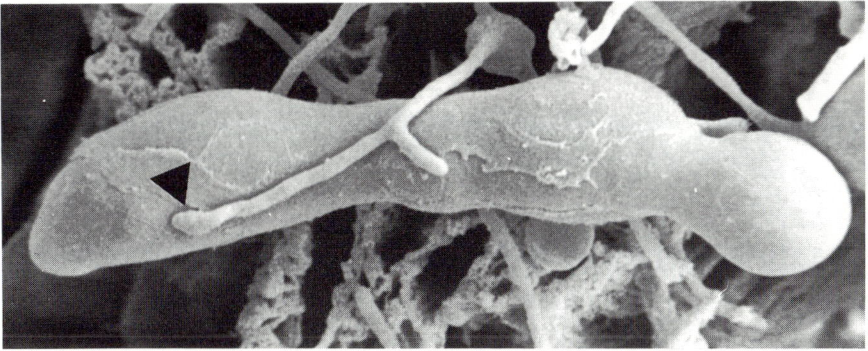

Fig. 4.3. Growth of a hypha of *Piptocephalis unispora* over the surface of a hypha of *Cokeromyes recurvatus*. An appressorium (arrow) has formed at the apex of the parasitic hypha (SEM: × 300).

but its host range within the Mortierellaceae is confined to the subgenus *Micromucor* (Cuthbert and Jeffries, 1984). Germ-tubes of the parasite do not apparently recognize the presence of hyphae of members of the subgenus *Mortierella*, directed growth does not occur and appressoria are not produced on contact. Some other species of *Piptocephalis*, however, such as *P. fimbriata*, have a wider host range and parasitize fungi from both subgenera of the Mortierellaceae. Caution must therefore be exhibited in using information relating to the host range as an aid to taxonomic distinctions between host fungi. In exceptional cases, notably *Piptocephalis xenophila*, *Dispira simplex* and possibly *D. parvispora* (Dobbs and English, 1954; Benjamin, 1961; Brunk and Barnett, 1966), hosts other than Mucorales are parasitized. Whether this reflects a major biochemical difference in the metabolism of these species or whether there is merely a difference in recognition phenomena remains to be determined. Such recognition phenomena are not absolute, however, as self-penetration of hyphae of *Dimargaris cristalligena* has also been observed, albeit rarely (Jeffries and Cuthbert, 1984), in which plasmalemmal invagination occurs in a manner similar to that noted during penetration of a conventional host, but invading hyphae fail to differentiate into a haustorial body. Germ-tubes of *Piptocephalis virginiana* will attach to cleaned hyphal wall fragments of host cells but not to those of non-host cells (Manocha, 1985). Attachment is inhibited by the addition of sugars, chitobiose and chitotriose, and by treatment with acid or alkali indicating the involvement of proteins or glycoproteins in recognizing sugar residues at the cell surface.

Once contact with the host is achieved, a germ-tube or mycelial hypha of the mycoparasite grows along the surface of a hypha or germinating spore of a potential host. When a suitable infection site is reached apical extension ceases and the tip enlarges forming an appressorium. Presumably some biochemical or

Fig. 4.4. Appressoria of *Piptocephalis unispora* (arrows) on a hypha of *Cokeromyces recurvatus* (SEM: × 1000).

tactile stimulus is responsible for triggering appressorial development. Appressoria are usually formed on the thinner part of the wall of the intact host either near the hyphal apex or where a germ-tube emerges from a spore or Y-phase cell (Figs 4.3 and 4.4). Attachment of the invading hypha to the wall of the host may involve specific chemical binding sites. In *Piptocephalis virginiana*, germ-tubes will attach to, and appressoria will form on, wall fragments of the potential host, indicating that the living host is not essential at this stage. Germ-tubes killed by heating or glutaraldehyde fail to attach, however, which suggests that some property of living cytoplasm of the parasite may be required for the germ-tube to be able to locate specific binding sites on the surface of the host. Tests of adhesion involving fluorescein-labelled sugar residue-binding lectins indicate that *N*-acetylglucosamine oligomers are specifically bound and may be involved in the

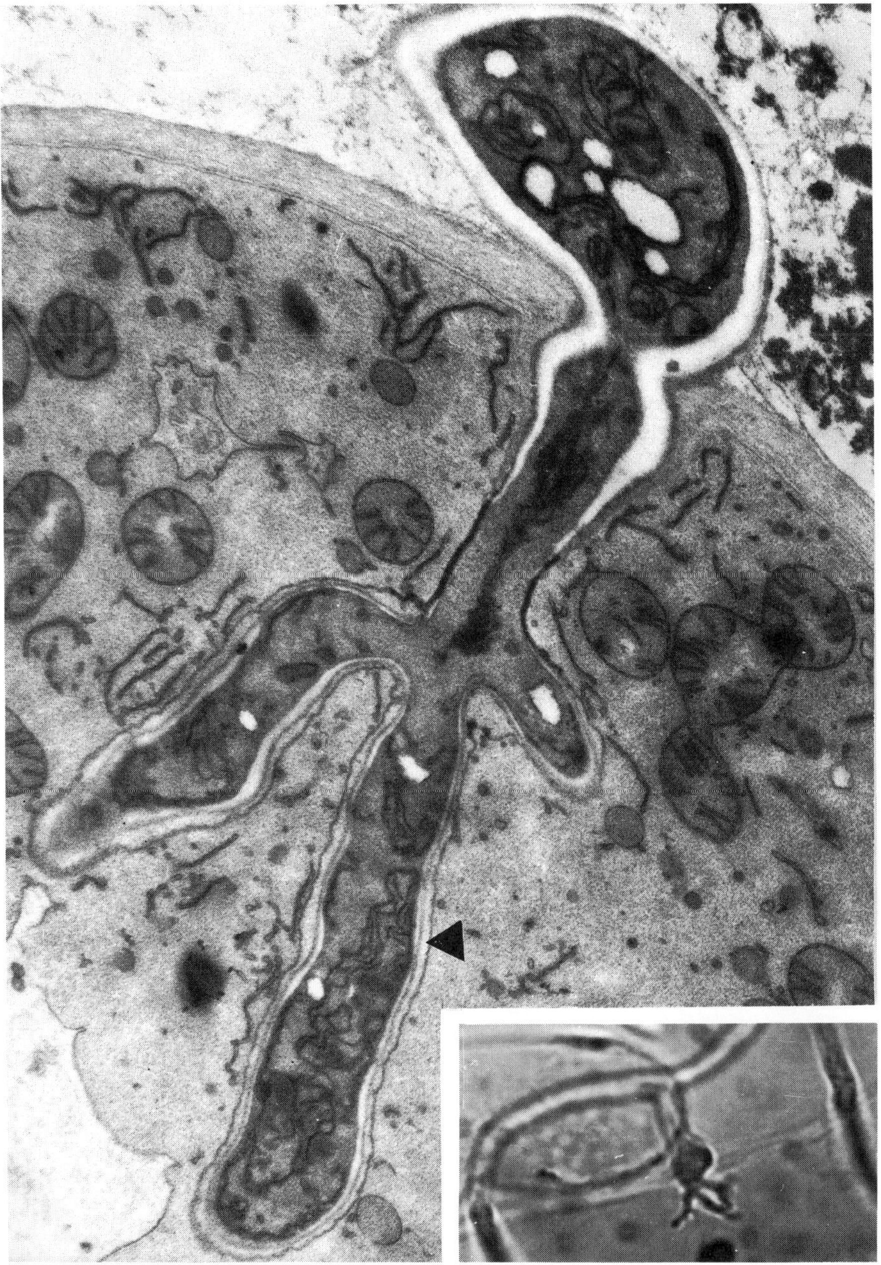

Fig. 4.5. Longitudinal section through a haustorium of *Piptocephalis unispora*. The extrahaustorial matrix (arrow) can be distinguished (TEM: × 5000). Inset shows a haustorium at lower magnification (LM: × 350).

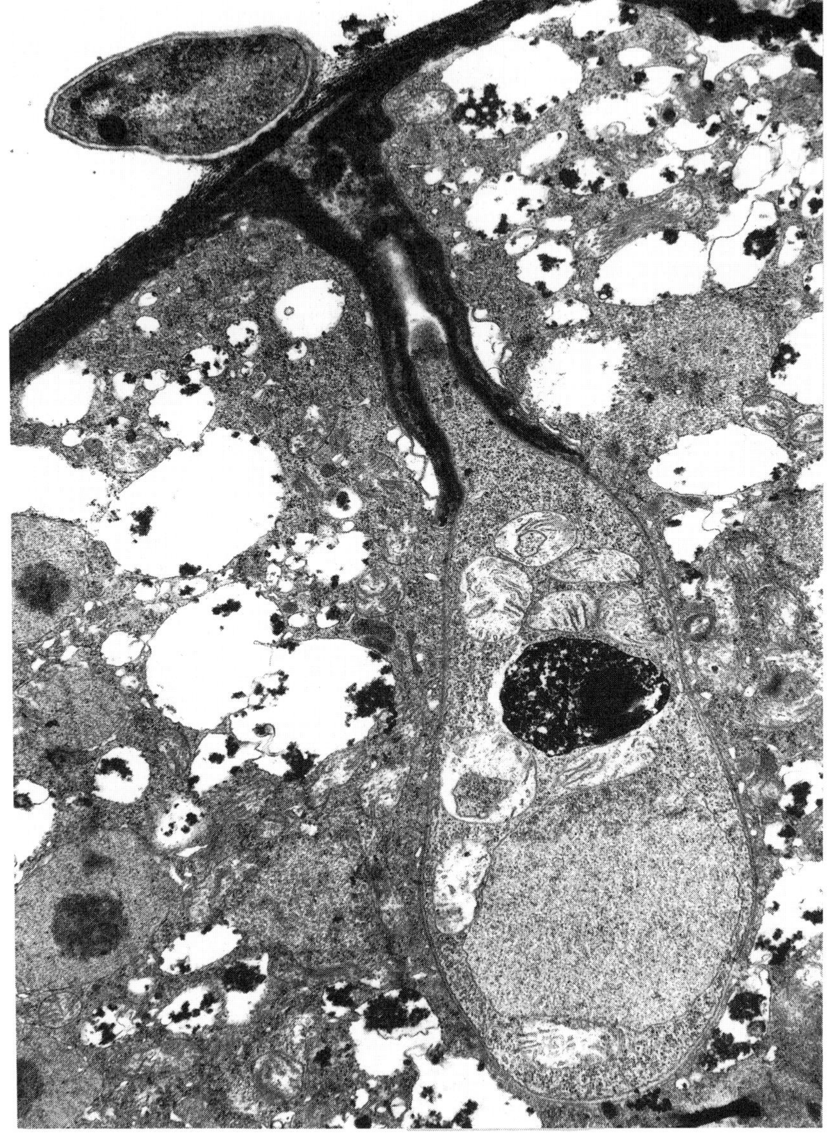

Fig. 4.6. Longitudinal section through a haustorium of *Dimargaris cristalligena* within a hypha of the host *Cokeromyces recurvatus* (TEM: × 3500).

mechanism of adhesion with fungal lectins (Manocha, 1985). Lectin-mediated adhesion has also been implicated in the capture of nematodes by nematode-trapping fungi (Nordbring-Hertz and Mattiason, 1979) where the trapped animals are held fast, unable to break free even with the most violent threshing movements. The removal of glycoproteins from isolated walls by means of treatment with acid or alkali also removes the binding sites and prevents the germ-tubes from attaching to wall fragments. This is taken to indicate the involvement of glycoproteins. Gel electrophoresis of cell surface proteins has revealed the presence of certain glycoprotein peaks in the potential host *Choanephora cucurbitarum* which are absent from those of the non-host *Linderina pennispora*, to which the germ-tubes do not adhere. This appears to lend further support to the view that glycoproteins are involved. *Linderina pennispora* may not be the most appropriate choice of test fungus, however, as it belongs in Kickxellales, not Mucorales, and wall structure in the Kickxellales appears structurally different from that of Mucorales (Young, 1970, 1973a,b, 1974, 1985).

In other studies (Manocha *et al.*, 1990), differences in the distribution patterns of glycosyl residues on the hyphal walls of host and non-host fungi have again been implicated in recognition phenomena between *Piptocephalis virginiana* and various mucoraceous fungi. Pretreatment of the mycoparasite with glucose and *N*-acetylglucosamine inhibited its attachment to the surface of the host, but had no obvious effect on appressorium formation. On the other hand, appressorium formation was inhibited by heat treatment of host wall fragments which still permitted attachment, thus indicating that the factors responsible for attachment and for appressorium formation are different (Manocha *et al.*, 1990). It has also been demonstrated that germ-tubes of the mycoparasite are able to bind to protoplasts of host fungi, and the percentage attachment is higher in protoplasts from the host fungi compared to those from non-hosts (Manocha *et al.*, 1990). These findings suggest that differences in hyphal wall biochemistry may not be the primary determinant of susceptibility. Protoplasts of host fungi incubated in the presence of germinating spores of *P. virginiana* became enlarged, highly vacuolate and finally collapsed, irrespective of attachment of the hyphae of the parasite (Sivakami Sundari and Manocha, 1991). Protoplasts of a non-host species of *Mortierella*, however, were not affected by the presence of the mycoparasite. This suggests that an extracellular metabolite produced by the mycoparasite may also play a role in the mycoparasitic process.

An appressorium adheres firmly and host penetration is achieved by growth of an infection peg through the wall of the host and the ultrastructural evidence indicates both enzymatic and mechanical involvement (e.g. Manocha and Lee, 1971; Jeffries and Young, 1976b, 1981; Evans *et al.*, 1984). After breaching the wall, the tip of the penetration peg enlarges and the lobe(s) of the haustorium develop, often in the vicinity of a nucleus of the host. The formation of such infection structures in response to the presence of the host has been reviewed (Staples and Macko, 1980; Manocha, 1987) and the main ideas are presented here. In ultrastructural studies involving thin sections of the structure of the host–parasite interface the interpretation of data is facilitated when the morphology of the interacting fungi is obviously different. Thus, in species of *Dimargaris*, *Piptocephalis* and *Tieghemiomyces* infecting the hyphae of *Cokeromyces*

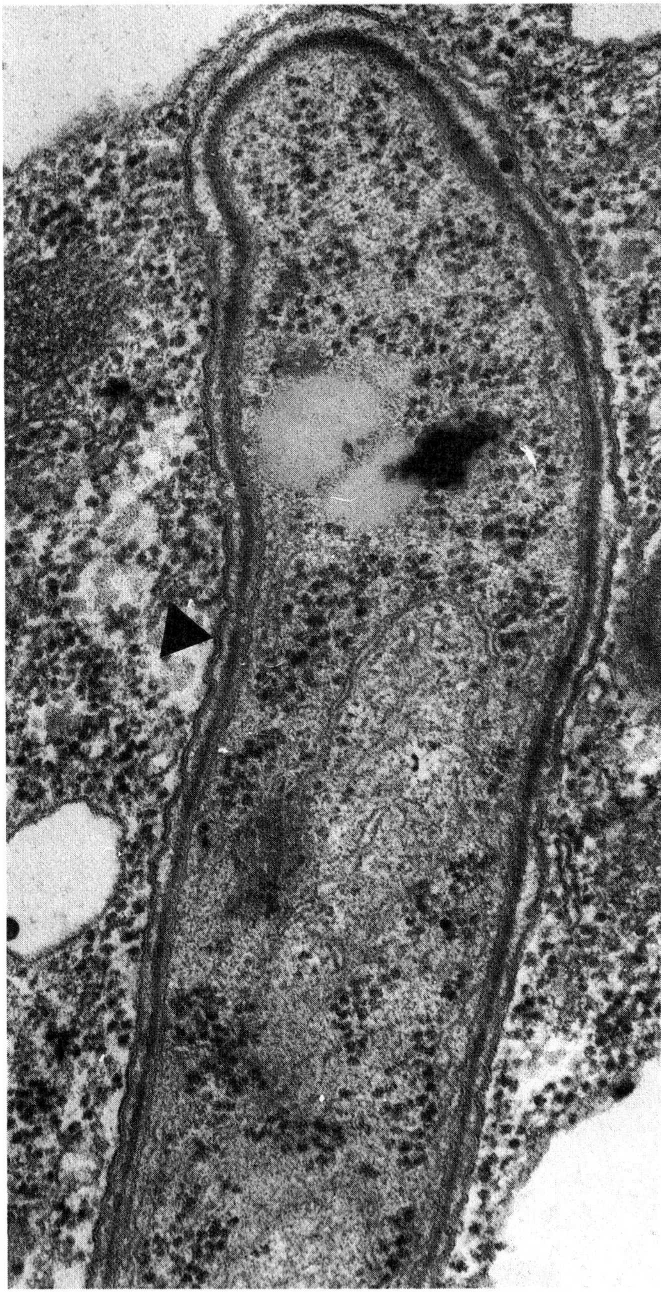

Fig. 4.7. A haustorial lobe of *Piptocephalis unispora* within a hypha of *Cokeromyces recurvatus*. The extrahaustorial membrane is arrowed (TEM: × 82,000).

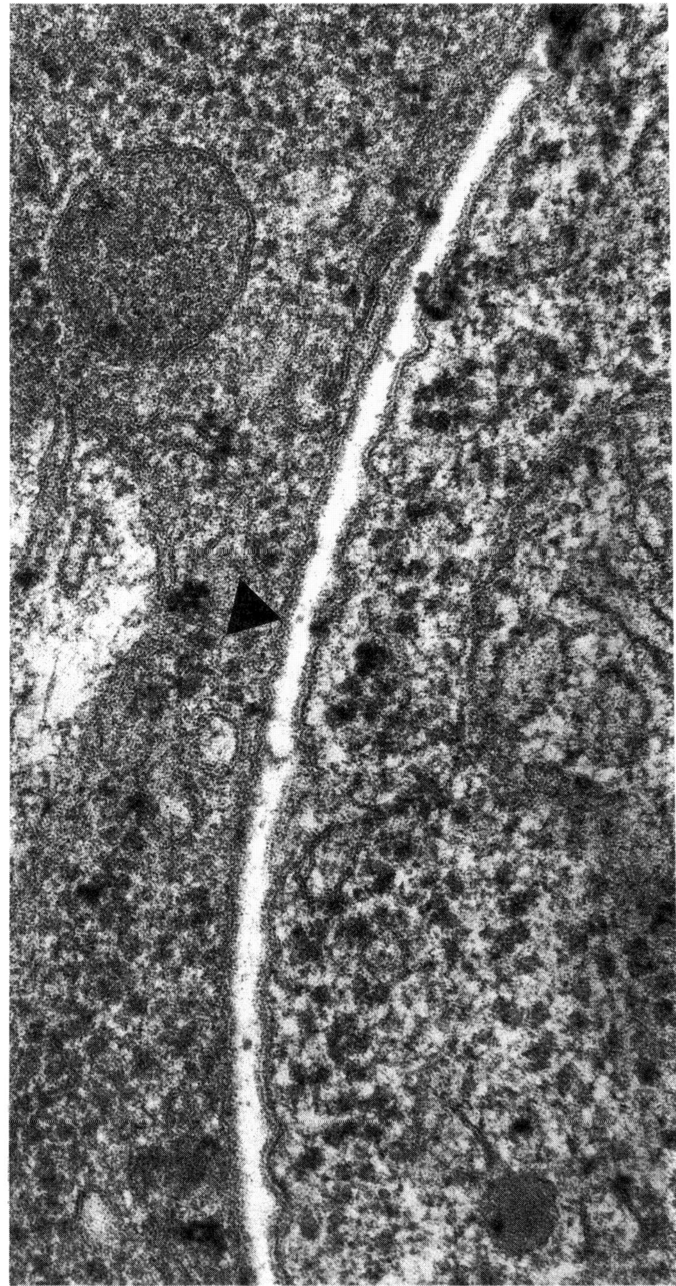

Fig. 4.8. A haustorial lobe of *Dimargaris cristalligena* within a hypha of *Cokeromyces recurvatus*. The extrahaustorial membrane is arrowed (TEM: × 1,000,000).

recurvatus the mycelial phase of the parasite is distinct from that of the host and there is unlikely to be confusion in the interpretation of electron micrographs. Septal structure may also be distinctive. In *Dimargaris* and *Tieghemiomyces* the septum of the vegetative hypha bears a central pore which contains a biumbonate plug with an electron dense, subglobose body attached to each side of the septum (Brain *et al.*, 1982). This structure, incidentally, is diagnostic for Dimargaritales, the order to which *Dimargaris* and *Tieghemiomyces* belong.

In a compatible reaction there is little evidence of extensive papilla formation at the penetration site (Jeffries, 1985), although a small apposition is often present which is synthesized as the infection peg passes through the host wall. Once penetration is achieved, the apex of the peg develops into the body of the haustorium, accompanied by the accumulation of host cytoplasm in the immediate vicinity. A **neckband** is present in the haustorial apparatus of *Piptocephalis unispora* (Jeffries and Young, 1976b) and presumably has a similar function to those found in other haustorial pathogens in sealing the **apoplastic** route for transport of materials (for a discussion of this aspect see Manners and Gay, 1983). The neckband resembles that observed in the haustorium of rust fungi in which this structure has been compared at the functional and ultrastructural level with the casparian strip of the roots of vascular plants (Heath, 1976). As the lobes of the haustorium develop, the mature host–parasite interface is formed and it is here that the major differences occur between the Dimargaritales and Piptocephalis. At first, the lobes of the haustorium of *Piptocephalis* are in contact with the extrahaustorial membrane, derived from the invaginated plasmalemma of the host, but as the haustorium matures an amorphous interstitial matrix develops, analogous to the extrahaustorial matrix of biotrophic plant pathogens (Figs 4.5 and 4.7). In *Piptocephalis* the haustorial lobes retain a wall which morphologically resembles that of the vegetative hypha. Any material to be transferred between host and parasite via the haustorium must thus traverse the extrahaustorial membrane, matrix, parasite wall and parasite plasmalemma. In contrast, the host–parasite interface in species from the Dimargaritales is less complex. Although the neck region of the haustorial complex is elaborate and includes a **penetration jacket** and a septal complex (Jeffries and Young, 1978, 1981), the wall of the haustorial lobe is substantially reduced. The plasmalemma of the host and that of the parasite are separated only by a narrow amorphous zone which is derived from the wall of the invading hypha in the region at the base of the haustorial neck (Figs 4.6 and 4.8). The haustorium of *Dimargaris cristalligena* and *Tieghemiomyces californicus* possesses a nucleus, a neck septum, extensive collar and a penetration matrix, yet lacks a neckband and a complex extrahaustorial matrix. In contrast, *Piptocephalis* haustoria are enucleate, aseptate, and a penetration jacket is not involved. A neckband is present, however, and the haustorial lobes also differ from those of the Dimargaritales in the presence of a wall. The extrahaustorial matrix is also more complex than that of the Dimargaritales. Thus, although the overall process of haustorial development is similar in all the haustorial Zygomycota, there are marked differences in the ultrastructure of the mature haustorial apparatus.

In both types of haustorial apparatus the final phase of the infection cycle involves the breakdown of the host–parasite interface. This is observed as a gradual disorganization of organelles of the host, resulting eventually in the disintegration of the extrahaustorial membrane. It is not clear how much of this change is related to natural hyphal senescence as the cytoplasm ages, or whether senescence is accelerated by the presence of the parasite, as these cytological changes are also apparent in uninfected hyphae in the zone of vacuolation. The functional life of the haustorium in *P. virginiana* infecting the mucoralean host *Mycotypha microspora* is probably 6–36 h as the cytoplasm begins to degenerate after 6 h and has necrosed after 36 h (Armentrout and Wilson, 1969). Infection of host hyphae by *Piptocephalis* may induce growth abnormalities such as 'witches brooms' (England, 1969) or the development of **Y-phase** (yeast-phase) cells (Evans *et al.*, 1978), or cause changes in the organization of marginal hyphae (Evans and Cooke, 1981). Depending on the host–parasite combination and temperature, several species of *Piptocephalis* can markedly affect the linear growth rate of their host fungi, the effects being either stimulatory or inhibitory (Curtis *et al.*, 1978).

Resistance to haustorial mycoparasitism, other than non-host resistance, or the resistance exhibited by older regions of the mycelium of susceptible hosts (England, 1969), is rare. One example is *Phascolomyces articulosus*, a thamnidiaceous fungus (Table 4.2) in which primary haustoria of *Piptocephalis* are formed, yet which does not support the secondary proliferation of the mycoparasite (Benny and Benjamin, 1976; Jeffries and Young, 1978). The germ-tubes of *P. unispora* exhibit directed growth towards the hyphae of the potential host but after the formation of appressoria, infection pegs and primary haustoria, however, the attack ceases and the mycelium of *Phascolomyces articulosus* rapidly overgrows that of *P. unispora.* The mature infection apparatus of *P. unispora* parasitizing *Phascolomyces articulosus* is composed of an appresssorium, penetration peg and a lobed haustorium surrounded by a sheath matrix enclosed in an extrahaustorial membrane, and a connecting neck region with a collar and a neckband. This is similar to observations made of haustoria formed within the highly susceptible host *Cokeromyces recurvatus* (Jeffries and Young, 1976b). Extensive papilla formation is often associated with the failure of the parasitic attack (Manocha and Graham, 1982), but the fact that degenerate haustoria have been observed in the absence of papilla formation (Jeffries and Young, 1978) suggests that papillae are not the primary cause of resistance. In a small proportion ($< 5\%$) of attacks, both *P. unispora* and *D. cristalligena* fail even to penetrate *Cokeromyces recurvatus*, a standard reference host for these fungi, and an extensive papilla is usually found at such sites (Jeffries and Young, 1981). It is not known whether the papilla forms either before or after the invading hypha ceases to attack. Wall reactions and papillae have also been observed in other mycoparasitic combinations such as that involving the necrotroph *Pythium acanthicum* (Hoch and Fuller, 1977). This suggests that papillae

Table 4.2. Host range of *Piptocephalis unispora* (Zoopagales). (Based on Jeffries and Young, 1978.)

Order Mucorales Family	Species	Development of *P. unispora*
Choanephoraceae	*Choanephora curbitarum*	+++
	Poitrasia circinans (= *Choanephera circinans*)	+++
Cunninghamellaceae	*Cunninghamella elegans*	+++
	C. vesiculosa	++
Mortierellaceae	*Micromucor ramannianus*	−
	Mortierella polycephala	+
	M. stylospora	−
	M. vinacea	++
Mucoraceae	*Absidia glauca*	+
	A. spinosa	++
	Gilbertella persicaria	++
	Mucor plumbeus	+
	M. racemosus	++
Pilobolaceae	*Pilaira anomala*	++
Radiomycetaceae	*Hesseltinella vesiculosa*	++
	Radiomyces spectabilis	+++
Syncephalastraceae	*Syncephalastrum racemosum*	+++
Thamnidiaceae	*Chaetocladium brefeldii*	+
	Cokeromyces recurvatus	+++
	Ellisomyces anomalus	++
	Phascolomyces articulosus	−

Note: −, no parasitism observed, only dwarf sporangiophores of *P. unispora* present; +, parasitism observed but only sparse growth of *P. unispora*; ++ dense mycelium of *P. unispora* develops at the point of inoculation but the host rapidly outgrows the parasite; +++, dense growth of *P. unispora* obscures the colony of the host.

may develop in response to contact and chemical stimulation by the parasite. It is generally assumed that wall apposition results from localized secretory activity of the host. This view is supported by studies of the haustoria of plant-pathogenic fungi which show cytoplasmic vesicles or dictyosomes of the host near to wall appositions. Vesicles and membranous inclusions present in the papilla are presumed to arise, at least in part, from the entrapment of cytoplasm and organelles of the host during papilla formation. Their presence and subsequent degeneration during papilla maturation commonly occurs where papilla development is extensive (Hoch and Fuller, 1977). Interestingly, wall ingrowths resembling those found in the vegetative septa of *C. recurvatus* have been observed following the secondary wall synthesis associated with papilla formation (Jeffries and Young, 1983). This suggests a common morphogenetic route for secondary wall synthesis whether triggered by internal (septation) or external (wall damage) stimuli. Other haustorial mycoparasites, such as *Dimargaris cristalligena* readily parasitize *Phascolomyces articulosus* and the morphology of this interaction differs little from that occurring during infection of the reference host *C. recurvatus* (Jeffries and Young, 1976b). A chitin synthase preparation from *Phas. articulosus* is activated by all the proteases secreted from *Piptocephalis virginiana* whereas only the acid proteases activate the chitin synthase of the susceptible host *Choanephora cucurbitarum.* This differential activation accounts for the more rapid response to invasion in the latter fungus resulting in the expression of resistance (Manocha and Begum, 1985).

Gross differences in hyphal wall composition and architecture have been proposed as determinants of host specificity for haustorial mycoparasites (Manocha and Golesorkhi, 1979; Manocha, 1981; Manocha and Graham, 1982; Manocha and Campbell, 1983). Evidence has been presented for the importance of a second fibrillar layer on the inner surface of the young hyphal wall of *Phas. articulosus*, and in older hyphae of *Choanephora cucurbitarum*, in the resistance to penetration by hyphae of *Piptocephalis virginiana.* Nevertheless, the hyphae of these resistant forms do not withstand initial penetration, yet fail to support secondary proliferation of the parasite (Jeffries and Young, 1978). It also seems that *Phas. articulosus* possesses a wall of one layer only (Cuthbert, 1984), and it may be that formation of a second wall layer reduces the degree of diffusion of components which induce directed growth of the hyphae of the parasite so that fewer challenges to these hyphae are made.

The presence of chitosan within the wall has also been suggested as a determinant of host specificity for *P. virginiana* (Manocha, 1981), as the Mucorales are distinctive in possessing high mural levels of this polymer (Bartnicki-Garcia, 1968). It is evident, though, that this could not be a general explanation for host range specificity if the parasitism of ascomycete hosts by *P. xenophila* and *D. simplex* is also considered. It is noteworthy that the hyphal walls of *P. unispora* and *D. cornuta* appear to lack chitosan (Cuthbert, 1984) which would greatly facilitate the involvement of a chitosanase during penetration of walls of host fungi. The addition of host cell wall fragments stimulates the release of chitinase and chitosanase from spores of *P. virginiana* germinating in a liquid medium (Balasubramanian and Manocha, 1986). Localized release of these enzymes at the infection site of an invaded hypha would facilitate penetration in the absence of extensive wall damage and this is consistent with ultrastructural observations of the infection process (Armentrout and Wilson, 1969; Manocha and Lee, 1972; Jeffries and Young, 1976b).

Most studies of biotrophic mycoparasitism have been made using dual cultures under laboratory conditions where growth of the host is favoured as the

Table 4.3. Non-mucoralean fungi tested for infection by *P. unispora*; none were infected (based on Jeffries and Young, 1978).* Hosts of *Piptocephalis xenophila* (Dobbs and English, 1954).

Oomycota	*Dictyuchus sterile*
	Phytophthora infestans
	Saprolegnia megasperma
Zygomycota	*Basidiobolus ranarum*
	Conidiobolus coronatus
	Linderina pennispora
Ascomycota	*Aspergillus reptans*
	**Penicillium chrysogenum* (= *P. notatum*)
	**P. waksmanii*
Basidiomycota	*Schizophyllum commune*
Deuteromycetes	*Fusarium culmorum*
	Trichoderma viride

environmental conditions can be readily manipulated. In the field, however, growth of the mycelium of the host could be suboptimal through nutrient deficiency, imbalance and competition and, in this situation, the effects of parasitism on the weakened host could be considerably enhanced. This theory is substantiated by laboratory experiments where the symbiotic balance is shifted in favour of the parasite. For example, if dual spore inocula are placed onto a nutrient medium with the spores of the parasite in excess, the germinating spores of the host are frequently attacked by several parasitic hyphae before the host can develop an extensive mycelium. Limited growth of the parasite occurs and the host dies. Suppression of growth of the host is occasionally seen in pure dual culture experiments (Jeffries, 1985).

The host range of most species of *Piptocephalis* appears restricted to Mucorales although not all of the Mucorales tested are susceptible to infection (Tables 4.2 and 4.3). A similar conclusion was reached for *Piptocephalis lepidula* which only parasitized Mucorales with the exception of *Mortierella candelabrum* (McDaniel and Hindal, 1979). Table 4.2 shows that the degree of growth of the parasite on the different mucoralean hosts varies widely. Hosts such as *Cokeromyces recurvatus* and *Cunninghamella elegans* are highly susceptible whereas species of *Mortierella* and *Phascolomyces articulosus* appear immune. Other species of *Piptocephalis* have also failed to parasitize *Phascolomyces articulosus* (Benny and Benjamin, 1976; Manocha and Golesorkhi, 1979) and *P. unispora* did not parasitize 13 of 18 species of *Mortierella* tested (Cuthbert and Jeffries, 1984). Of the susceptible species of *Mortierella*, all belong to the subgenus *Micromucor* which indicates biochemical specificity at the molecular level. A suggestion is that the ability of *Piptocephalis virginiana* to parasitize *Micromucor isabellinus* (= *Mortierella isabellina*) could be due to glycoproteins found in the walls of susceptible species of *Mortierella.* Such glycoproteins are, apparently, also found in extracts of the walls of *Phascolomyces articulosus.* As *Phascolomyces articulosus* does not support sustained infection by *Piptocephalis* species but is susceptible to *D. cristalligena* the glycoprotein hypothesis requires further exploration.

The mechanism of recognition of a potential host by the parasite at the molecular level is a difficult problem to solve and the evidence to date is largely circumstantial. As the hyphal wall is the first physical barrier to penetration encountered by the mycoparasite, the analysis of wall chemistry is likely to provide important clues to the processes involved in recognition of the host.

4.1.2 Other haustorial biotrophs

Among the other haustorial mycoparasites that are known to attack the mycelia of their hosts are members of the Basidiomycota. Many mycoparasitic species have been described from the Auriculariales and Tremellales and often haustorial structures have been observed, whereas in others the nature of the relationship has not yet been determined. Some *Tremella* species are also lichenicolous. The tremelloid haustorial apparatus comprises a basal clamp connexion bearing a small subglobose element from which thread-like irregularly curved outgrowths develop which can penetrate the hyphae of the host (Fig. 4.9). Many species of the genus *Tremella* are parasites of other fungi including other basidiomycetes. Members of the Tremellales which parasitize the mycelium of *Stereum* and produce basidiomes comprising hyphae of both the host and the parasite are known throughout much of the world, including North and South America, Asia and Europe. The basidiome characteristically consists of a white-coloured fibrous to fleshy core, or the lobes may be hollow, enveloped by a gelatinous layer consisting mainly of elements of the mycoparasite (Bandoni and Zang, 1990). The gelatinous *Tremella aurantia* (= *T. spectabilis*), *T. spectabilis* (= *T. aurantia*, Bandoni and Oberwinkler, 1983), *T. steidleri* and *T. tremelloidea* are found growing on the resupinate basidiomes of *Stereum hirsutum*, and *T. encephala* on *S. sanguinolentum.* Recently *Tremella aurantialba* which parasitizes *Stereum hirsutum* and related fungi which grow on species of *Quercus* and *Betula* has been described from China (Bandoni and Zang, 1990). The corticiaceous *Aleurodiscus amorphus* is parasitized by *T. mycophaga*, *A. tsugae* by *T. simplex*, *Corticium* species by *T. versicolor* and members of the fairy clubs by *T. parasitica.* Two members of the Tremellales also reported to be fungicolous are *Sirotrema translucens* (= *Pseudostipella translucens*) and *Xenolachne longicornis* (Ellis and Ellis, 1988). *Sirotrema translucens* parasitizes the apothecia of *Lophodermium conigenum*, *L. pinastri* and *L. seditiosum* and *Hypodermella* species which are found on the dead needles of *Abies* and *Pinus.* Other species of *Sirotrema* are *S. parvula* which grows in the ascomata of *L. pinastri* on the fallen needles of *Pinus contorta* and *S. pusilla* which is found in the ascomata of *Hypoderma pacificensis*

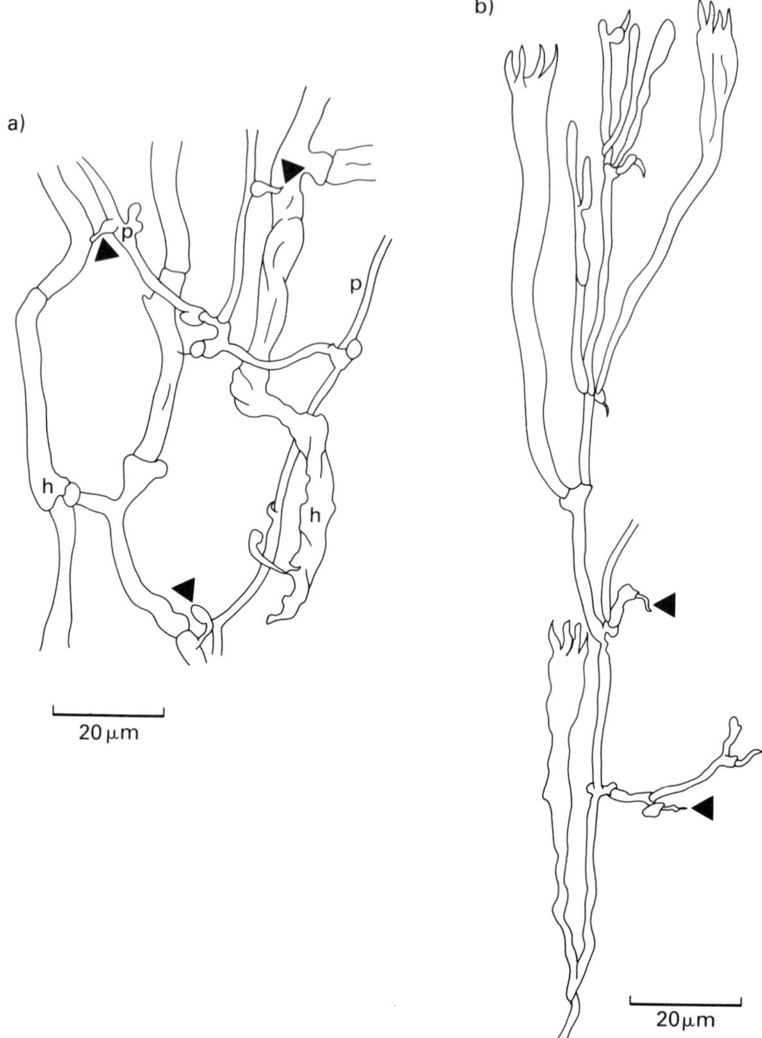

Fig. 4.9. The tremellaceous mycoparasite *Syzygospora alba* (redrawn from Oberwinkler and Bandoni, 1982). (a) Parasite hyphae attached to host hyphae with haustoria (arrows). (b) Hymenial section showing basidia and hyphae with haustorium-like outgrowths (arrows).

on the fallen petioles of *Acer macrophyllum*. The tremellaceous basidiome of *Xenolachne longicornis* develops on the apothecia of *Cudoniella clavus*, *Discinella margarita* and *Hymenoscyphus vernus* where it replaces the hymenium of the host. *Rhynchogastrema coronata*, isolated from soil, produces tremelloid haustoria in axenic culture and an implication is that this fungus is probably mycoparasitic although further study is required to support this conclusion (Metzler *et al.*, 1989). A conidial anamorph referred to Tremellales is *Tremellina pyrenophila* which grows slowly on agar in axenic culture and produces tremelloid haustoria (Bandoni, 1986). *Tremellina pyrenophila* was isolated from a species of *Ceratocystis* growing in the decaying litter of a palm tree where the incompletely developed ascomata were enclosed by the sporulating mycelium of the mycoparasite. Possibly the mycoparasite inhibits development of the ascomata of the host (Bandoni, 1986). There are also intrahymenial parasites with tremelloid affinities, such as *Tremella obscura* in *Dacrymyces* species (Olive, 1946; Christiansen, 1959; McNabb, 1965), *T. rhytidhysterii* in the lichenized ascomycete *Rhytidhysterium rufulum* (= *Rhytidhesteron rufulum*) (Oberwinkler and Bandoni, 1982), and *Sebacina penetrans* (Hauerslev, 1969), also in *Dacrymyces* species.

Haustoria have also been described from *Filobasidiella neoformans*, *Filobasidium floriforme*, *Rhyncogastrema coronata*, *Syzygospora alba* (= *Christiansenia alba*) and *Trimorphomyces papilionaceus* (Oberwinkler and Bandoni, 1982, 1983; Oberwinkler *et al.*, 1984; Metzler *et al.*, 1989). The host–parasite interaction of all these species is by haustoria which apparently penetrate the host cells. Morphologically the haustorial type is highly characteristic: small subglobose cells subtended by basal clamps form apical, thread-like and irregularly curved outgrowths which are capable of penetrating the host cells (Fig. 4.9). For example, *Syzygospora* contains several species described from the basidiomes of other members of the Basidiomycota (Ginns, 1986). *Syzygospora alba*, the type species, was described from a corticoid member of the Aphyllophorales (Oberwinkler and Lowy, 1981), *S. norvegica* from the surface of **hypertrophied** lamellae in the pileus of *Collybia butyracea*, and *S. marasmoidea*, *S. solida*, and *S. subsolida* on hypertrophied pilei of *Marasmius pallidocephalus*. Some of these relationships have been examined in more detail, and for example in *Tremella rhytidhysterii* (Bezerra and Kimbrough, 1978), a parasite of *Rhytidhysterium rufulum*, the close interaction with the host has been demonstrated by electron microscopy. In interactions such as these it appears that a comparatively narrow penetration neck is formed and that inside the host cell the parasitic hypha expands again. Furthermore, the host plasmalemma surrounds the intracellular part of the haustorium. A papilla is sometimes formed around the invading haustorium. Degeneration of host cytoplasm has been reported around the mature haustoria (Oberwinkler *et al.*, 1984) and it has been assumed that these structures have an absorptive role like that of similar structures in other host–parasite interactions. These fungi are unusual, however, as structures identical to the haustoria can also be regularly developed from hyphae of the parasite that are not in contact with a host cell, and can be formed in pure cultures in the absence of a host (Oberwinkler and Bandoni, 1982). Similar structures have also been seen in pure cultures of related species, such as *Tremella mesenterica*, which have not been recorded as fungicolous. The specialized cells with the tremelloid haustorium morphology may function likewise in other species where they are present. It can be seen, nevertheless, that many haustoria are not attached to host cells and, additionally, that haustoria are formed in pure cultures without the host. It is believed that species such as *T. mesenterica* with these pseudohaustoria may at least be potentially parasitic, even though a host is not known (Oberwinkler and Bandoni, 1982). The *Tremella* haustorial type is of great taxonomic significance and can be used for suprageneric classification (Oberwinkler and Bandoni, 1982).

A similar intramycelial structure has also been described in interactions of *Carcinomyces* with other fungi, but details are lacking. The Carcinomycetaceae, a family of the heterobasidiomycetes apparently allied to Cryptococcales, Exobasidiales and Tremellales (Oberwinkler and Bandoni, 1982), comprises the entirely mycoparasitic genera *Carcinomyces*, *Christiansenia* and *Syzygospora*. According to Ginns (1986), however, *Carcinomyces* and *Christiansenia* are synonyms of *Syzygospora* and all should be placed in the Syzygosporaceae. All grow and sporulate on homobasidiomycetes although a discrete basidiome may not be produced by the parasite. It is possible that further intrahymenial mycoparasites which do not produce recognizable basidiomes remain to be discovered (Oberwinkler and Bandoni, 1982) as the presence of such fungi in a basidiome clearly requires patient investigation and careful interpretation. As an example, the parasitic interactions of *Carcinomyces* species with *Collybia dryophila* are not well understood. Ginns and Sunhede (1978) found no evidence that the parasites penetrate the host cells. In *Carcinomyces effibulatus* (= *Syzygospora mycetophila* = *S. effibulata* = *Christiansenia effibulata*), narrow, ramified parasite hyphae with pointed ends were found attached to the host cells (Oberwinkler and Bandoni, 1982). However, these structures did not show the tremelloid haustorial morphology. In *Carcinomyces mycetophilus* a clear morphological interaction of species could also be seen in spite of the fact that it was old herbarium material that was examined.

With the aid of phase-contrast light microscopy a haustorial apparatus was detected in *Syzygospora tumifaciens* (= *Christiansenia tumefaciens*) and its structure resembles somewhat the tremelloid haustorial type (Oberwinkler and Bandoni, 1982). The fine structure, however, reveals a more intimate relationship, that of a fusion biotroph described in more detail in section 4.2. More detailed studies are certainly necessary to understand whether the haustorial structures of the *Christiansenia-Syzygospora* group and those of the *Carcinomyces* species are homologous. Indeed, much more information regarding the physiology and inter-relationships of these fungi needs to be obtained before firm conclusions can be made about the physiological similarity of these structures with the haustoria of other groups. For example, a very unusual interface type has been described (see section 4.2) in the interaction of the tremelloid fungus *Tetragoniomyces uliginosus*, the type and only species of the genus (Tetragoniomycetaceae; Tremellales) with sclerotia of its host *Rhizoctonia* (Bauer and Oberwinkler, 1990a). This was thought to be a simple haustorium as described above but, when the interface was examined in detail at an ultrastructural level, it was revealed that a micropore between the juxtaposed cells allows direct cytoplasm–cytoplasm connection to be made. The pore membrane is continuous with the plasmalemma of both elements. This interface type is also similar to those found in the fusion biotrophs and more details are given in section 4.2. It could be that several of the other interfaces described here as haustoria may also be found to be fusion interfaces on detailed examination as was the case in *Christiansenia pallida* (= *Syzygospora pallida*) (Bauer and Oberwinkler, 1990b). Regarding the taxonomy of *Christiansenia pallida*, Oberwinkler and Bandoni (1982) included *Christiansenia* in the Carcinomycetaceae, together with the type genus *Carcinomyces*, and *Syzygospora*. They used characteristics of basidia, conidial stages, septal pore ultrastructure, and mycoparasitism for a comprehensive circumscription of the family. *Christiansenia* appears to be a genus in that taxon. *Carcinomyces* species lack zygoconidia and, in addition, seem to be confined to the agaric host, *Collybia dryophila*. *Christiansenia pallida* is certainly closely related to *Syzygospora alba*, another species with zygoconidia. The Carcinomycetaceae characteristically, although not invariably, induce gall formation on the basidiomes of the hosts (which are either Agaricales or Aphyllophorales). Some species, e.g. *Carcinomyces effibulatus*, *C. mycetophilus* and *Christiansenia pallida* have also been grown in axenic culture when both a yeast-phase and a mycelial phase may be produced (Ginns, 1986). *Carcinomyces effibulata* and *C. mycetophilus* are known only to infect the basidiome of the agaric *Collybia dryophila* in which tumours due to hypertrophic growth of the host are produced although the host range may be wider than this. *Syzygospora norvegica* infects the basidiome of *Collybia butyracea* where the mycoparasite appears to be restricted, however, to growth on the lamellae of the host. The basidiomes of *Marasmius pallidocephalus*, infected by *Syzygospora marasmoidea*, *S. solida* and *S. subsolida*, also undergo hypertrophy and the pileus becomes deformed and may become unrecognizable through surface growth of the mycoparasite. *Syzygospora marasmoidea* induces the development of subcerebriform galls on the surface of the pileus. In *Syzygospora alba*, known only from Panama and Mexico, the macroscopic white-coloured tremelloid basidiomes (5–10 cm diameter) are initially soft and gelatinous but become hardened and dark brown on drying. *Syzygospora alba* produces tremelloid haustoria and apparently parasitizes the hyphae of corticoid Aphyllophorales which are intermixed with those of the lacunose basidiome of *S. alba* (Oberwinkler and Lowy, 1981; Oberwinkler and Bandoni, 1982). In the closely related *Christiansenia pallida* the mycelium develops both on and within the hymenium of *Phanerochaete sordida*(= *P. cremea*, Corticiaceae). The basidiome in *C. pallida*, which develops on the host, also comprises minute, soft gelatinous pustules (0.1–1 mm diameter) which shrink and toughen on drying and the host of this species does not undergo hypertrophy.

An apparently widespread, but infrequently collected tremellaceous fungus is *Trimorphomyces papilionaceus* which constantly associates with the dematiaceous *Arthrinium sphaerospermum* found on members of the Gramineae, e.g. *Dactylis glomerata* and *Phalaris arundinacea* (Oberwinkler and Bandoni, 1983).

This tremellaceous fungus develops minute (1–2 mm) gelatinous, tuberculate sporomes which produce conidia initially, then basidia. Prior to release from the sporomes, the conidia conjugate in pairs appearing H-shaped and are dikaryotic. The conidia undergo germination to produce pairs of buds, initiating a yeast-phase in the life cycle. The pairs of yeast-phase cells also conjugate to produce dikaryotic yeast cells. In culture, in the presence of hyphae of the potential host species *A. sphaerospermum*, both the dikaryotic conidia and the yeast cell dikaryons germinate producing hyphae and typical tremelloid haustoria (Oberwinkler and Bandoni, 1983).

Interactions of the extracellular mycoparasitic Chytridiomycota and their hosts have not been studied extensively with the electron microscope, with the exception of *Caulochytrium protostelioides* (Powell, 1981). This fungus invades its host *Cladosporium cladosporioides* with multiple small unbranched haustoria (Figs 4.10 and 4.11). The initial infection peg of the parasite is surrounded by a thin collar continuous with the inner layer of the host wall and covered by the host plasma membrane. The infection peg penetrates this collar and develops into a haustorium with a tubular neck and bulbous end. An extrahaustorial membrane surrounds the walled haustorium and an extrahaustorial matrix sepa-

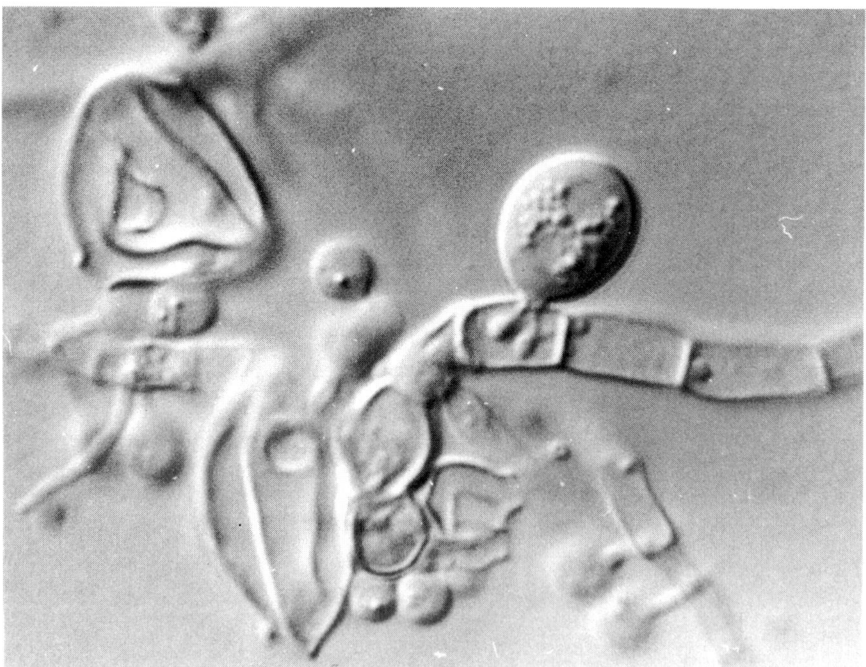

Fig. 4.10. (see also Fig. 4.11) Attachment of *Caulochytrium protostelioides* to a hypha of *Cladosporium cladosporioides*. Two haustoria (LM: × 200). (Courtesy of Professor M.J. Powell.)

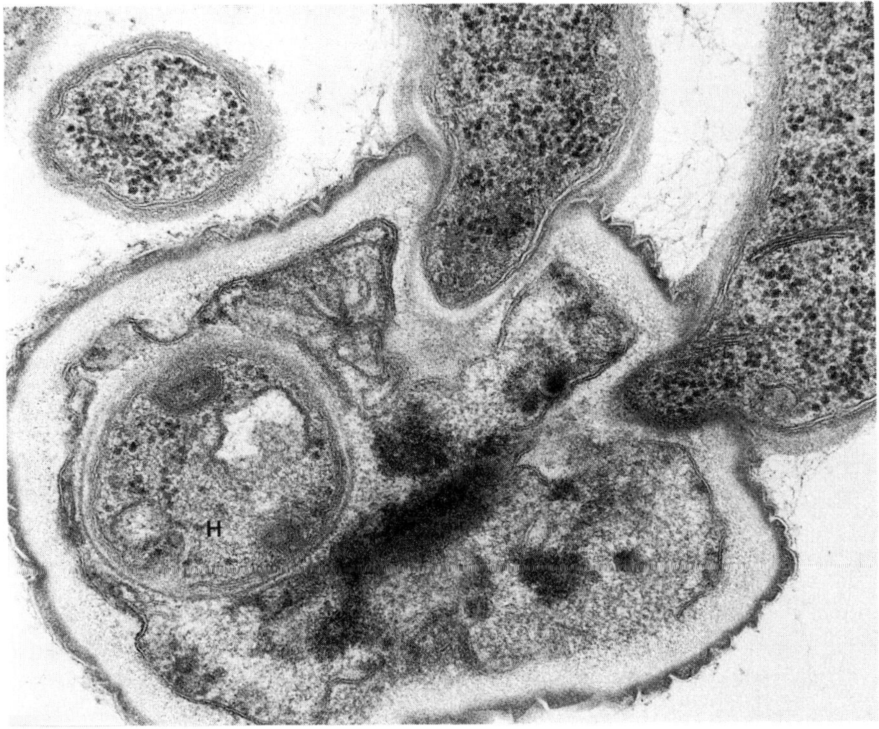

Fig. 4.11. (see also Fig. 4.10) Attachment of *Caulochytrium protostelioides* to a hypha of *Cladosporium cladosporioides*. Two infecting hyphae and a haustorium (H) in cross section (TEM: × 76,000). (Courtesy of Professor M.J. Powell.)

rates it from the cytoplasm of the host. The ultrastructural features of the haustorium thus closely resemble those of the haustoria of *Piptocephalis* species. Papilla formation is also a common phenomenon and haustoria may become encased by electron-dense material. The presence of the haustorium of *C. protostelioides* appears to damage the host little if infection of a single cell is not heavy, but eventually the cytoplasm appears degenerate and becomes disrupted (Powell, 1981) There are many other examples of epibiotic mycoparasites in the Chytridiomycota and Oomycota, but as these are detailed by Karling (1977) only examples are provided here. Sometimes only the spores of the host are attacked. An example is *Sparrowia* with two species *S. parasitica* and *S. subcrusiformis* which parasitize the oogonia and oospores of *Pythium debaryanum* (Peronosporales) and *Aphanomyces laevis* (Saprolegniales), and the zoosporangia of *Chytriomyces annulatus* and *C. poculatus*, *Karlingea rosea*, and *Rhizophydium coronum* (Chytridiales). The mature thallus of *Sparrowia* spp. comprises a zoosporangium attached to the host wall by a very fine unbranched or branched rhizoidal hypha which penetrates the host and enters the cytoplasm from which it is presumed that nutrients are absorbed. In many other epibiotic mycoparasitic chytrids, for example *Rhizidiomycopsis* spp., thallus development takes place on the surface of the host and the cytoplasm is invaded by fine branched rhizoidal hyphae which are presumed to behave like haustoria. Whether these structures are ultrastructurally similar to those of *Caulochytrium* has not been investigated. Most other chytridiaceous mycoparasites are endobiotic and are discussed in section 4.3.

One further example of a mycoparasite which produces a haustorium-like infection structure is *Sporidesmium sclerotivorum*. This fungus produces haustoria in the cortical and medullary hyphae of sclerotia of *Sclerotinia minor* (Bullock *et al.*, 1986). The haustorial structures, which may be spheroidal, clavate or extensively lobed, do not penetrate the plasmalemma of the infected element of the host and, also typical of the fungal haustorium, an electronlucent zone develops between the wall of the haustorium and the plasmalemma of the host. Cytoplasm of the haustorium appears more electron-dense than that of the host. Conidium production in *S. sclerotivorum*, which occurs 14–35 days after infection, seems to be correlated with the production of haustoria which suggests a nutrient absorptive function of these structures. The density of the cytoplasm of the haustoria is taken to indicate intense metabolic activity which also coincides with the period during which the need to absorb nutrients by the mycoparasite during spore production is greatest. After 40 days, when sporulation has ceased, the haustorial cytoplasm and that of the host appears degenerate and is presumed to have become functionless.

Another sclerotial mycoparasite, *Teratosperma oligocladum* may also produce similar haustoria, but the microscopical evidence is at present lacking. This mycoparasite is known to infect sclerotia in much the same way as *Sporidesmium sclerotivorum* (Ayers and Adams, 1981) and shares several other characteristics with this fungus. Both produce distinctive, dark-pigmented conidia, and a similar *Selenosporella* anamorphic state, and both form microsclerotia during their parasitic phases (Ayers and Adams, 1981). In recognition of these similarities, Hughes (1979) has transferred *S. sclerotivorum* to the genus *Teratosperma*. Both fungi have a restricted host range and parasitize sclerotia of *Sclerotinia minor*, *S. sclerotiorum*, *S. trifoliorum*, *Sclerotium cepivorum* and *Botrytis cinerea*, but not those of *Corticium rolfsii* and *Macrophomina phaseolina*. *Teratosperma oligocladum* grows better on living sclerotia than dead ones. Taken together, these characteristics suggest that *T. oligocladum* may not be a destructive, necrotrophic mycoparasite of sclerotia, but may also form a transient biotrophic infection of the host in a similar fashion to *S. sclerotivorum*. It is significant that the haustoria of *S. sclerotivorum* were not originally detected, but were discovered only during an ultrastructural study of the infection process (Bullock *et al.*, 1986).

4.2 Fusion Biotrophs

The fusion biotrophs establish physical contact with their hosts but no penetration of the host is believed to occur. Instead, the juxtaposed walls undergo mutual dissolution. Before the ultrastructural studies by Hoch (1977a,b, 1978) it was not known that direct channels of communication existed between host and parasite and these relationships were referred to as the contact mycoparasites (e.g. Barnett and Binder, 1973; Jordan and Barnett, 1978). However, ultrastructural studies have given a much clearer understanding of the host–parasite interface and shown that fusion of host and parasite protoplasts is an integral feature of these relationships. For this reason the term fusion biotrophs has been introduced.

Possible effects of these parasites on their hosts appear to be a reduction in growth rate and the dry weight of the mycelium and an increase in the rate of autolysis of the cytoplasm. The conidiospores of two fusion biotrophs *Gonatobotrys simplex* and *Dicyma parasitica* (= *Calcarisporum parasiticum* = *Hansfordia parasitica*), which require the growth-stimulating compound **mycotrophein** to promote germination (see section 5.1.2), produce germ-tubes which apparently secrete an attractant chemical which causes the hyphae of a potential host to grow towards them. This unusual ability to direct growth of the hyphae of the potential host seems to aid survival of the mycoparasite as the germ-tubes are only capable of limited axenic growth before contact. Apart from reducing the growth rate of their hosts, the fusion biotrophs often, but not invariably, appear to cause them little harm. No toxin production has been recorded and the reduced growth rates could be due to increased competition for nutrients.

Fusion biotrophs are mainly anamorphs from the hyphomycetes, and teleomorphic states in the Ascomycota are also known. The hosts of the fusion biotrophs are other hyphomycetes and Ascomycota. Several of the fusion biotrophs are closely related and belong to three genera; *Nematogonum*, *Gonatobotrys* and *Gonatobotryum* (Walker and Minter, 1981). *Nematogonum ferrugineum* (= *Gonatorrhodiella highlei*) has often been recorded from dead wood and bark, and is now known to be a fusion biotroph often associated in nature with *Nectria coccinea* (often with the *Cylindrocarpon* anamorph of this species) the causal agent of beech bark disease. It also infects *Chaetomella*, *Cladosporium*, *Graphium*, *Tritirachium* and *Verticillium* (Gain and Barnett, 1970) in laboratory studies but is not known to parasitize them in nature (Walker *et al.*, 1982). The related *N. mycophilum* has been isolated from a coelomycete genus, *Eleutheromyces*, itself growing on a decaying agaric (Walker and Minter, 1981) but this fungus has not been firmly established to have a fusion biotrophic relationship with its host. Other fusion biotrophs have also been described as common associates of other fungi in nature. For example, *Gonatobotryum fuscum* has been observed on field collections of *Ceratocystis fagacearum*, the causal agent of oak wilt disease (Shigo, 1960a), and has also been recorded from bark and wood of various trees including *Fagus* and *Quercus*, and as a parasite of a variety of fungi including *Chalaropsis*, *Graphium* and *Leptographium* from Europe (UK, Italy) and North America (Walker and Minter, 1981). *Gonatobotryum parasiticum* is also parasitic on other fungi and although the mode of parasitism is unspecified it is likely to resemble that of *G. fuscum*. It has been recorded from *Ganoderma*, *Hypocrea*, *Hypomyces*, *Polyporus*, *Poria*, *Tremella* and *Trichoderma* in Europe and North America. It appears that *G. parasiticum* parasitizes members of the Hypocreaceae (Walker and Miner, 1981) themselves growing on other decaying fungi (see section 2.1).

Melanospora zamiae, also a fusion biotroph, has a wide host range (*Ceratocystis fagacearum*, *Ophiostoma ulmi* (= *O. ovo-ulmi* = *Ceratocystis ulmi*), *Cladosporium* sp., *Stenocarpella maydis* (= *Diplodia zeae* = *Diplodia maydis*), *Fusarium oxysporum* f. sp. *lycopersici*, *Fusarium* sp., *F. roseum*, *Sphaeropsis malorum*, *Tritirachium* sp., *Verticillium albo-atrum*, and *V. dahliae*). However, contact branches do not form with the following which are presumed to be non-

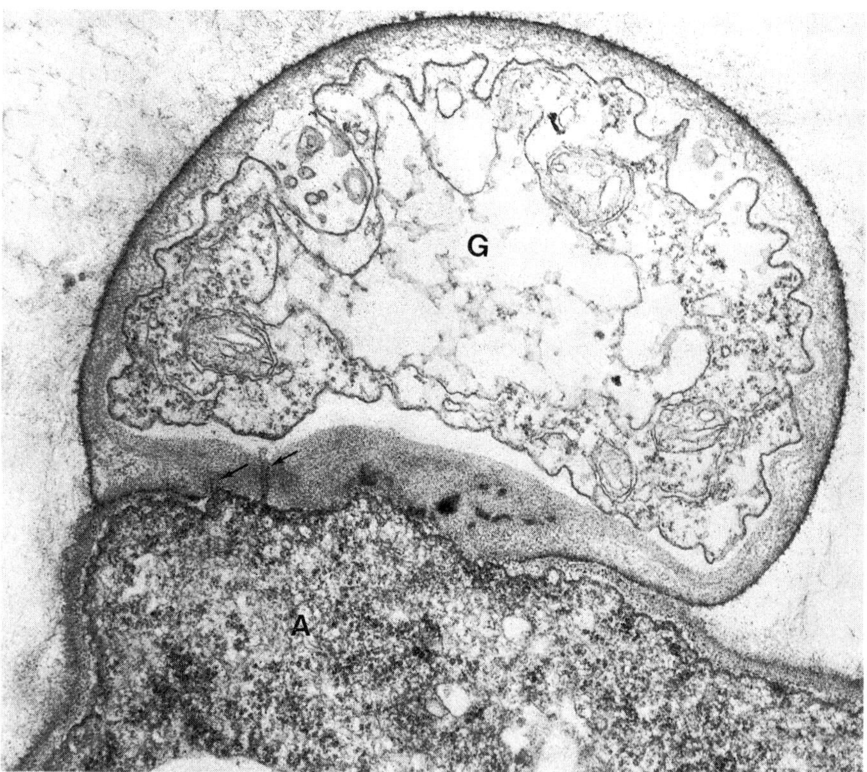

Fig. 4.12. Contact zone between the mycoparasite *Gonatobotrys simplex* (G) and its host *Alternaria tenuis* (A). Fine pores have developed (arrows) through the walls of the adjacent hyphae (TEM: × 78,600). (Courtesy of Dr H.C. Hoch.)

hosts: *Armillaria mellea, Arthrobotrys musiformis, Aspergillus carneus, A. rugulovalvus, Chaetomium globosum* (= *C. cochliodes* = *C. olivaceum*), *Penicillium vulpinum, Bjerkandera adusta* (= *Polyporus adustus*), *Rhizoctonia solani, Corticium rolfsii, Chondrostereum purpureum* (= *Stereum purpureum*), *Zygorhynchus moelleri* (Jordan and Barnett, 1978). Microconidiophores of other mycoparasitic *Melanospora* species have been described under the anamorphic state *Harzia.* A similar *Harzia* form occurs in other fusion biotrophic hyphomycetes such as *Acladium tenellum* which suggests that these fungi may also be anamorphs of Centrostomataceae. *Gonatobotrys simplex* is the anamorph of *Melanospora damnosa* which is itself a contact mycopathogen of species of *Fusarium* (Vakili, 1989). *Fusarium sacchari* (= *F. subglutinans*) is the causal agent of stalk rot of *Zea mays* (corn) and perithecia of the teleomorph *M. damnosa* are found in association with the *Fusarium* host growing in the pith of the stem at the origin of the leaves. Individual ascospore isolates of *M. damnosa* germinated on potato dextrose agar (PDA) produce mycelium of the anamorph *G. simplex.* Although the majority of the ascospores in a sample germinate and produce a mycelium capable of axenic growth on PDA, about 6% require supplementation with thiamine and 2% with biotin. Nutritional dependence (**auxotrophy**) in some form appears to be commonly associated with biotrophic behaviour. Pathogenicity of the mycelia derived from isolated ascospores is variable, ranging from no visible effect to highly pathogenic. The mycoparasitic mycelia produce hook cells which partially encircle the hyphae of the host. Where such contacts are made the

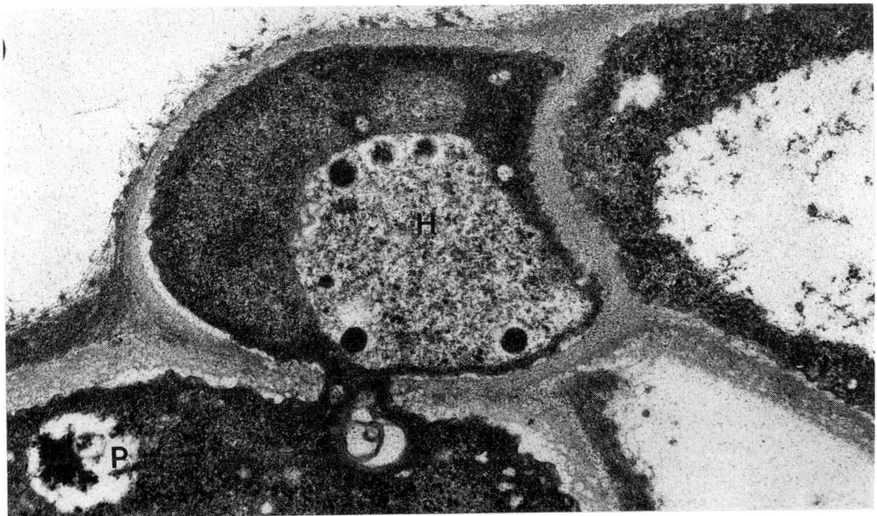

Fig. 4.13 Contact zone between the fusion mycoparasite *Dicyma parasitica* (H) and its host *Physalospora obtusa* (P). A pore is evident at the interface of the two fungi (TEM: × 45,000). (Courtesy of Dr H.C. Hoch.)

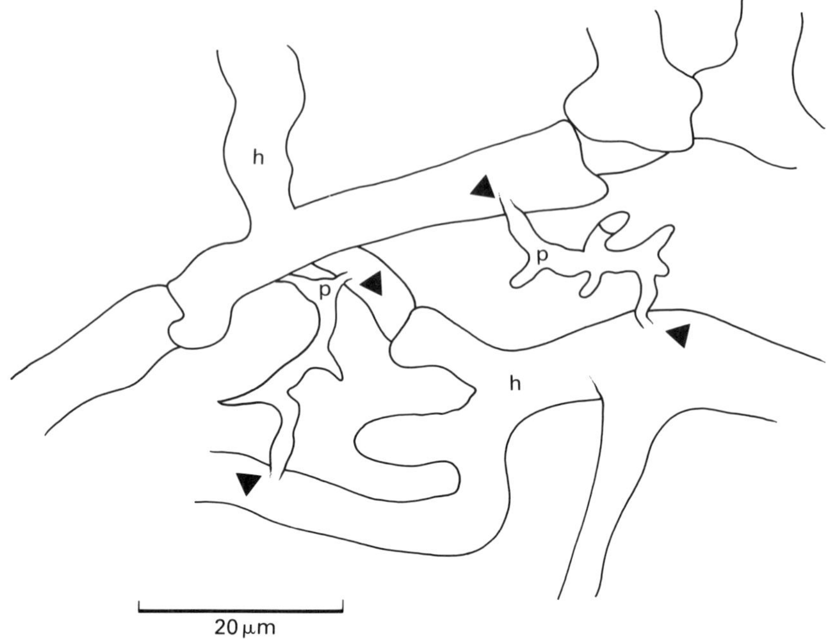

Fig. 4.14. Line drawing of the interaction (arrows) of a tremellaceous mycoparasite (p) and its host (h). (Redrawn from Oberwinkler and Bandoni, 1982.)

cytoplasm of the host dies and growth ceases. Six hyphomycetes, *Acladium tenellum*, *Dicyma parasitica*, *Gonatobotryum fuscum*, *Gonatobotrys simplex*, *Nematogonum ferrugineum* and *Stephanoma phaeospora* have all been reported as contact biotrophs of *Fusarium* spp. and at least four *Melanospora* species have been reported similarly (Vakili, 1989). *Acladium tenellum*, for example, was observed growing in association with *Fusarium moniliforme* on ears of corn (Kuykendall *et al.*, 1983). *Harzia acremonioides* which parasitizes *Stemphylium botryosum* also produces unicellular, occasionally multicellular, hooked or lobed contact cells (Urbasch, 1986).

Calcarisporium species have been described from other mycoparasitic niches. The genus was erected to accommodate *C. arbuscula*, a fungicolous associate of Basidiomycota and Ascomycota. *Calcarisporium ovalisporum* (= *Cladobotryum ovalisporum*), on the other hand, is usually found as a mycoparasite of entomopathogenic fungi, and was originally thought to be entomopathogenic itself (and was described as such as *Cladobotryum ovalisporum*) until the true nature of its relationship was known. It has recently been described growing within the mycelium and synnemata of *Hirsutella citriformis*, an important entomopathogen of the brown plant-hopper (Rombach and Roberts, 1987).

One of the most complete accounts of the behaviour of a fusion biotroph is

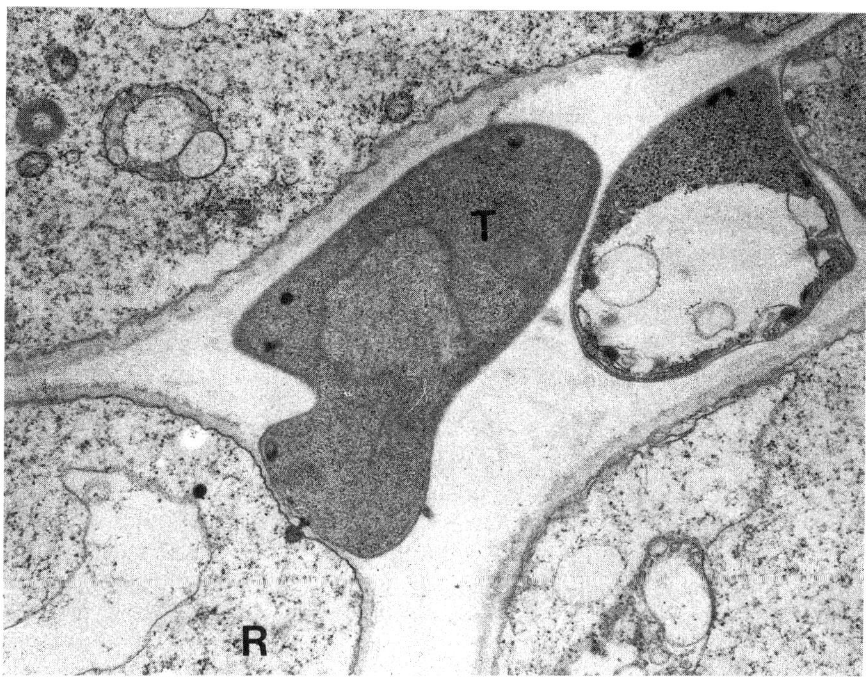

Fig. 4.15. (see also Figs 4.16 and 4.17) Longitudinal section through haustoria of *Tetragoniomyces uliginosus* (T) within a hypha of *Rhizoctonia solani* (R). The haustorium enters the host hypha through a membrane-lined micropore. TEM: × 11,300. (Courtesy of Dr R. Bauer and Professor Dr F. Oberwinkler.)

that of *Dicyma parasitica* which parasitizes species of *Physalospora* and related fungi (Barnett and Lily, 1958). Contact between hyphae of *D. parasitica* and those of the host is established through the production of lateral branches from each fungus which touch apically. A specialized buffer zone approximately 2.0 μm wide normally develops which serves to provide the interface between host and parasite.

An aquatic fungus, *Sphaerulomyces coralloides*, which infects the conidia of the hyphomycetes *Anguillospora crassa* and *Tricladium splendens*, was described as a fusion biotroph (Marvanova, 1977). There is, however, no experimental evidence for biotrophy, nor has any dissolution of the depressed walls been observed. The cytoplasm in the spores of the host dies in the region of contact, and this fungus might be better placed at the present time with the contact necrotrophs.

Ultrastructurally, at the contact interface between *Gonatobotrys simplex* and the hypha of *Alternaria tenuis* (= *A. alternata*), the cytoplasm of the two fungi is in direct connection through minute channels which traverse the interface

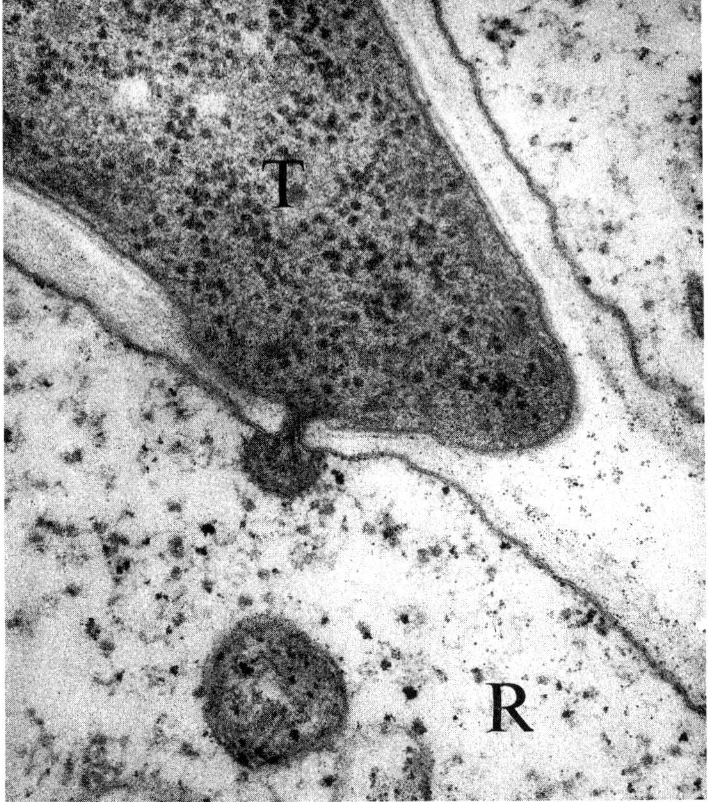

Fig. 4.16. (see also Figs 4.15 and 4.17) Longitudinal section through a haustorium of *Tetragoniomyces uliginosus* (T) within a hypha of *Rhizoctonia solani* (R). TEM: × 81,000. (Courtesy of Dr R. Bauer and Professor Dr F. Oberwinkler.)

(Hoch, 1977a). Structurally, these fine cytoplasmic connections resemble the plasmodesmata of the cells of plants and their occurrence provides strong circumstantial evidence that this is likely to be the route through which the transfer of nutrients to the mycoparasite occurs (Fig. 4.12). Similar plasmodesmata have also been demonstrated at the interface of *Stephanoma phaeospora* which parasitizes *Fusarium* sp. (Hoch, 1978) and in the mycoparasitic heterobasidiomycete *Christiansenia pallida* (Oberwinkler and Bauer, 1990). A much larger fusion zone has been described for *Dicyma parasitica*, in which a single pore, 0.2–1.0 μm in diameter, is formed between the contacting hypha and the wall of the hypha of the host (Fig. 4.13) through which free cytoplasmic exchange could presumably occur (Hoch, 1977b). A suggestion is that this type of interaction might allow fungal viruses (mycoviruses) to be transferred from one fungus to another. As the aggressiveness of certain plant-pathogenic fungi is known to be attenuated when

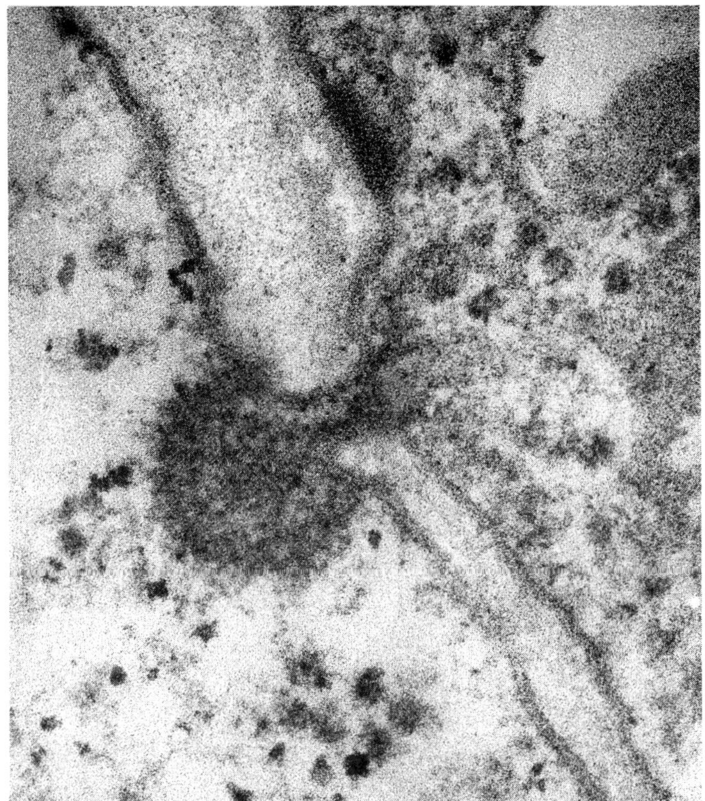

Fig. 4.17. (see also Figs 4.15 and 4.16) Longitudinal section through a haustorium of *Tetragoniomyces uliginosus* (T) within a hypha of *Rhizoctonia solani* (R). TEM: × 229,000. (Courtesy of Dr R. Bauer and Professor Dr F. Oberwinkler.)

they are infected with mycoviruses, this phenomenon suggests a further role of the mycoparasites in the biological control of plant pathogens. A single micropore is also found in the interface of the tremelloid mycoparasite *Tetragoniomyces uliginosus* and the host *Rhizoctonia* sp. (Bauer and Oberwinkler, 1990a). As described in section 4.1.2, more detailed examination of the interfaces between the 'haustoria' produced by some of the heterobasidiomycetaceous fungi has revealed direct cytoplasm–cytoplasm connections. These examples thus illustrate an interface type that shares characteristics of the haustorial biotrophs, yet also exhibits cytoplasmic fusion between partners (Fig. 4.14).

When *Tetragoniomyces uliginosus* infects *Rhizoctonia* sp. several stages (Figs 4.15–4.17) in the development of cellular interactions can be recognized:

1. contact of the haustorial apex with the host cell;

2. penetration of the haustorial apex into the host cell wall;
3. development of a body close to the haustorial plasmalemma;
4. fusion of the haustorial and host cell protoplasts beneath the body into the pore channel and partially into the host side (Bauer and Oberwinkler, 1990a).

The micropore is approximately 14–19 nm in diameter, and the pore membrane is continuous with the plasmalemma of both hyphae. The pore clearly allows the movement of a homogeneously-structured body of medium electron-density to move partially into the host cell from the parasite (Figs 4.16 and 4.17). This body, originally conical and approximately 160–190 nm diam. and 155–190 nm long, does not contain ribosomes or other organelles. During the later stages of the interaction, it becomes located mainly on the host side of the pore and an organelle-free zone often surrounds the side of the pore within the cytoplasm of the parasite.

In a parallel study, the interaction of *Christiansenia pallida* and *Phanerochaete sordida* (= *P. cremea*) was restudied by transmission electron microscopy (Bauer and Oberwinkler, 1990b). Several stages could again be recognized, some of which are different in detail from those described above for *T. uliginosus*:

1. contact of the haustorial apex with the host hypha;
2. nesting of the haustorial apex into the wall of the hypha of the host;
3. development of a digitate protrusion and production of vesicle-like particles;
4. penetration of the hypha of the host;
5. fusion of the protoplasts of hyphae of the host and the parasite through micropores.

The distinctive features of the infection apparatus of *C. pallida* are the digitate protrusion, the vesicle-like particles and the micropores. Similar protrusions and vesicle-like particles have not been described for host–parasite interactions of other basidiomycetous mycoparasites. In the extrahaustorial matrix of *Urocystis* spp. vesicle-like bodies have been observed (Nagler and Oberwinkler 1989), but their function and origin are unknown. In *C. pallida* the vesicle-like particles are of parasitic origin, derived from the digitate protrusion. The protrusion apparently develops rapidly during early developmental stages of parasitic interaction. After penetration several micropores are produced between the haustorial filament and the host cell. The plasmalemma is continuous through the pore channels. Up to 15 micropores, approximately 14–18 nm in diameter, have been seen in serial sections of a single invasive protrusion. This contrasts with the situation for *T. uliginosus*, where only a single micropore is formed, and where there is no invasion of the host by a digitate protrusion. However, early developmental stages of parasitic interactions in *T. uliginosus* and *C. pallida* are similar. In both species the haustorial apex nests into the wall of the hypha of the host and a homogeneously-structured body appears in the centre of the haustorial apex in close association with haustorial plasmalemma. The similar types of parasitic interaction in *C. pallida* and *T. uliginosus* can be interpreted as a character of common phylogenetic origin. The '*Christiansenia* haustorial type' appears to be more advanced in comparison with the '*Tetragoniomyces* interaction type' (Bauer and Oberwinkler, 1990b). Whether micropores are present in all tremelloid host–parasite interactions is unknown as morphologically similar haustoria are present in a variety of *Tremella* species (Olive, 1946; Bandoni 1961, 1987; Bezerra and Kimbrough, 1978), in *Syzygospora alba* (Oberwinkler and Lowy, 1981), in *Trimorphomyces papilionaceus* (Oberwinkler and Bandoni, 1983), in *Filobasidium floriforme* (Olive, 1968), and in *Filobasidiella neoformans* (Kwon-Chung, 1976). Micropores with cytoplasmic connections have not yet been described from these fungi. Additional ultrastructural investigations are therefore needed to clarify the host-parasite interactions in this group.

Another type of host–mycoparasite interaction has been described from *Phragmoxenidium mycophilum*, a basidiomycetous parasite of *Uthatobasidium fusisporum* (Oberwinkler *et al.*, 1990). Micropores protrude through the abutted walls of the hyphae of the parasite and the host but, in this example, cytoplasmic connections through the pores have not been found.

One group of fungal interactions which have not been investigated in detail at an ultrastructural level are the gall-forming members of the Mucorales. Species of *Chaetocladium*, for example, parasitize other mucoralean fungi by inducing galls comprising intermingled outgrowths of hyphae of both the host and the parasite. When grown in the vicinity of a potential host colony, directed growth

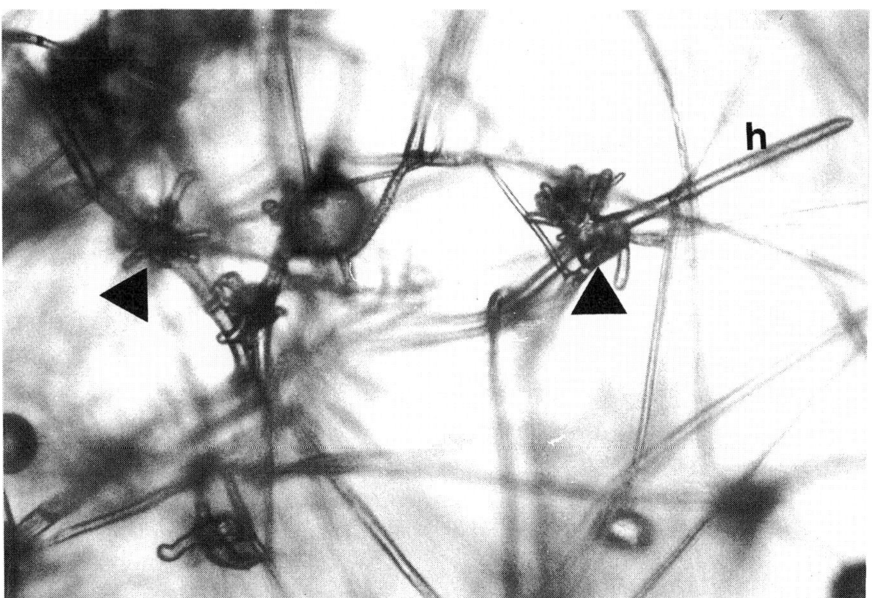

Fig. 4.18. Gall formation (arrows) between a hypha of *Chaetocladium brefeldii* and the host (h) *Cokeromyces recurvatus* (LM: × 200).

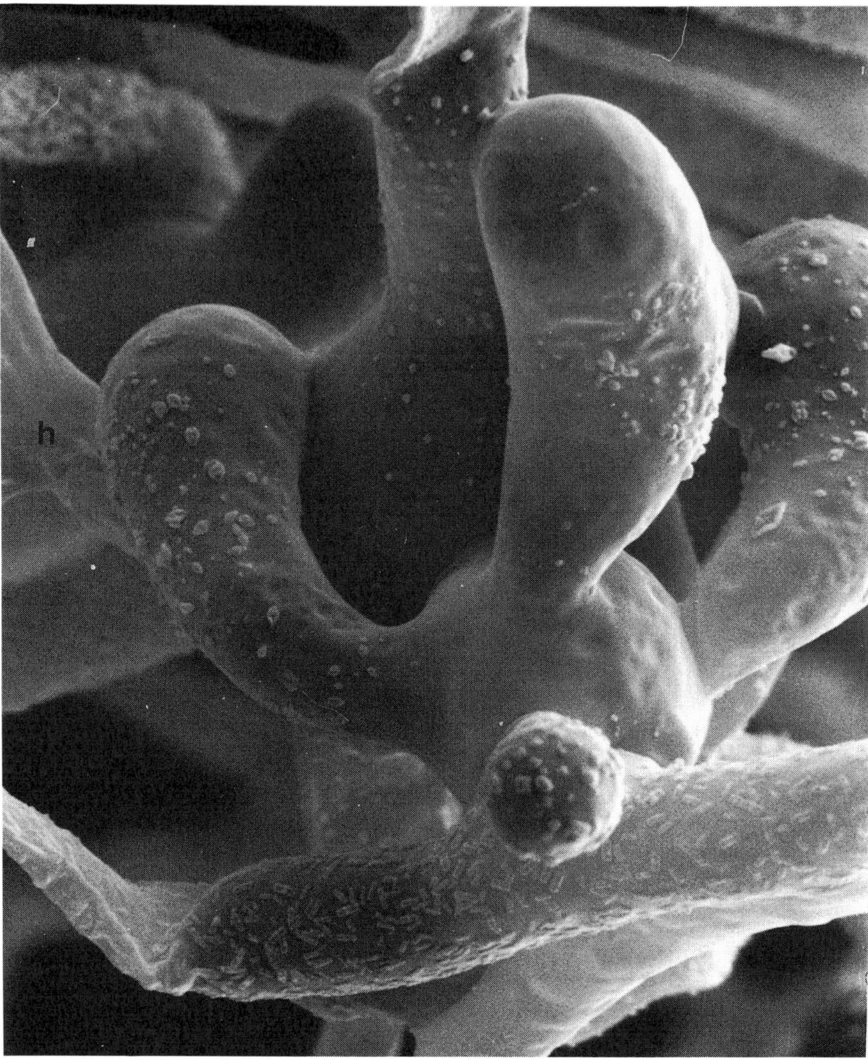

Fig. 4.19. Gall formation between *Parasitella parasitica* and the host (h) *Cokeromyces recurvatus* (SEM: × 3000).

of the hyphae occurs towards those of the host and galls form in the region of contact (Fig. 4.18). The galls are said to contain nuclei derived from both partners which indicates that wall breakdown and cytoplasmic coalescence occur. Although there is no experimental information to show that the galls are absorptive, circumstantial evidence of mycoparasitism is observed, as, following gall formation, there is stimulation of the growth of *Chaetocladium* accompanied by

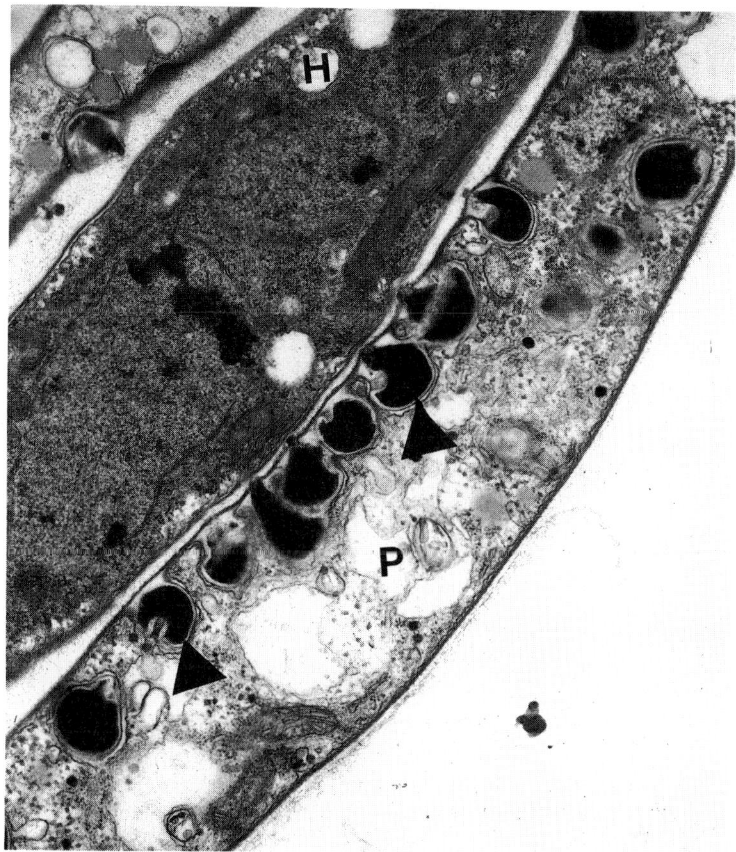

Fig. 4.20. Interaction of hyphae of the parasite *Colacogloea peniophorae* (P) and the host *Hyphoderma praetermissum* (H) showing colacosomes (arrowed). (TEM: × 19,100). (Courtesy of Dr R. Bauer and Professor Dr F. Oberwinkler.)

suppression of growth of the host. This situation contrasts with that in a related fungus, *Parasitella*, also a gall-former (Fig. 4.19) which does not, however, exhibit differential growth effects following gall formation. Nevertheless, there are similarities with *Chaetocladium* as the hyphae show directed growth towards a potential 'host' and there are specific localized zones where galls are formed. It is likely that gall formation in *Parasitella* represents abortive sexual conjugation as it is known to be a sex-linked phenomenon. *Parasitella* is heterothallic, with two mating strains appearing morphologically similar being designated '+' and '−'. The plus mating strains of *Parasitella* will only form galls on mycelia of opposite mating type and, in addition, the morphology of gall formation closely resembles the initial stages in the development of zygosporangia, the sexual

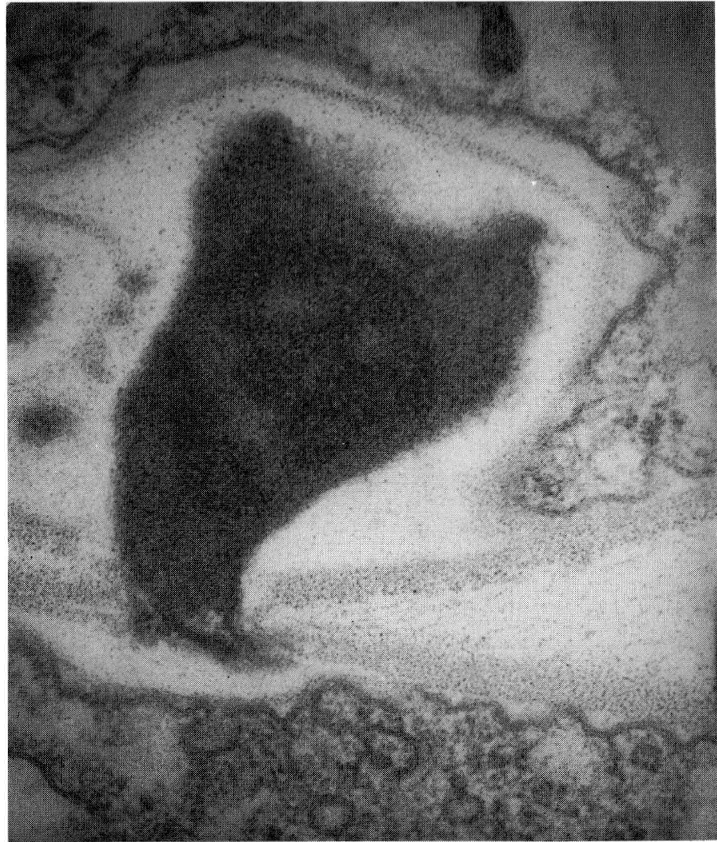

Fig. 4.21. High magnification of a colacosome similar to those shown in Fig. 4.20 (TEM: × 137,000). (Courtesy of Dr R. Bauer and Professor Dr F. Oberwinkler.)

stage. Galls are also known to form between opposite mating strains of *Parasitella*. In *Chaetocladium*, however, the interaction is not linked with a pseudosexual process and appears truly mycoparasitic. Thus it seems that the morphological similarities between gall formation in *Chaetocladium* and *Parasitella* are not reflected in their physiological behaviour (see also section 1.4).

The final interface to be described here concerns a unique structure formed at the interface between certain heterobasidiomycete fungi and their hosts. It involves the formation of a distinct organelle, called a colacosome, within the periplasmic region of the hypha of the mycoparasite where it is in close contact with a hypha of the host fungus (Bauer and Oberwinkler, 1991). This structure eventually penetrates the walls of the parasite and host such that its tip is in contact with the invaginated plasmalemma of the host. The physiological role of

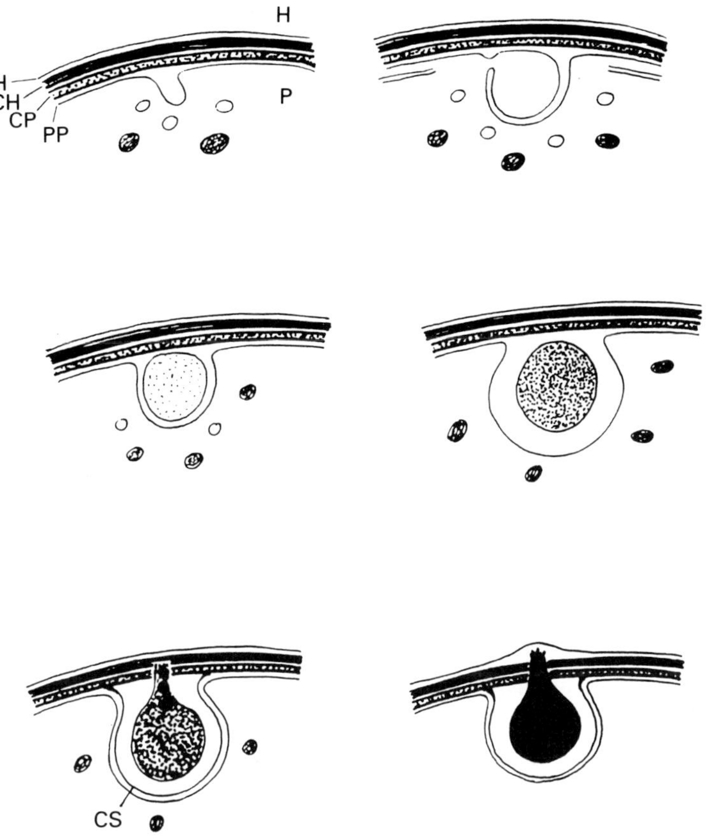

Fig. 4.22. Diagram of colacosome development, based on serial sections, and with reference to the original transmission electron micrographs of this chapter. Abbreviations and symbols: CH, cell wall of the host; CP, cell wall of the parasite; CS, secondary cell wall layer; H, cell of the host fungus *Hyphoderma praetermissum*; P, cell of the parasite *Colacogloea peniophorae*; PH, plasmalemma of the host; PP, plasmalemma of the parasite. (i) Initial stage of invagination of the plasmalemma of the parasite. (ii) The plasmalemma of the parasite recurves. (iii) Delimitation of the young colacosome from the cytoplasm. (iv) The central part of the colacosome becomes homogeneous and more and more electron-dense. The electronlucent sheath of the colacosome increases in thickness. (v) The electron-opaque core material penetrates the cell wall of the parasite and begins to intrude into the cell wall of the host. (vi) Final developmental stage with colacosome penetration through host cell wall. (Modified from Bauer and Oberwinkler, 1991.)

this organelle is unknown, but it is presumed to aid transfer of nutrients between the two fungi. If the parasite and/or the host hypha continues to grow, additional colacosomes are rapidly developed such that the number of connections between the partner fungi is continually increased and the hyphae continue to grow in close contact with one another.

The colacosome was described in detail from the interaction of the parasite *Colacogloea* (= *Platygloea*) *peniophorae* (Auriculariales) and the host *Hyphoderma praetermissum*. *Colacogloea peniophorae* hyphae grow internally in fructifications of their hosts and develop basidiomes on the surface of the host hymenia. Colacosomes develop in the contact area as illustrated in Figs 4.20, 4.21 and 4.22. They are globular, subglobular or beaked. The central part of the colacosome is electron-dense, 0.3–0.4 μm diam., enclosed by a membrane and surrounded by an electronlucent, unstructured sheath of approximately 0.05 μm diam. (Fig. 4.21). The colacosome is covered by the plasmalemma. A thin secondary cell wall layer is often present along the plasmalemma covering the colacosome. During their formation, the plasmalemma of the parasite is folded into the cytoplasm, then recurves, and finally fuses with itself at a distance of 0.2–0.3 μm from the original outgrowth. Thus, a more or less globose compartment is formed. Consequently, it is surrounded by a membrane derived from the plasmalemma. The globose compartment is now separated from the cytoplasm, the vesicular core becomes homogeneous and finally more and more electron-dense. Simultaneously, the intermembranous space between the central part of the colacosome and the cytoplasm increases slightly in thickness. Interaction of the juxtaposed hyphae starts with intrusion of electron-dense core material of the colacosome into the hypha of the parasite. The wall close to the intrusion peg becomes electronlucent and indistinct in substructure. Intrusion then continues through the closely attached hyphal wall of the host into an electronlucent protuberance formed between the hyphal wall and the plasmalemma of the host.

The majority of the other species included in the Auriculariales are parasites of fungi, mosses, ferns, and flowering plants (Bandoni, 1984). Among auricularioid taxa, the genera *Cystobasidium*, *Mycogloea*, *Colacogloea* and *Phragmoxenidium* contain mycoparasitic species (Lagerheim, 1898; Olive, 1951; Bandoni, 1956, 1984; McNabb, 1965; Oberwinkler *et al.*, 1990). *Colacogloea peniophorae* grows in and on fructifications of basidiomycetes, mainly *Hyphoderma praetermissum* (Corticiaceae), but it is also reported as a parasite from species of the genera *Dacrymyces*, *Odontia*, *Phanerochaete*, and *Poria*. However, it is unclear whether all of these reports definitely refer to *C. peniophorae* (Bandoni, 1956). The colacosome has also been found in another heterobasidiomycete, *Cryptomycocolax abnorme* (Cryptomycocolacales). This fungus was described as a mycoparasite of the sclerotioid bodies of an unnamed ascomycete growing on the rotten culms of *Cirsium subcoriaceaum* in a pasture in Costa Rica (Oberwinkler and Bauer, 1990). The thallus of *C. abnorme* comprises a septate, dikaryotic mycelium with clamp connections which enters the sclerotia and associates with the hyphal elements of the host. In the contact zone, the clamped hyphae of the parasite are penetrated by colacosomes from the unclamped hyphae of the host. In this particular interaction, a second type of colacosome was formed which had a more electronlucent core and a pore, 7–14 nm diam., through the region where the two hyphal walls were closely adpressed. The latter type of colacosome was found along cytoplasmic intrusions of the host formed in the hypha of the parasite. There is clearly a complex reaction between these two fungi. Colacosomes also form, apparently, when hyphae of *C. abnorme* are also closely adpressed to one another. *Cryptomycocolax abnorme* produces thick-walled intercellular chlamydospores within the sclerotia and naked elongate holobasidia each bearing a tetrad of sessile basidiospores on the surface. One of the most outstanding features of this mycoparasite, however, are the simple hyphal septa each with a central pore with Woronin bodies on each side which were formerly regarded as diagnostic of the Ascomycota. Perhaps *Cryptomycocolax* represents, a previously missing evolutionary link between Ascomycota and Basidiomycota and it is fitting that the link should appear to be a mycoparasite!

4.3 Intracellular Biotrophs

The Oomycota and Chytridiomycota contain fungi characterized by the formation of a dispersive motile spore, the zoospore. In parasitic forms, the zoospore is the infective agent which is attracted to host tissues, on which encystment takes place and penetration occurs. The penetration process of endobiotic species involves the movement of the complete protoplast of the parasite into the underlying cytoplasm of the host. This sequence of events is shared by a number of mycoparasitic species.

Catenaria, *Olpidium*, *Phlyctidium* and *Rozella* are examples of Chytridiales, an order characterized by a posteriorly uniflagellate zoospore. *Olpidiopsis* and *Rozellopsis*, on the other hand, are placed in the Lagenidales with biflagellate zoospores. During the infection process in these examples the entire thallus becomes attached to the host, which may be the thallus of another chytrid or the mycelium or sporulating stage of another filamentous fungus. Zoospores of the parasite are attracted chemotactically by a diffusible substance produced by the living host. On contact, the zoospores bind strongly to the hyphae of the host and penetration by the protoplast occurs rapidly to form the intracellular biotrophic infection. Two distinct types of infection are known to occur (Held, 1981). The first is typified by species of *Rozella*. When *Rozella* infects the mycelial chytrid *Allomyces*, for example, the invasive protoplast which fills part of the host hypha stimulates the development of septa which isolate the parasite from adjacent uninfected regions (Held, 1972, 1981). The hyphal wall of the host thus surrounds its own cytoplasm and that of the parasite which remains bounded by its own plasmalemma (Fig. 4.23). The cytology of the interaction has been well illustrated for *Rozella polyphagi* infecting *Polyphagus euglenae* (Powell, 1984).

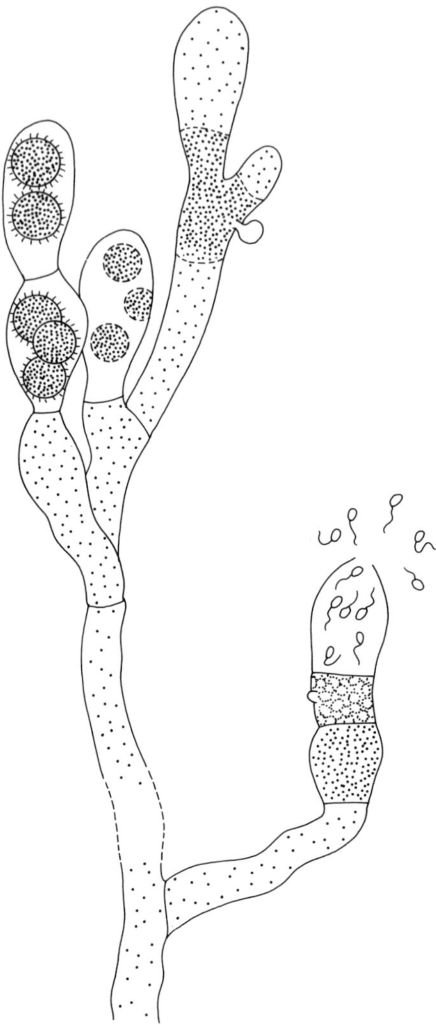

Fig. 4.23. A hypha of *Allomyces* parasitized by the chytrid *Rozella* showing several stages of the infection cycle.

Initially, the mycoparasite appears to cause little damage and this stage is presumed to be biotrophic. This structural interface is interesting in that the unwalled thallus of the endoparasite is in direct contact with the protoplasm of the host with no intervening host membranous envelope (Fig. 4.24). This type of interface was considered structurally unusual for endoparasites (Bracker and Littlefield, 1973), but similar interfaces have now been found in a number of other endoparasites (Powell, 1984). Later, however, cytoplasm of the host is

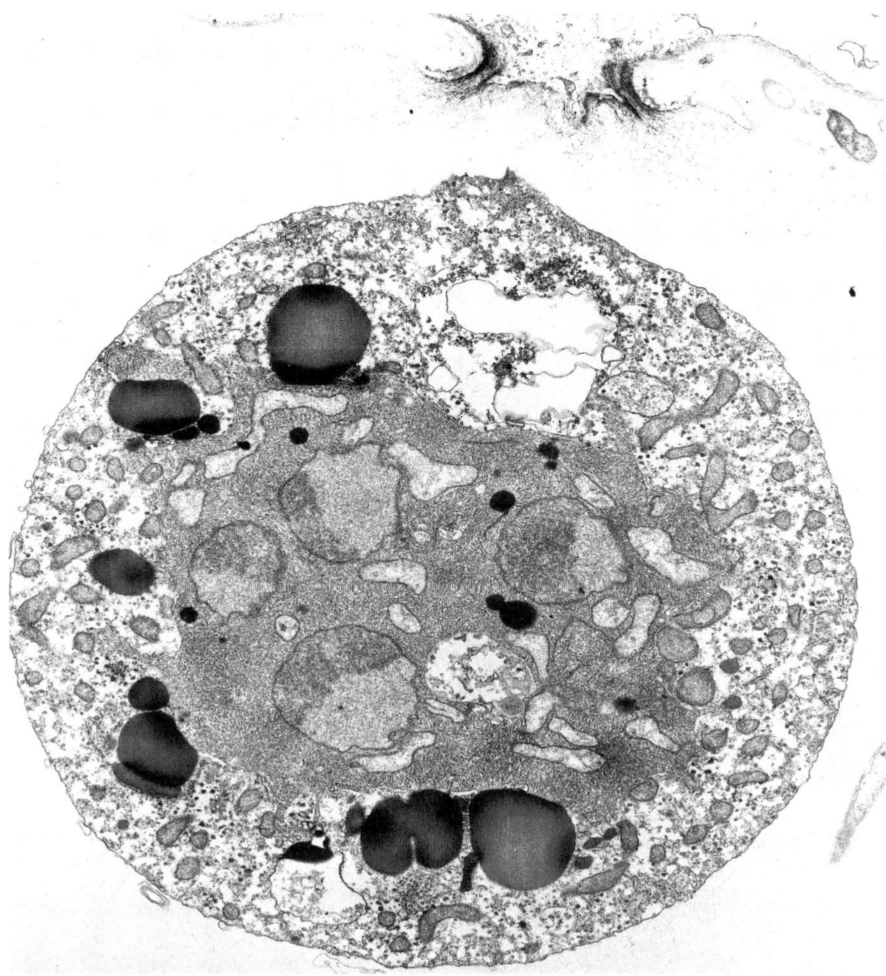

Fig. 4.24. Naked thallus of the endobiotic mycoparasite *Rozella polyphagi* inside a sporangium of the host *Polyphagus euglenae* (TEM: × 21,500). (Courtesy of Professor M.J. Powell.)

destroyed as that of the parasite differentiates into another generation of zoospores. Destruction of the cytoplasm of the host is apparently through phagocytosis by the endoparasite, which engulfs and digests host material. Once the complete protoplast of the host has been removed, sporulation of the parasite takes place. In contrast, in all other intracellular, or endobiotic biotrophic mycoparasites, the invasive element does not stimulate septum formation and several mycoparasitic individuals may share common cytoplasm. The protoplasts of these latter fungi, for example mycoparasitic species of *Catenaria*, *Olpidium*, and

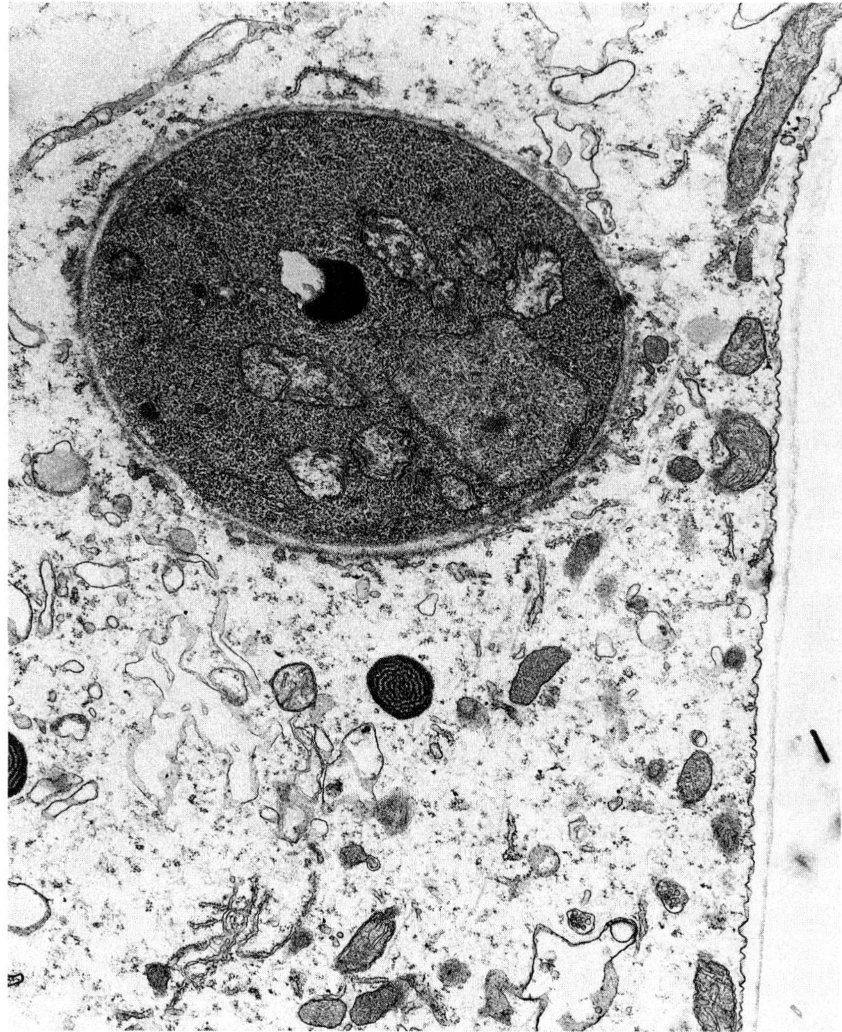

Fig. 4.25. Intracellular thallus of *Catenaria allomycis* inside a hypha of *Allomyces javanicus*. (TEM: × 25,000). (Courtesy of Professor M.J. Powell.)

Olpidiopsis, retain distinct walls during the parasitic phase (Figs 4.25 and 4.26). It is not always clear if the parasite has penetrated the plasmalemma of the host or whether this membrane remains as part of the host-parasite interface. The nature of this interface may change during the intracellular phase of the life-cycle, and this is illustrated by *Catenaria allomycis* in the cytoplasm of its host, *Allomyces javanicus* (Sykes and Porter, 1980; Powell, 1982). Zoospores attach to the hyphae of the host, round up and encyst by the secretion of a wall. The encysted

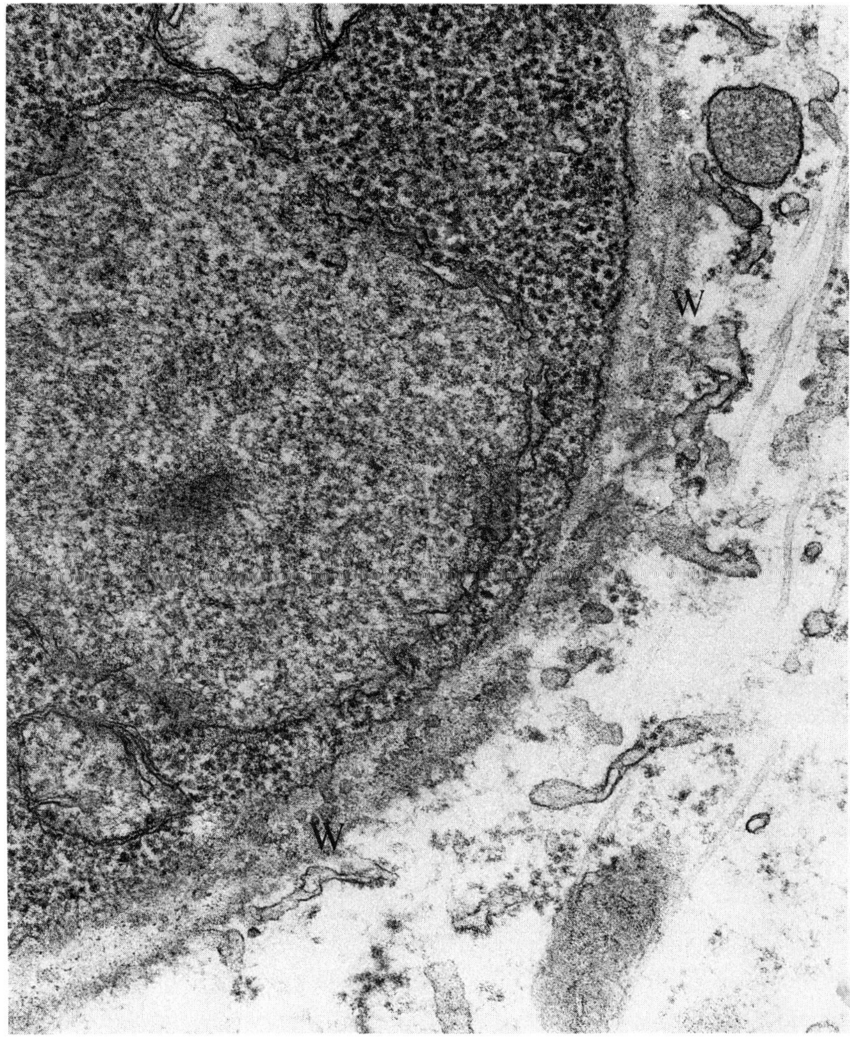

Fig. 4.26. Enlarged area from Fig. 4.25. The interface includes the parasite wall (w) and the endoplasmic reticulum of the host. (TEM: × 75,000). (Courtesy of Professor M.J. Powell.)

zoospore produces a short germ-tube which attaches to the underlying hyphal wall of the host, swells and develops into an appressorium (Sykes and Porter, 1980). An infection tube arises from the appressorium and penetrates the host wall. A double-layered papilla is produced around the invading hypha, and this structure is morphologically very similar to that observed during penetration of cabbage cotyledons by *Albugo candida* (Coffey, 1975) and also resembles the electron-dense deposits occurring when the mycoparasite *Dimargaris cristalligena* either infects *Cokeromyces recurvatus* (Jeffries and Young, 1981) or penetrates its own hyphae (Jeffries and Cuthbert, 1984). The papilla eventually ceases growth, and remains as a tubular collar around the neck of the penetration pore. The plasmalemma of the host is breached during the penetration process such that the wall of the parasite interfaces directly with the cytoplasm of the host (Fig. 4.25). As the parasite develops, host organelles show a progressive sequence of preferential association with the parasite. During the early stages of the interaction (4–5 h postpenetration), host endoplasmic reticulum proliferates and ensheaths the parasitic thallus (Fig. 4.26). In the later stages, the endoplasmic reticulum becomes organized into membranous labyrinths continuous with whorls of single cisternae, and lipid droplets of the host cluster around the parasite. When the sporangia of the parasite begin to mature (12–14 h postpenetration), the cytoplasm of the host begins to degenerate (Powell, 1982). These events occurring during the early stages of infection by *C. allomycis* are unusual, and in other chytridiaceous endoparasites of fungi, the thallus is generally invested in a specialized envelope, presumed to comprise either a thickened, unlayered membrane, or the juxtaposed plasmalemmas of the host and parasite.

During infection of *Pythium* and *Phytophthora* species by *Olpidiopsis gracilis*, zoospores of the parasite become encysted on the hyphae of the host. Transference of the bulk of the cytoplasm of the parasite to the host hyphae occurs within about 1 h, followed by the development of a pyriform swelling on the side of the host hypha opposite the zoospore cyst (Pemberton *et al.*, 1990). Detailed information on the nature of the host–parasite interface is not available at the present time.

5 Physiological Aspects of Mycoparasitism

5.1 Physiology of Mycoparasitism

5.1.1 Spore germination

If the mycoparasitic fungus is not an obligate parasite, spore germination studies may be performed using chemically defined liquid media and agar to which known nutrients are added. Using such techniques with a range of mycoparasites has shown that the rate of germination is affected by a range of physical factors such as the oxygen tension, hydrogen ion concentration of the growth medium, temperature of incubation and the age of the spores themselves. The free-living phase during germination of the spores is also influenced by the composition of the growth medium. The spores of some mycoparasites, such as *Gonatobotryum fuscum* germinate in distilled water, but most require specific factors for germination.

For example, investigation of the conditions required for germination of the sporangiospores of the haustorial mycoparasite *Piptocephalis unispora* has shown that they will germinate in either the presence or absence of a potential host (Jeffries and Young, 1976a). Anaerobic conditions inhibit germination but the same spores can germinate when exposed to an atmosphere which contains oxygen. On 2% malt extract agar at 25°C, cytoplasm of viable sporangiospores stains blue in dilute methylene blue when the spores begin to swell and absorb material from the growth medium. This change is detectable after incubation for about 4 h. The ungerminated, ovoid sporangiospores do not stain. The sporangiospores undergo a spherical phase of growth (spore swelling) and the volume increases dramatically. Germ-tubes may be detected after 12 h, by which time the sporangiospores cease to enlarge. Although up to six germ-tubes may grow from a single sporangiospore, between one and three normally develop on a

Table 5.1. Effect of the carbon and nitrogen source on germination of the sporangiospores of *Piptocephalis unispora* after 20 h at 25°. Nutrients were supplied at 1% (w/v) strength.

Medium (liquid)	Spore diam. (μm)	Germ-tube length (μm)
Distilled water	4.0	–
Casein hydrolysate (CH)	8.0	15
Glucose	5.5	<5
Maltose	5.5	<5
Glycerol	6.0	<5
Glucose + CH	8.5	25
Maltose + CH	8.5	26
Glycerol + CH	8.5	45

medium containing an adequate source of nitrogen. The average length of the germ-tube is influenced by the composition of the growth medium (Table 5.1). In media containing appropriate carbon and nitrogen sources, maximum swelling and germ-tube production occur within a 24 h incubation period. The rate of growth of the germ-tubes is higher when glycerol, as compared with carbohydrates, is provided as the source of carbon with casein hydrolysate as the nitrogen source and the likelihood is that infectivity is higher than from sporangiospores germinated on a depauperate food base.

Germination, determined by the production of germ-tubes, occurs at pH 3–10, extension growth at pH 5–9 with the optimum at pH 6. Thus different aspects of growth during germination are affected differently by the hydrogen ion concentration. The degree of spore swelling and rate of germ-tube extension appear to be related to the ambient temperature (Table 5.2). Maximum swelling and the emergence of germ-tubes ensues at 15–40°C whereas the maximum rate of germ-tube growth is restricted to about 30°C. The cytoplasm of the sporangiospores is also sensitive to temperature as they are no longer viable after incubation at 45°C for 24 h. On the other hand, sporangiospores exposed to −25°C for 24 h remain viable and can still germinate within the temperature range 15–40°C. Sporangiospore samples allowed to age at room temperature (about 20°C) lose viability after six months whereas up to 30% of those stored at temperatures down to −25°C may remain viable after this period. It is notable that the sporangiospores of *Cokeromyces recurvatus*, the reference host of *P. unispora*, germinate readily on standard laboratory nutrient media and germ-tubes, which are well developed after incubation for 10–12 h at 25°C, are produced in the range 15–30°C. It is possible that the temperature and pH range over which *P. unispora* is capable of growing enables the fungus to mirror those conditions under which a potential host may also grow.

The physiology of germination of several other species of *Piptocephalis* has

Table 5.2. Effect of incubation of sporangiospores for 24 h at different temperatures on germination in *Piptocephalis unispora*.

	Temperature (°C)									
	−25	5	10	15	20	25	30	35	40	45
% germination	0	0	0	0	95	98	97	97	0	0
germ-tube length (μm)	0	0	0	0	6	26	84	19	0	0
% germination on transfer to 25°C	95	96	94	98	–	–	–	–	96	0

— = all viable spores already germinated.

also been studied in some detail. Germination of the sporangiospores of *P. virginiana*, for example, does not occur in distilled water or on water agar alone (Berry and Barnett, 1957). Although over 90% of the spores germinate when incubated on malt extract agar for up to 24 h, growth rarely proceeds beyond the production of germ-tubes. Spores can also swell on media that do not support germ-tube formation (McDaniel and Hindal, 1982). Of a wide range of amino acids, sugars, vitamins and other nutrients tested as potentially stimulatory for sporangiospore germination in *Piptocephalis* species, methionine and other metabolically related amino acids are effective. Methionine also reverses the inhibitory effects of compounds such as ethionine and norleucine. Dithiocarbamic acid and other metal chelators also inhibit the process of infection of the host but the effect is reversed by addition of cupric ions to the growth medium. A suggestion is that copper metalloenzyme(s) may be involved in the initial growth of the parasite in addition to various amino acids which appear to be at their greatest concentrations during the initial stages of the process of parasitism. The spores of *Piptocephalis lepidula* have been shown to be capable of swelling and germinating in the absence of exogenous nutrients (McDaniel and Hindal, 1982), although best germ-tube elongation occurred on media containing selected nitrogen sources. This observation contrasts with that of other *Piptocephalis* species, demonstrating that generalizations about the physiology of spore germination require caution.

The sporangiospores of *P. xenophila*, the only species of *Piptocephalis* known to infect certain Ascomycota in addition to Zygomycota, appear to require the presence of a potential host in order to be stimulated to germinate (Dobbs and English, 1954). Thus although this fungus is capable of limited axenic development, growth seems to be entirely host-dependent. Spore germination in several other mycoparasitic fungi has also been studied in depth. Macroconidia of *Sporidesmium sclerotivorum* germinate after 3 days burial in soil adjacent to sclerotia of *Sclerotinia minor*, and on membrane filters placed on soil containing sclerotia. No macroconidia placed 1 cm or more away from sclerotia germinate (Ayers and Adams, 1979). Germination of macroconidia is also induced on tapwater agar containing sclerotial homogenates, and occurs over a pH range of 5–7.

5.1.2 Axenic growth of mycoparasites

The *in vitro* study of parasitic fungi is facilitated if the fungus can be grown on laboratory media in the absence of a host. When a parasitic fungus is able to grow in the absence of a host, even though the growth it makes may be sparse, this phase of growth is referred to as axenic. This is not always possible, however, and mycoparasitic fungi, like their plant-pathogenic counterparts, differ in their nutritional flexibility with respect to being able to grow saprotrophically. Those parasitic fungi which cannot be grown in the absence of the host are usually termed 'obligate' parasites. This is an artificial term, however, as once their nutritional requirements have been determined, they can be grown in axenic culture. Plant-pathogenic fungi have been particularly studied in this respect. For example, the first successful axenic culture of a rust fungus from uredospores was reported for *Puccinia graminis* f.sp. *tritici* by Williams *et al.* (1966, 1967) and since then many other rust fungi have been cultured on artificial media. A similar situation occurs in mycoparasitic fungi, and a number of contact biotrophs require the presence of host extracts for axenic culture to be successful.

Laboratory studies of mycoparasitic fungi constitute a convenient way of learning about the principles of the processes associated with the infection of host organisms. Thus the factors relating to the breaking of dormancy in fungal spores, the directed growth of germ-tubes of parasites towards their hosts, the mechanism of penetration of the wall of the host, processes related to host/parasite recognition, and nutritional aspects of the development of infection may all be studied in Petri dish cultures. These processes may be parallelled in the infection of higher plants by parasitic fungi and it sometimes happens that the study of one group of organisms may shed light on the structure and function of an apparently dissimilar group. For example, the fine structure of haustoria appears to be remarkably similar whether they are developed by plant parasites such as the black stem rust of wheat, *Puccinia graminis*, and the powdery mildew of wheat, *Erysiphe graminis*, or by mycoparasites such as *Piptocephalis unispora* and *Dimargaris cristalligena*. The likelihood is that the various types of haustoria found in these parasitic fungi are also similar in function. Although it is not a simple matter to demonstrate unequivocally that fungal haustoria absorb nutrients, it seems that mycoparasitic systems offer considerable potential in this respect because of the ease with which they may be handled within the laboratory.

When the mycoparasite can be grown *in vitro* in axenic culture its nutritional requirements for mycelial development may be determined using a chemically defined liquid nutrient medium containing materials of analytical grade in known concentrations. Test cultures, incubated under standard conditions of temperature, hydrogen ion concentration and light intensity, are usually also shaken mechanically to assist aeration of the medium. The uptake of specific nutrients, alteration in pH, nutrient concentration, growth in weight and volume can be determined by sampling replicates at standard intervals. It should be emphasized, however, that the nutrient requirements of a mycoparasite grown *in vitro* may not be identical to those of the same fungus growing on a host. Probably a more natural system is represented when pure dual cultures are used. Complications with pure dual cultures are that the nutrient requirements of the host must also be determined in order to distinguish those of the host from those of the parasite. Second, a parasitized host may elaborate or modify the nutrients in the growth medium such that they are passed onto the parasite in a form different from that in which they were originally absorbed. This is likely to be the case with biotrophic mycoparasites, many of which only grow well when parasitizing the host. Third, the nutrient requirements of the parasitized host may differ from those of the unparasitized host. Notwithstanding the obvious problems associated with nutritional experiments, some of the growth requirements of several mycoparasites are known. The interaction of physiological factors may also affect the aggressiveness of the parasite and the susceptibility of the host.

There are few instances where a nutritional advantage has been positively demonstrated for any fungus existing mycoparasitically. One particular example involves *Pythium oligandrum* growing in association with cellulolytic hosts. *Pythium oligandrum* cannot grow on cellulose because it is unable to break down this organic polymer. It can, however, grow successfully on a cellulose nutrient source such as filter paper or cellophane, if in association with a fungus such as *Botryotrichum piluliferum* which is cellulolytic and breaks down the cellulose extracellularly and absorbs the resultant sugar monomers. The hyphae of *Pythium* penetrate those of *Botryotrichum* and presumably take up the simpler molecules from within. The parasitic relationship is further demonstrated when the reproductive capacity of the mycoparasite is examined. Although *P. oligandrum* will grow on a simple glucose medium in the absence of a host, it does not reproduce sexually unless certain sterols are added to the medium. If, however, a potential host fungus is present on the sterol-free medium the mycoparasite is again able to reproduce sexually as it presumably obtains sterols or their precursors from the mycelium of the host. A nutritional effect has also been reported for *Fusarium udum* which parasitizes the hyphae and sporangiophores of the soil-inhabiting fungus *Syncephalastrum racemosum* and grows within them as long as nutrients are available (Upadhyay *et al.*, 1981). When active growth ceases, however, owing to depletion of the nutrients available from the host, intercalary chlamydospores are formed. Chlamydospores are the propagules responsible for infection of growing hyphae of the host.

Necrotrophic mycoparasites such as *Trichoderma* and *Gliocladium* species may grow rapidly and sporulate readily in pure culture on ordinary nutrient media. They are capable of existing saprotrophically on a wide variety of substrates for an unlimited period in the absence of a host and do not appear to be exacting in their nutrient requirements. Few necrotrophs have been intensively studied from a nutritional viewpoint and further research could provide useful information on mycoparasitic associations. In general, the nutritional requirements of most necrotrophic fungi do not differ greatly from those of saprotrophic species. For example, in a study of the effect of substrate C:N ratio, iron, and vitamins on biomass production by *Gliocladium virens*, *Trichoderma pseudokoningii*, and two strains of *T. viride* (Jackson *et al.*, 1991a) it was found that in a glucose–alanine medium, optimum dry weight production per gram of carbon provided occurred with a C:N ratio of 15:1 and this approximates to the C:N value for carbon utilization for other fungi. Addition of iron to a glucose–alanine basal medium increased biomass production of all isolates, although a high concentration was toxic to the *Trichoderma* spp. Lack of Mg, P, K, or S in the medium decreased growth, but the degree of reduction differed between isolates. Sporulation on agar was reduced in the absence of Mg, P, K, and N. Addition of the vitamins biotin, *p*-aminobenzoic acid, and thiamine HCl increased biomass production, but the increases were small. Glucose–alanine basal medium supported more growth of all isolates compared with a potentially commercial molasses–yeast medium. Conidial production was greater in molasses–yeast medium than in glucose–alanine medium. Chlamydospores were produced by all isolates, but the numbers varied depending on the cultural conditions used.

A detailed examination has been undertaken of the growth of and interactions of the glasshouse pathogens, *Rhizoctonia solani*, *Fusarium oxysporum*, *Pyrenochaeta lycopersici*, *Phomopsis sclerotioides*, *Sclerotinia sclerotiorum* and *Botrytis cinerea* and the antagonistic fungi, *Trichoderma harzianum*, *T. viride*, *Coniothyrium minitans*, *Gliocladium roseum* and *Pythium oligandrum* on tapwater agar, soil extract agar and potato dextrose agar (Whipps, 1987a). The medium used had significant effects on the growth rates and morphology of the fungi, the production of volatile and non-volatile antibiotics by the antagonists and the responses of the pathogens to these antibiotics. It also affected competition between the pathogen and antagonists and mycelial interactions. Antagonists were able to produce volatile and non-volatile antibiotics which are active against the pathogens tested but each pathogen showed a different range of responses depending on the media. Several interrelated factors are involved. First, if the antagonist is under stress, for instance from lack of nutrients or low water availability, it may not be able to grow normally and may be unable to produce large quantities of antibiotics. Second, if the pathogen is under stress it likewise may be unable to grow normally but its response to the effects of antibiotics may change becoming either more or less susceptible. Observations of the interaction plates reflected all these factors. For example, the growth of *S. sclerotiorum* was significantly inhibited by *C. minitans* on tapwater agar and soil extract agar but not on potato dextrose agar where growth of both fungi was maximal. These results have implications for the use of such *in vitro* tests as part of a general screen for efficacy of action antagonists against pathogens. It is obvious that no single medium can reflect all the possible antibiotic–growth–mycoparasitic interactions that can occur. Similarly, antagonist–pathogen interactions have also been shown to be dependent upon temperature and water potential. Growth rates gradually declined for most fungi with decreasing water potential but the rate of decline was different for each fungus and was dependent upon both medium and osmoticum used. Colony morphology and pathogen–antagonist interaction also changed with decreasing water potential and media. The media used in these studies were chosen partly because of their widespread adoption in studies of this kind, however, they are subject to inherent variability depending, for instance, on soil or water type or agar batch, and thus they are not strictly defined. A standard defined medium for use in the future has been advocated (Whipps, 1987a). The interactions that may be observed also depend upon the isolates obtained. In the experiments described above, the *T. harzianum* isolates differed in their antagonistic abilities. Variation in isolate effectiveness is a well documented phenomenon. Similarly some isolates of pathogens may be more resistant than others, producing antibiotics or acting as mycoparasites themselves. However, a range of these *in vitro* tests on a variety of defined rich and poor media using different environmental variables would still have a place within a screening system. They would indicate the potential of any organism to produce antagonistic chemicals, or act as a mycoparasite, and the likely outcome of any specific confrontation under a range of known nutritional and environmental conditions. Differences in outcome of interactions have also been reported for the mycoparasite *Verticillium biguttatum*, a potential biocontrol agent against *Rhizoctonia solani* in potatoes (Jager and Velvis, 1988). Differences in growth response between *R. solani* and *V. biguttatum* hinder successful biocontrol in the field, as the mycoparasite needs higher temperatures for growth and proliferates more slowly than its host fungus (Boogert and Jager, 1984). In a worldwide search for isolates with more appropriate temperature requirements the minimum growth temperature of 57 *V. biguttatum* isolates was found to be in the range from 10–13°C, irrespective of their geographical origin. A non-linear logistic growth model was used to describe the radial growth in *Rhizoctonia* mycelium on nutrient agar plates. At near-minimum temperature the maximum colony radii varied considerably; they were up to 3.8 times that of the reference isolate. Fast- and slow-growing isolates could be distinguished. Although the growth properties of *V. biguttatum* isolates from different locations varied, the presence of fast- and slow-growing isolates was not restricted to particular areas and both types could be found in the same field. However, bioassays with selected fast- and slow-growing isolates did not support the assumption that growth at near-minimum temperatures is a relevant criterion for screening isolates of *V. biguttatum* in terms of effectiveness for biocontrol of *R. solani.*

Although biotrophic mycoparasites, such as *Gonatobotrys simplex*, a fusion biotroph parasitizing certain Ascomycota, may grow well in culture in the absence of a host, others such as *Piptocephalis* species may make only feeble growth axenically and the ability of the fungus to sporulate is limited. Several haustorial biotrophic fungi can, however, be grown successfully in axenic culture although it would seem unlikely that they could compete saprotrophically in their natural habitat in soil or on the dung of herbivorous animals. Strains of *Ampelomyces quisqualis*, a biotroph in the initial stages of infection of powdery mildews and later a necrotroph (Barnett and Binder, 1973) produce *in vitro* a range of extracellular enzymes including β-glucosidase, *N*-acetylglucosaminidase and β-1–4-glucanase. Small amounts of phospholipase are produced and proteases are either not detected or are secreted in traces. During the initial stages of infection, the plasmalemma of the host remains intact, possibly through the lack of production of protease and phospholipase by the mycoparasite, thus extending the biotrophic phase. It is suggested that the β-1–4-glucanase enables the mycoparasite to mobilize the glycogen reserve and utilize the product glucose as a carbon source. Chitin synthesis is affected by *N*-acetylglucosaminidase, and β-glucosidase acts on carbohydrate moieties. The breakdown products are believed to serve as nutrients. If these enzymes also bind with the plasmalemma of the host, the diffusion and leakage of nutrients may be reduced thus prolonging the nutrient source for the mycoparasite. The extracellular enzymes are likely also to be important in nutrition of *A. quisqualis* in the necrotrophic phase of growth (Philipp, 1985). This is an example of the interference of metabolism of the host to the advantage of the mycoparasite.

The importance of enzymes has also been demonstrated in a mycoparasitic association of a soil fungus with an aquatic fungus (Willoughby, 1988). Species of *Saprolegnia* found in bodies of freshwater such as lakes and ponds grow saprotrophically on organic detritus and also parasitize fish. They can be baited in water with boiled hemp seeds or cereal grains on which they grow saprotrophically. An isolate of *Saprolegnia* was found to be parasitized by *Mortierella alpina* in the water although the latter species is normally regarded as a soil fungus. Experimental infection of *S. unispora* and *S. hypogyna* grown on hemp seed showed that the mycelium of *M. alpina* could penetrate the hyphae and cause disintegration of the oospores. *Mortierella alpina* also antagonizes *Phytophthora cactorum*, a parasite of apple trees. Both *Phytophthora* and *Saprolegnia* species are known to have a proportion of cellulose in the hyphal walls whereas this compound is absent from the hyphal walls of Mucorales to which *Mortierella* belongs. It appears that Mucorales do not normally produce cellulose-digesting enzymes although some species of *Zygorhynchus* (Taiwo *et al.*, 1987a,b) and *Mortierella* are believed to do so. Whether *Mortierella alpina* can produce cellulases requires investigation but such an enzymic capability could constitute part of the parasitic armoury of this fungus. *Mortierella alpina* has been isolated from fields irrigated with sewage (Domsch *et al.*, 1980). *Mortierella spinosa*, although not tested as a mycoparasite against Saprolegniales, has been isolated from organic material collected from streams and incubated in water (Park, 1972). A suggestion is that these species of *Mortierella* may play a role in the aquatic environment, that of *M. alpina* as a parasite of Saprolegniales.

During parasitism of *Choanephora cucurbitarum* by *Piptocephalis virginiana*, especially early on in the process of infection, it appears that electrolytes, sugars and amino acids are caused to leak from the infected hyphae which implies that an effect of mycoparasitism is to alter the permeability of the cytoplasm of the host (Manocha and Maharaj, 1981). A similar phenomenon is frequently reported as one of the early stages in the infection of plant hosts by plant-pathogenic fungi. Limited growth of the biotrophic parasite in axenic culture may result from the inability or poor ability of glucose and other carbon sources to traverse the plasmalemma. Possibly, however, in the parasitic phase the permeability of the plasmalemma of the haustorium is altered, enabling glucose and other nutrients to be absorbed from the cytoplasm of the host. Water extracts from germinating spores of *Trichoderma* species and *Gliocladium virens* cause the leakage of soluble protein, carbohydrate, amino acids and salts from the mycelium of *Rhizoctonia solani* (Lewis and Papavizas, 1987). The effect on the host is presumably twofold in bringing about rapid death of the cytoplasm of the host and providing a utilizable source of nutrients for the necrotrophs. The nutrients released are utilized by the necrotroph. The more dramatic physiological effect of the necrotrophs on the host highlights the delicate nutritional and physiological balance which has to be maintained between the host and the parasite in biotrophic haustorial associations.

Natural media or extracts of complex naturally occurring substrates have been used to determine whether haustorial biotrophic fungi such as species of *Dispira* are able to grow in the absence of a potential host. Strong axenic growth of *Dispira cornuta* occurred on autoclaved agar media containing egg, beef or swordfish (Ayers, 1933, 1935). The mycoparasite also grew axenically although it did not sporulate in liquid media containing such compounds as peptone, proteose–peptone, nutrient broth and beef infusion. Various carbohydrate and vegetable oil liquid media did not support growth of the mycoparasite and it was concluded that growth would only occur in solutions of derived proteins. It was also found that nutrition of the host influenced the degree of parasitism which was much greater on media containing a high proportion of nitrogen. The host range of this mycoparasite was found to be restricted to Mucorales. *Dispira cornuta* does not normally utilize glucose and other sugars but axenic growth occurs when glycerol and acetate are used as single carbon sources when either casein hydrolysate or yeast extract are provided as the source of nitrogen. Dry weight determination of the quantity of axenic growth compared with that which occurs on the host shows that far less growth takes place in axenic culture. A deficiency of one or more vitamins in the growth medium could account for the poorer growth in axenic culture.

Species of *Syncephalis*, also mycoparasites of Mucorales, were originally thought to be obligate parasites as they could not be grown in pure culture in the laboratory. Five species of *Syncephalis* were eventually grown axenically on a natural medium prepared by autoclaving small cubes of calf liver and pouring an agar containing tryptone and potassium phosphate over them (Ellis, 1966). The liver supplied the essential nutrient(s) required for axenic growth. Homogenation of the medium followed by centrifugation divided the medium into a precipitate and a supernatant solution. As growth occurred only on the precipitated material, it was concluded that *Syncephalis* species probably need a water-insoluble material for growth which was provided by the liver. A wide variety of carbon sources tested were not utilized indicating the possibility that specific carbon sources may be required for growth. *Syncephalis sphaerica* has also been grown on a medium similar to that used by Ellis (1966) but was not capable of axenic growth on a variety of host extracts or killed mycelium (Baker *et al.*, 1977).

Within the genus *Pythium*, marked nutritional differences between the mycoparasitic species, e.g. *P. oligandrum* and *P. acanthicum* and non-mycoparasitic species, e.g. *P. ultimum* and *P. mamillatum* have been found (Foley and Deacon, 1986). On the basis of dry weight and linear growth determinations it was shown that the mycoparasitic strains require an organic source of nitrogen such as asparagine, and thiamine. Non-mycoparasitic isolates, on the other hand, can utilize inorganic sources of nitrogen such as sodium nitrate or ammonium sulphate and most are self-sufficient for the vitamin thiamine. The mycoparasitic pythia also lack cellulolytic ability whereas the non-mycoparasites possess this enzymic capability. As the non-mycoparasitic pythia are also aggressive plant-pathogens, it would seem that the ability to colonize cellulosic substrata could be of importance in this respect. Although the mycoparasitic isolates may depend on the presence of potential host fungi in the soil, the nutritional dependence could also be advantageous as the potential hosts will have already colonized cellulosic substrata in the soil which would otherwise probably not be available to them for growth.

Fusion biotrophic parasites such as *Gonatobotrys simplex*, *Gonatobotyrum fuscum*, *Dicyma parasitica*, *Nematogonum ferrugineum*, and *Stephanoma phaeospora*, require an unidentified nutrient found in hot water extracts of the host termed 'mycotrophein' (Whaley and Barnett,1963; Calderone and Barnett, 1972). Mycotrophein is effective in the promotion of growth even after high dilution to one part per million. Although efforts have been made to identify the compound its identity remains unknown (Hwang *et al.*, 1985). Some non-host fungi also contain mycotrophein which indicates that susceptibility to parasitism is also related to other factors. The nutrient requirements of these fungi when grown on mycotrophein-containing media are varied, with no common factors. Some, such as *D. parasitica* and *G. fuscum*, will utilize single amino acids as nitrogen sources, whereas others such as *Gonatobotrys simplex* have more complex nitrogen requirements. *G. fuscum* also requires abnormally high levels of thiamine and biotin. *Stephanoma phaeospora*, formerly considered to be a typical contact biotrophic mycoparasite, has also been shown to require host extracts for growth in axenic culture (Rakvidhyasastra and Butler, 1973). Some fungi, especially in the genus *Aspergillus*, released this factor into the substrate,

Table 5.3. Effect of host extracts on the sporulation ability of dwarf sporangiophores of *Piptocephalis unispora*.

Medium	No. spores/sporangiophore
Nutrient alone	4
Cokeromyces extract	5
Cunninghamella extract	9

thus supporting saprotrophic growth of *S. phaeospora* in pure dual culture of these two fungi. *Stephanoma phaeospora* is now recognized as a fusion biotroph and presumably obtains the necessary growth factors from host fungi by direct transfer through the cytoplasmic channels in the host–parasite interface. In a study of nutrition of another contact biotroph *Melanospora zamiae*, Jordan and Barnett (1978) found that the parasite was able to obtain all nutrients necessary for growth and sporulation from washed living mycelia of host fungi separated from the culture medium. The required nutrients were held within the host mycelium and were not secreted into the culture medium. The host range included eleven species of Ascomycota and deuteromycetes. *Melanospora zamiae* was deficient for thiamine and biotin but not the growth factor mycotrophein, which it synthesized. The addition of zinc to a glucose–casein hydrolysate liquid medium supplemented with thiamine and biotin was required for maximum axenic growth and sporulation. D-Glucose, D-fructose, D-mannose and cellobiose were the best sources of carbon. The nitrogen sources casein hydrolysate, L-asparagine, L-aspartic acid, L-alanine, L-glutamic acid and urea were utilized to the greatest degree.

In the absence of a suitable host, most species of haustorial zygomycetes are capable of limited axenic growth. Sometimes, e.g. *Piptocephalis xenophila*, germ-tube outgrowth does not occur in the absence of a host mycelium. This is not so with most species, however, and on malt extract agar, for example in *P. virginiana*, the spores germinate, producing a mycelium of sparse growth which forms depauperate (dwarf) sporangiophores (Leadbeater and Mercer, 1957). Table 5.3 gives estimates of the number of sporangiospores produced by the dwarf sporangiophores developing on the axenically grown mycelium of *Piptocephalis unispora* on various media containing extracts of potential hosts. Table 5.3 shows that the number of sporangiospores produced on each sporangiophore varies with the type of host extract included in the nutrient medium provided. The potential host *Cunninghamella elegans* stimulates the greatest development of the sporangiospores.

Analysis of the secondary sporangiospores of *P. fimbriata* shows that almost 13% are enucleate compared with less than 1% from the parasitic mycelium. An implication of such behaviour is that the host fungus is responsible for providing the stimulus to further growth, possibly nutritional, which is not furnished by nutrient agar alone. Inclusion in the growth medium of the alternate carbon source glycerol and the complex undefined nitrogen source casein hydrolysate tends to promote axenic growth and sporulation of *P. unispora* (Jeffries, 1976). However, the viability of axenically produced spores is less than 40% and the mycelium they produce is not itself capable of sporulating. Thus sustained axenic propagation of *Piptocephalis* has not yet been obtained beyond the first generation of sporangiospores. In contrast, several species of dimargaritaceous mycoparasites, such as *Dispira cornuta* and *Tieghemiomyces parasiticus*, can be maintained indefinitely in axenic culture when provided with either glycerol or acetate as a source of carbon and casein hydrolysate as a nitrogen source. Glucose and other sugars do not normally appear to sustain extensive axenic growth (Kurtzman, 1968; Barnett, 1970), although interisolate differences may occur as exemplified by the discovery that one strain of *D. cornuta* could be grown on glucose (Singh, 1975). *Dispira parvispora* has also been grown successfully using glucose as the source of carbon (Barnett and Binder, 1973). Most of these fungi also require the addition of biotin and thiamine to the culture medium in order to obtain appreciable axenic development (Barnett, 1970), although an isolate of *D. cornuta* has been shown to be exceptional in not requiring addition of these vitamins (Singh, 1975). As glucose is readily absorbed by most growing fungal mycelia, suggestions are that the plasmalemma of haustorial biotrophs may be less permeable to glucose or that the fungi lack the enzymes required to metabolize glucose. It has since been discovered that the apparent inability of *T. parasiticus* to absorb glucose is partly overcome by addition of the surfactant Tween 80 to the culture medium which increases the permeability of the hyphae (Barnett, 1970; Binder, 1974). Glycerol is apparently accumulated via a specific transport system (Binder and Pierce, 1976) whereas glucose enters only slowly. Furthermore, cell-free extracts of glycerol-grown cultures contain all the enzymes necessary for the metabolism of glucose, and measurements of the rate of liberation of carbon dioxide when cultures are grown on either ^{14}C-labelled glycerol or glucose indicate that glucose is metabolized but at a much lower rate than glycerol (Binder and Barnett, 1973; Binder, 1974). These results substantiate the hypothesis that it is a reduction in the ability to take up glucose directly from the environment which lies behind this phenomenon. The parasite may, however, absorb alternative carbon compounds such as acetate from the cytoplasm of the host. Casein hydrolysate, although not apparently used as a carbon source, provides the amino acids necessary for growth. The replacement of casein hydrolysate with L-alanine, L-aspartic acid or L-glutamic acid enables *D. cornuta* to grow well axenically (Barker and Barnett, 1973). *Tieghemiomyces parasiticus*, however, cannot utilize these amino acids when they are provided singly (Barnett, 1970; Binder and Barnett, 1974). In addition, *T. parasiticus* has an absolute growth requirement for either cysteine or methionine, and leucine and valine are also important for sustained axenic growth. Inorganic nitrogen supplies such as nitrate cannot be used as a substitute for the essential amino acids by either of the two mycoparasites. Gas–liquid chromatographic analyses of the free amino acid pool in the mycelium of *Choanephora cucurbitarum*, a host of *T. parasiticus* and *P. virginiana*, reveals high concentrations of glutamic acid and, to a lesser extent, other amino acids when nitrogen sources or C:N ratios favourable to parasitism are present in the growth medium. Apparently, glutamic and other amino acids in the pool from the mycelium of the host are a primary determinant of the degree of susceptibility of the mycelium of the host to parasitism (Phipps and Barnett, 1975).

Tests of the enzymic capability of *Piptocephalis* species indicate a poor ability to digest lipid in axenic culture and it is possible that the fungus may be unable to produce sufficient of the enzymes required for sustained axenic growth on basic lipid, and other organic nutrients. Theoretically, the stimulation of growth should be possible by supplying the mycoparasite with breakdown products of nutrients such as lipid in the growth medium, assuming that the mechanism of absorption in the absence of the host is not impaired.

In spite of supplying axenically growing species of *Piptocephalis* with a wide range of test nutrients (Berry, 1959; Richardson, 1963; Morris, 1965) it has not been possible to obtain sustained axenic growth. As it seems unlikely that limited axenic growth is due to the absence of appropriate nutrients or the inability to degrade them into forms suitable for absorption and assimilation, other explanations may have to be sought. A correlation between the degree of parasitism of *C. cucurbitarum* by *P. virginiana* and the level of γ-linolenic acid (γ-18:3:cis, cis, cis, 6, 9, 12-octadecatrienoic acid) in the mycelium of the host has been demonstrated (Manocha, 1975; Manocha and Deven, 1975). Sporangiospores of axenically grown cultures differ from those grown parasitically on malt extract–yeast extract agar in lacking polar lipid bodies and γ-linolenic acid. Of the mucoralean fungi tested, all appear to contain γ-linolenic acid (Shaw, 1965, 1966) in the mycelium although the quantity detected is influenced by cultural conditions such as temperature, light, age of the culture and composition of the growth medium. These conditions may also affect growth of the parasite and a suggestion is that *P. virginiana* is confined to parasitism of fungi which contain γ-linolenic acid. Tests of the culture filtrate from *C. cucurbitarum* for leakage of fatty acids from the mycelium show the presence of many fatty acids but not γ-linolenic. As γ-linolenic acid probably plays a role in the permeability and possibly maintenance or the establishment of the plasmalemma, its absence from axenically grown *P. virginiana* may result in the development of defective membranes with altered permeability which could oblige the fungus to grow parasitically. Certainly, the permeability of the mycelium of the host *Choanephora cucurbitarum* is affected by the mycoparasite because as early as 24 h after infection, electrolytes, amino acids and sugars can be detected as leachates from the mycelium of the host (Manocha and Maharaj, 1981). Several fungi outside the Mucorales, however, also contain this substance (Weete, 1980) and are not hosts of this species, and the atypical host range of *P. xenophila* remains unexplained.

Thus it seems unlikely that γ-linolenic acid is a prerequisite for axenic growth of *Piptocephalis* species.

Investigation of metabolism in relation to nitrogen source shows that quantitative aspects of the free amino acid pool within the mycelium of the host constitute the primary determinant of the degree of parasitism of *C. cucurbitarum* by either *P. virginiana* or *T. parasiticus* (Phipps and Barnett, 1975). On the other hand, the importance of carbon metabolism within the mycelium has been emphasized. Of five hosts and 13 *Piptocephalis*–host combinations, there were no qualitative differences in soluble carbohydrates between host and mycoparasite–host mycelia (Evans *et al.*, 1981). This observation makes it difficult to postulate mechanisms by which *Piptocephalis* species may obtain carbon compounds from their hosts. Trehalose was shown to be the major soluble carbohydrate of the host, and probably also of the parasite. Quantitative changes in trehalose and glucose during mycoparasitism indicated that the physiologies of host and parasite might be linked in some way via this disaccharide (Evans *et al.*, 1981). Thus there are indications that differences in carbohydrate, nitrogen, and lipid metabolism occur between axenic and parasitic mycelia, but none of them fully explains why axenic growth cannot be maintained *in vitro* for more than a single generation. Biochemical studies such as these are hindered as alterations in one pathway of the primary metabolic processes may lead to far-reaching effects throughout metabolism in general. Thus it is difficult to separate causes from observed effects. It is clear, however, that further research is needed before the biochemical basis of the nutritional relationship between these mycoparasites and their hosts is fully understood.

5.2 Fungicolous Relationships

Nutritional studies have also been carried out on a range of fungicolous fungi, particularly those that grow on basidiomes of agarics. Mycelium of *Asterophora lycoperdoides* and the closely related *A. parasitica* grows well in pure culture when the substrate contains a high ratio of an organic nitrogenous nutrient such as asparagine to a carbohydrate such as glucose. Why mycoparasites should have a regular association with hymenomycetes has been investigated (Singh and Plunkett, 1967). Apparently, several hymenomycetes are known to produce hydrocyanic acid in the mycelium. Cyanide in high dilution normally inhibits respiratory enzymes, preventing growth, although it may be utilized as a source of nitrogen by *A. lycoperdoides.* The growth of several toadstool-inhabiting fungi tested in the presence of M/650 KCN, however, was either unaffected or slightly affected whereas that of non-fungicolous cultures was completely inhibited. Tolerance of cyanide has been advanced as a partial explanation for the association of these fungi with the basidiomes of hymenomycetes. A similar phenomenon is noted for the mycelium of *Calcarisporium arbuscula* which grows within the basidiomes of species of *Russula* and *Lactarius* without apparently affecting their development.

The mycelium and germinating conidia of *Calcarisporium arbuscula* are tolerant of high concentrations of isovelleral, an antifungal compound. Isovelleral is produced by the basidiomes of Russulales in response to injury and it has been suggested that its production is part of a chemical defence mechanism which repels attack by other microorganisms. *Calcarisporium arbuscula*, however, is able to withstand the potentially toxic effect of isovelleral and it is also capable of detoxifying the compound through transformation to isovellerol (Anke and Sterner, 1988). The mycelium of *Spinellus macrocarpus*, a mucoralean fungus found on basidiomes of agarics, especially *Mycena* and *Collybia*, grows in axenic culture on media containing glucose, fructose, mannitol or trehalose as sources of carbon with mixtures of amino acids, casein hydrolysate or yeast extract as nitrogen sources (Watson, 1962). The mycelium, which appears sensitive to temperature, growing best at 15°C but not above 23°C, is unable to utilize gluconate as a carbon source. Sporangiospores, on the other hand, which have long been regarded as difficult to germinate in culture require chemical stimulation before they will do so. Various substances including ascorbic and gluconic acids are capable of stimulating the processes which lead to the breaking of dormancy in the sporangiospores of this fungus. Possibly such chemicals are present on the caps and gills of those toadstools which species of *Spinellus* are able to parasitize. Species of *Spinellus* appear to have a limited host range and may be species specific. Incompatibility with a toadstool other than the specific host species could indicate the necessity of specific chemicals for growth. The related fungicolous zygomycete, *Syzygites megalocarpus*, on the other hand, grows vigorously as a saprotroph on ordinary nutrient agars in the laboratory and is found to grow upon the moribund basidiomes of a range of agarics. If *Syzygites* is a mycoparasite, it is probably weakly so.

5.3 Effects of Mycoparasitism on the Host

5.3.1 Necrotrophic mycoparasites

The effects of necrotrophic mycoparasites on their hosts are by definition much more devastating than those of the biotrophs. In order to test the potential antagonistic effects of a fungus, pure dual cultures are prepared by placing agar plugs containing the mycelia of the antagonists on opposite sides of the agar plate. Control plates are inoculated with only one of the antagonists but on opposite sides of each individual plate in order to simulate comparable growth conditions to those in the test plates. Replicates comprising a minimum of three plates per combination of each pair of the fungi under test are usually set up. A cover slip placed over the zone of contact of the opposing mycelia enables the interaction to be examined microscopically and photographed, if necessary. Translucent agar such as tapwater agar facilitates observation of the interacting hyphae because the mycelium of the antagonists is less dense than that grown on nutrient media and light is readily transmitted through it.

Interference in growth of the antagonized fungus due to the production of fungistatic or fungitoxic metabolites by the antagonist will then be evident, which may precede an apparently overt parasitic reaction such as hyphal coiling, the production of specialized contact cells or penetration. The fungitoxic metabolites may be enzymes, organic acids or antibiotics which, after release from the mycelium of the antagonist, act on the antagonized fungus usually, although not invariably, at some distance from the originator mycelium. Zones of the inhibition of growth of the antagonized fungus may be observed on the plates, indicating the effect of the antibiotic or the production of other inhibitory compounds such as organic acids. The antibiotic effect enables the antagonist to out-compete the antagonized fungus for nutrients and space and to colonize a substratum to the detriment of the latter. In addition to causing a reduction in growth rate of the mycelium, inhibitory metabolites may also be responsible for the prevention of formation of vegetative propagative and survival structures such as sclerotia and sporulation. Overgrowth of the antagonized fungus by the antagonist may also occur and it is in this stage of the process that physical contact and penetration of the host may happen. Penetration which probably occurs through the agency of wall-digesting enzymes which may include chitinases, glucanases and proteases (Whipps *et al.*, 1988) coupled perhaps with mechanical pressure at the points of contact, usually but not invariably precedes death of the cytoplasm.

A technique used to study the interaction of the destructive necrotroph *Nectria inventa* with 14 fungi associated with rapeseed involved incubating an equivolume mixture of spores of the mycoparasite and the potential host spread evenly over the surface of an 8 cm cellophane disc on tapwater agar (Tsuneda and Skoropad, 1980). After incubation, the cellophane was cut into strips and mounted in water or cotton blue in lactophenol and examined microscopically for hyphal interference. This technique also enabled the course of the reaction to be observed throughout the period of incubation. Some fungi produce volatile antibiotics which may not be detected using this dual culture technique. For example, *Trichoderma harzianum* produces volatile broad-spectrum antifungal alkyl pyrones responsible for the inhibition of growth of the mycelium of competing fungi (Claydon *et al.*, 1987).

In paired cultures, a zone of inhibition also develops between the advancing fronts of the mycelia of *Fusarium solani* and *Sclerotinia sclerotiorum* where the hyphae from neither fungus penetrate. Hyphae of *S. sclerotiorum* are antagonized, however, the effect being manifested by short, thin apical hyphae which branch extensively producing a ‘witches broom’ effect. Effects caused by other antagonists include necrosis and granulation of the cytoplasm in areas of contact and lysis of hyphal tips allowing the cytoplasm to escape. No zone of inhibition, however, develops with *Trichoderma harzianum* and *T. viride* which are both powerful antagonists of *S. sclerotiorum* and are frequently responsible for killing the sclerotia. The morphological effects on hyphae of *S. sclerotiorum* are probably due initially to antibiosis then enzyme activity resulting in the lysis of hyphal walls. Different isolates of the same species of fungus may vary in their ability to antagonize a potential host and their behaviour in culture may not necessarily correspond with that occurring in soil.

A simple cultural technique in which pieces of cellulose film laid on sand saturated with mineral salt solution were inoculated with pairs of antagonistic fungi to determine their antagonistic properties was devised by Tribe (1966). Cellulolytic species such as *Botryotrichum piluliferum*, *Rhizoctonia solani* and *Trichoderma viride* were allowed to interact with each other and with the non-cellulolytic *Pythium oligandrum*. *Pythium oligandrum* did not grow with *R. solani*, grew very poorly with *T. viride* and grew well in association with *B. piluliferum*. Where the growth of a non-cellulolytic species is stimulated by the growth of an antagonist, an implication is that some extracellular product of the antagonist is responsible for the improved growth. The observation that mycoparasitic *Pythium* species cannot use cellulose or nitrate as nutrient sources (see section 6.1.2) has enabled the investigation into the direct effects of these mycoparasites on their hosts. Non-mycoparasitic *Pythium* species were grown on cellophane films overlying agar containing nitrate as a source of nitrogen. A range of mycoparasitic *Pythium* species were compared in relation to their effects on the host fungi (Laing and Deacon, 1990). The various isolates behaved differently as evidenced by their ability to reduce cellulolytic activity of their hosts. *Pythium oligandrum* was a more aggressive mycoparasite than *P. mycoparasiticum* which, in turn, was usually more aggressive than *P. nunn*. Both *P.oligandrum* and *P. mycoparasiticum* produced oogonia in the presence of susceptible host fungi, thus deriving carbon, organic nitrogen and sterols from the host. *Pythium nunn* which did not form oogonia with any host, was the only mycoparasite that could utilize nitrate as sole nitrogen source, suggesting that it may grow mainly as a competitor for substrates rather than a mycoparasite in nature (Laing and Deacon, 1990). This is supported by a study of interactions of *P. nunn* and *P. ultimum* on bean leaves (Paulitz and Baker, 1988). *Pythium nunn* was able to occupy bean leaves previously colonized by *P. ultimum*, and can displace the latter fungus. It was suggested that mycoparasitism could be important for secondary resource capture, but it could also be due to competition between the fungi for primary colonization of the substrate. *Pythium mycoparasiticum* and *P. oligandrum*, on the other hand, are likely to be mycoparasitic in nature. The fact that *P. mycoparasiticum* was less aggressive in these tests might be explained by the fact that this fungus grows more slowly than *P. oligandrum*. Thus host hyphae might outgrow those of *P. mycoparasiticum* resulting in 'disease-escape' equivalent to that recorded for plant–fungus interactions (Garrett, 1970). On water agar, where growth rates were more similar, *P. mycoparasiticum* was equally aggressive as *P. oligandrum*. Other growth experiments *in vitro* have been used to show a marked reduction in cellulolytic ability and sclerotium production by *Rhizoctonia solani* when it is grown on nutrient-soaked filter paper

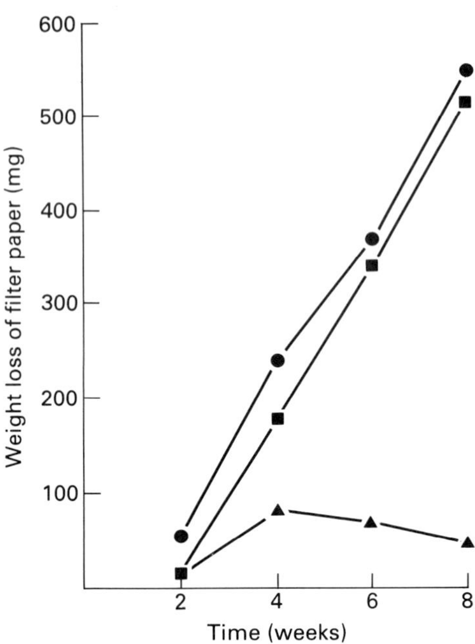

Fig. 5.1. Effect of *Pythium* species on cellulolysis by *R. solani* alone, -•-; with *P. oligandrum*, -▲-; with *P. ultimum*, -■-. (Modified from Al-Hamdani and Cooke, 1983.)

wads in the presence of *Pythium oligandrum* (Al-Hamdani and Cooke, 1983). Even when well established, *R. solani* showed little capacity to resist the influence of *P. oligandrum*. In this regard, the relative mycelial extension rates of the two species may again be of significance, these being 1.6 and 4.0 cm per 24 h for *R. solani* and *P. oligandrum* respectively at 25°C on cornmeal agar. The strain of *P. oligandrum* used by Al-Hamdani and Cooke (1983) was particularly aggressive and had a drastic effect on the ability of *R. solani* to degrade cellulose (Fig. 5.1). This result differs from that of Deacon (1976) where cellulolysis was reduced by only 3%. However, this difference might be due to the use of a less aggressive strain of the mycoparasite in Deacon's study, or by differences in cellulolytic activity between the two strains of *R. solani* studied by the respective authors.

Fungal interactions may also have different outcomes dependent on the environmental conditions. For example, *Rhizoctonia solani*, discussed elsewhere as a host of mycoparasites (section 7.2) is itself an aggressive parasite of fungi especially Mucorales and Peronosporales (Butler, 1957). When *R. solani* and a potential host grow together on an agar plate, a zone of dense mycelium of the parasite forms initially where the advancing edges of the colonies contact. The dense zone then extends throughout the colony of the host and the latter is parasitized either

Table 5.4. Hyphal interaction of *Nectria inventa* with selected fungi isolated from rapeseed. (Based on Tsuneda and Skoropad, 1980.)

	Test fungus		
Reaction	*Alternaria brassicae*	*Sclerotinia sclerotiorum*	*Trichoderma harzianum*
Nectria overgrows colony	yes	no	no
Nectria colony overgrown	no	yes	yes
Hyphae susceptible	very susceptible	susceptible	immune
Spores susceptible	very susceptible	not tested	immune
Hyphae show directed growth	yes	yes	no
Appressorium formation	yes	yes	no
Hyphal coiling	yes	yes	no
Hyphal penetration	yes	yes	no
Granulation of cytoplasm	yes	yes	no
Vacuolation of cytoplasm	yes	yes	no
Loss of cytoplasm	yes	yes	no

by hyphal coiling, for example in susceptible *Pythium* species, or by coiling and penetration as in Mucorales. The behaviour of *Pythium oligandrum*, recorded as immune to attack by *R. solani* (Butler, 1957) varies according the constitution of the medium on which it grows (Whipps, 1987a). On tapwater agar and soil extract agar the mycelium of *P. oligandrum* overgrows that of *R. solani* but does not prevent the growth of the latter. On potato dextrose agar, however, the situation is reversed as the mycelium of *P. oligandrum* is overgrown and inhibited by that of *R. solani.* A suggestion is that the constitution of the growth medium affects the ability of a fungus to produce antifungal antibiotics responsible for inhibiting growth of one of the pair of interacting fungi. Other factors such as variation in the temperature of growth and the water potential of the growth medium may also play a part in the type of hyphal interaction which occurs. In species of *Rhizopus* and *Mucor* the aerial sporangiophores and sporangia may be penetrated resulting in the production of abnormal sporangiospores which are smaller than normal and polymorphic as opposed to ellipsoid. Although much of the aerial mycelium is destroyed the host is not killed. A response to penetration is seen in the development of wall material of the host around the infection peg which may serve to isolate the infection (Butler, 1957).

The effects of the same antagonist on different hosts can also vary. With

Nectria inventa, the hyphae tend to grow preferentially towards those of susceptible fungi and the mode of parasitism has been categorized according to type of hyphal reaction observed (Tsuneda and Skoropad, 1980). Table 5.4 outlines the main reactions of *N. inventa* with *Alternaria brassicae*, *Sclerotinia sclerotiorum* and *Trichoderma harzianum* isolated from rapeseed and shows that the fungi tested react in different ways, the mycelium of *T. harzianum* being unaffected by *N. inventa.* On the other hand, the hyphae of *A. brassicae* appear more susceptible than those of *S. sclerotiorum*, as judged by the larger number of cytoplasmic abnormalities observed in the tests, although the range of reactions in the cytoplasm of the two potential hosts is similar. From results of this kind it was suggested that directed growth appears to be a prerequisite of mycoparasitism, the hyphae of susceptible species producing specific microenvironments which those of the parasite are capable of detecting at some distance from the potential host and respond to through directed growth. In the *N. inventa–A. brassicae* reaction the potential host most commonly influences the growth of the parasite when hyphae of the two are within 15 μm. The destructive necrotrophic action implies the involvement of cytotoxic compounds produced by the mycoparasite, and a number of such substances including antibiotics, ethylene and hydrogen cyanide are known to be produced by many fungi, including mycoparasites.

There are, however, some examples of effects that are not always so obvious, especially in the early stages of infection. Sclerotia of *Sclerotinia* species isolated from the field, for example, infected by *Coniothyrium minitans* are usually hard and rough and may not show external symptoms of infection (Turner and Tribe, 1976). In humid conditions of culture, however, the surface of the sclerotium may become covered with dark-brown droplets or a viscous blackish exudate containing spores from the numerous pycnia of *C. minitans* which develop on the rind. By then the infected sclerotia are usually soft, the medulla discolours brown to black and the cytoplasm has been killed. In sclerotia collected from the field, the pycnia may develop under the rind and in the medulla and the surface does not appear to produce droplets. Use is made of the appearance of the exudate containing pycniospores or the development of erumpment pycnia in the assessment of experimental infections of sclerotia in culture as these symptoms provide a visual means of determining whether infection has occurred. For one isolate of *S. sclerotiorum*, the optimum concentration of pycniospores required for the infection of sclerotia was 1×10^6 ml^{-1}, with 100% infection occurring after 4–5 weeks. Susceptibility of various strains of *S. sclerotiorum* to infection varies from resistant to highly susceptible and the infective ability of strains of *C. minitans* also varies which suggests that the mycoparasitic relationship is more complex than would appear at first sight. Variability in susceptibility of the sclerotia of *Claviceps purpurea* colonized by the mycelium of *Fusarium heterosporum* has also been recorded. Whether intact sclerotia on wet sand sprayed with conidial suspensions of *F. heterosporum* become infected depends on the source of the sclerotia. Ergots (sclerotia) obtained from the ryegrass *Lolium perenne* were infected by *F. heterosporum* whereas those originating from other grasses were not (Hornok and Walcz, 1983). Infected sclerotia permeated by the mycelium of *F. heterosporum* develop light-brown lesions mainly towards the base and these symptoms are regarded as evidence of infection for experimental purposes.

A mutually beneficial result from mycoparasitism has been described in the interaction of the contact mycoparasite *Gonatobotryum fuscum* and a species of *Graphium* (Barnett, 1968). *Gonatobotryum fuscum* is heterotrophic for biotin but can synthesize pyridoxine, whereas *Graphium* can synthesize biotin but not pyridoxine. If both fungi are grown together on vitamin-free media, the rate of growth is at first slow but gradually increases as each species supplements the vitamin needs of the other. This type of mutualistic nutritional relationship has not been demonstrated in any other hyphal interactions.

5.3.2 Biotrophic mycoparasites

Although biotrophic mycoparasites may affect the morphology, physiology, biochemical processes and fine structure of their hosts, few scientific assessments of their precise effects have been made. Much of the effort expended in the study of mycoparasitism has been directed towards the use of mycoparasitic fungi in the biocontrol of plant-pathogenic fungi (see Chapter 7). Biotrophic mycoparasites are probably of limited use as biocontrol agents and investigations of these interfungal symbioses have tended to concentrate on the extent of the host range, nutrient requirements and the mechanism of infection. Effects upon the host have been observed but such results are often mentioned incidentally. An advantage of using haustorial mycoparasites to determine how they may affect behaviour of the host lies in the ease with which they can be cultured in pure dual culture and they may be easily studied in the laboratory under conditions which can be carefully controlled. Colonies of *Piptocephalis* species, for example, produce well-defined, often heavily sporulating aerial growth over the surface of the host colony. The mycelium of the parasite keeps pace with that of the host through multiple penetrations of the wall of the mycelium of the latter. The apical region of the hypha of the host appears to be most susceptible to attack and the walls of older hyphae are probably not penetrated (England, 1969). The need to attack the younger parts of the mycelium may explain why, in agar plate cultures, the margin of the colony of the parasite is usually found close to the edge of the colony of the host. This does not occur in all species, however, and some species of *Piptocephalis*, such as *P. benjaminii*, occupy a relatively restricted region within the host colony, being surrounded by a zone of host mycelium up to 2 cm wide (Evans and Cooke, 1981).

It might be thought that the effect of a mycoparasite would always be to slow growth of the mycelium of the host. With various host–parasite combinations involving *Piptocephalis* species grown under different temperature regimes, it appears that growth of the mycelium of the host may be inhibited, unaffected, or even stimulated when the diameter of the infected colony exceeds that of the uninfected control (Curtis *et al.*, 1978). Whether the parasite is capable of stimulating growth of the host merits further investigation because this could be an adaptation to the parasitic mode of life which enables *Piptocephalis* to continue to sporulate. A suggested explanation of this observation is that when the spores of the parasite and the host are inoculated together onto nutrient agar plates the lag phase of germination which precedes hyphal growth is shortened, that of host spores alone being longer. Thus if a single measurement is taken after a specified time interval it appears as though the rate of growth of the mycelium of the host in dual culture exceeds that of the host grown alone. A reduction in the rate of growth and sporulation of the coprophilous *Pilaira anomala* infected with *Piptocephalis* species in culture has been noted over the temperature range 15–30°C (Wood and Cooke, 1986). In the *Pilaira anomala–Piptocephalis freseniana* combination grown at 20 and 25°C, however, radial growth occurred faster than in colonies of the host alone. The temperature of growth appears to be a determining factor in extension growth of the mycelium of the host when infected by this mycoparasite. Comparison of the dry mycelial weights of the host *Mycotypha microspora* grown in a liquid glucose–glutamic acid medium with those of dual combination with *Piptocephalis virginiana* show a decrease over control cultures of more than fivefold, indicating a significant effect of the parasite in the reduction of mycelial growth (Berry, 1959). On other media and with other host–parasite combinations, however, such marked inhibition of growth does not occur.

Piptocephalis species induce the development of morphological abnormalities in hyphae of the host which take the form of swellings and unusual branching patterns (England, 1969; Wood and Cooke, 1986). Infection of the mycelium of *Mycotypha microspora* by two species of *Piptocephalis* (*P. benjaminii* and *P. unispora*) resulted in significantly increased branching of the leading hyphae of the host, and a decrease in leading hypha density within the colony margin (Evans and Cooke, 1981). However, total hyphal density within the margin was unchanged so that, although neighbouring leader hyphae were further apart, increased branching produced potential leader hyphae which then occupied the available extra space. This effect was not induced, however, following infection by *P. cylindrospora*. The induction of additional branches in the mycelium of the host provides a continuous supply of susceptible hyphal tips and the phenomenon seems to represent an interesting adaptation to parasitism. *Syncephalis californica* has also been shown to produce hyphal abnormalities on infection of its mucoralean hosts, a feature that has been used to demonstrate mycoparasitism in field soils (see section 6.2). Growth effects due to mycoparasitism can also reflect natural behaviour of the affected host. For example, spores of *Mycotypha microspora* germinated axenically at 15°C give rise to a transient yeast-phase before the typical hyphae are developed. If *Piptocephalis fimbriata* is present, however, the yeast-phase persists and few hyphae are formed (Evans *et al.*, 1978). Those hyphae that did form either reverted to the yeast-phase, became severely attenuated, gave rise to depauperate sporangiophores or died. Such mycoparasite-mediated effects did not occur at 20°C.

Nutritional requirements for the growth of haustorial mycoparasites may be complex. In general, media containing a high concentration of nitrogen relative to carbon are favourable to parasitism compared to those with a high ratio of carbon to nitrogen. This can be demonstrated with the mycoparasite *Piptocephalis unispora* growing on the mucoralean host *Cokeromyces recurvatus.* The degree of mycoparasitism is estimated by measurement of the diameters of the colonies of the host and the parasite, and by recording the relative density of hyphal growth of the mycoparasite as judged visually. The effect of varying the ratio of carbon to nitrogen in the growth medium is marked. At the higher C:N ratios, the mycelium of the mycoparasite is restricted to the central region of the colony of the host and is barely visible to the naked eye, whereas at the lower C:N ratios the mycelium of the parasite grows densely over the entire colony of the host. A general observation is that a high concentration of utilizable nitrogen relative to carbon in the growth medium appears to favour mycoparasitism. More recently (Kurtzman, 1967, 1968) it has been shown that the carbon and nitrogen requirements of the host *Cokeromyces recurvatus* affect the degree of parasitism by *Dispira cornuta.* A low C:N ratio favours development of the parasite but sparse development occurs on media with a high C:N ratio. The development of *Olpidiopsis incrassata* on Saprolegniaceae is also affected by the carbon and nitrogen nutrition of the host and, in addition, the pH of the growth medium which has to be within the range 5–8 for growth to occur (Slifkin, 1963).

In media containing glucose to glutamic acid in the ratio 3:1 parasitism of *Mycotypha microspora* by *Piptocephalis virginiana* was rated as fair but this improved with a ratio of 3:8 (Berry, 1958, 1959). *Piptocephalis virginiana* fails to colonize the potential host *Mortierella pusilla* grown on a medium containing ammonium sulphate as the source of nitrogen whereas replacement of the inorganic source of nitrogen with glutamic acid enables the parasite to make excellent growth. Determination of the quantity of soluble nitrogen in the mycelium of *M. pusilla* after growth on media containing various ratios of carbon to nitrogen has shown that as the concentration of yeast extract is increased in the growth medium so the concentration of trichloroacetic acid (TCA) soluble nitrogen in the mycelium also increases (Shigo, 1960b; Shigo *et al.*, 1961). With high glucose concentrations, however, the concentration of TCA soluble nitrogen decreases sharply and it seems that the growth of *P. virginiana* on *M. pusilla* is correlated with the production of high concentrations of TCA-soluble nitrogen in the mycelium of the host. With *P. xenophila*, a biotrophic parasite of non-mucoralean hosts growth of the mycoparasite is further increased in liquid media by the inclusion of microelements, especially manganese. In a qualitative survey of soluble carbohydrates in a wide range of *Piptocephalis*-host combinations, no differences were found between host and mycoparasite-host mycelia (Evans *et al.*, 1981). During growth in liquid culture, endogenous glucose declined more rapidly in *Piptocephalis unispora–Circinella mucoroides* mycelia than in *C. mucoroides* mycelium alone, resulting in increased trehalose:glucose ratios in dual mycelia. Similarly enhanced ratios were found during development of mixed germinating spores of *Piptocephalis fimbriata–Mortierella vinacea* and *P. fimbriata–Mycotypha microspora* on agar. Proximity of a host to *P. fimbriata* spores induced increased mycoparasite trehalose activity. It was suggested that access to host trehalose is an important feature of haustorial mycoparasitism.

Mycoparasitism of *Melampsora larici-populina*, a rust of poplar leaves, by *Cladosporium tenuissimum* can be assessed from the experimental infection of leaf discs with the rust in the presence and absence of the mycoparasite. The numbers of uredia produced on the leaves can be counted and the reduction in the presence of the mycoparasite is taken as a direct measure of the extent of infection. The parasite may reduce the production of uredia by 50% (Sharma and Heather, 1988) and, by implication, the number of uredospores is also significantly reduced by the activity of the mycoparasite. *Cladosporium tenuissimum* is a necrotroph capable of destroying both ungerminated and germinated uredospores. Penetration of the spore wall occurs, the contents are lysed and the uredospore collapses. The mycelium of *C. tenuissimum* in the parasitized uredia sporulates, producing conidia which may adhere to the echinulate uredospores. It has been suggested that attachment of the conidia to the uredospores might assist aerial dispersal of the former and also serve to reduce the potential increase in the rust population. Uredospores of the bean rust *Uromyces appendiculatus* (= *U. phaseoli*) are similarly parasitized by *Verticillium lecanii* although lysis of the spores does not occur (Allen, 1982).

6

Ecological Aspects of Mycoparasitism

6.1 Isolation of Mycoparasitic Fungi

Although many mycoparasitic fungi can be isolated from soil, and it can be demonstrated that they are parasitic in laboratory culture, it is usually a presumption that they behave as mycoparasites in the natural soil environment. The fungi form a major component of the biomass in soil, and together with the bacteria they far outweigh all the other soil flora and fauna put together (Campbell, 1985). However, it is frequently difficult to show by direct microscopic examination of soil that there are any fungi present at all, irrespective of their mycoparasitic abilities. Nevertheless, such soil samples may yield considerable numbers of various types of fungi when a few milligrams are placed onto nutrient agar and incubated at 25°C for a few days. Mycoparasitic fungi may subsequently develop from the incubated samples, and can then be isolated by the methods described below. *Trichoderma* and *Gliocladium* species are very commonly isolated in this way, yet may also possess incidental mycoparasitic abilities. In other cases, specific isolation methods have been described for particular necrotrophic mycoparasites using selective media. Selectivity is often obtained by using a combination of antibiotics and other inhibitory agents. For example, a *Trichoderma*-selective medium (Elad *et al.*, 1981) and a *Pythium*-selective medium (Schmitthenner, 1962) have both been used to isolate these genera from soil samples. A combination of fungicides has been used in a similar way to isolate and enumerate *Laetisaria arvalis*, an antagonist of *Pythium ultimum* and *Rhizoctonia solani* (Papavizas *et al.*, 1983). Such selective media can then be used to study the occurrence of the antagonistic fungi in natural substrates. For example, *Athelia bombacina*, an antagonist of the apple scab pathogen *Venturia inaequalis*, was recovered from 75% of non-inoculated apple leaves and 96% of inoculated leaves over a six-month period (Young and Andrews, 1990).

Some mycoparasitic fungi are certainly common in the natural environment. The haustorial zygomycetes, for example, can easily be isolated from soil and from the dung of herbivorous animals. Freshly defaecated pellets of dung can be placed onto sterile moist filter paper in a Petri dish and examined daily for mycoparasitic fungi. Soil samples can also be sprinkled onto the surface of a nutrient medium such as potato carrot agar and incubated in the laboratory at 20–22°C. If the mycoparasite sporulates and is known to grow well axenically, sporophores can be picked off with a sterile loop, needle or forceps. The spores can be streaked onto a plate of nutrient agar which is then incubated, usually at 25°C, for two or three days, to allow them to germinate and produce mycelia from which subcultures may be made. A single sample of dung or soil may yield several different mycoparasites thus the samples should continue to be examined until it is clear that no further species can be isolated from the sample. It is important to use fresh dung pellets because the mycoparasites of mucoralean fungi cannot be isolated after the dung is about ten days old as the hosts are replaced in the ecological succession by other fungi.

On the other hand, if the mycoparasite makes only poor axenic growth, a simple technique may be used to isolate the mycoparasite in pure dual culture by using the host fungus as a selective bait. For example, yeast-phase cells of the susceptible host *Cokeromyces recurvatus* streaked onto a plate of agar can be used to isolate species of the mycoparasites *Piptocephalis* and *Syncephalis*. The yeast-phase cells are incubated for six hours until many germ-tubes have been produced and use is now made of the fact that germ-tube apices have particularly thin hyphal walls and are especially susceptible to attack by mycoparasites. Consequently, if germinating spores of the mycoparasite are present, they rapidly invade the germ-tubes of the host and a vigorous growth of the mycoparasite may develop (Jeffries and Kirk, 1976). *Piptocephalis* species have been isolated as parasites of other fungi commonly occurring on dung, soil and occasionally rotting fruit. Although *Piptocephalis* and *Syncephalis* species have usually been considered to be associated with hosts growing on the dung of herbivorous animals they can frequently be isolated from incubated leaf litter and A-horizon soil from woodland and pastures (Richardson and Leadbeater, 1972; Jeffries and Kirk, 1976). Possibly host fungi in the dung are invaded by the mycoparasites which are already established on hosts in the litter and the soil. Many experimental studies involve pure dual culture of the parasite and the host where observations may be made more readily than in the natural habitat. In this respect, it is useful to have a reference or standard host which enables researchers throughout the world to compare their results more readily than in those mycoparasitic combinations in which the host varies but the same mycoparasite is used. One such reference host used to study parasitism by species of *Piptocephalis* is *Cokeromyces recurvatus*, a mucoralean fungus which is readily cultured on ordinary nutrient media, the mycelium growing flush with the surface of the agar. The mycelium is very susceptible to infection by *P. unispora*, other species of *Piptocephalis* and several other haustorial mycoparasites. Owing to the low mycelial growth, the parasite is not obscured and thus observation of the latter is facilitated. Furthermore, asexual sporulation of the host can be prevented by growth in continuous dark thus facilitating collection of sporangiospores of the mycoparasite free from contamination by those of the host. This enables the behaviour of relatively large volumes of sporangiospores of the mycoparasite alone to be studied if required. *Cokeromyces recurvatus* is also morphologically distinct from species of *Piptocephalis* which reduces the possibility of confusion in identity of structures at the ultrastructural level of observation.

Several mycoparasitic species of *Pythium* have also been isolated by using potential host fungi as bait. In this case, agar plates are completely precolonized by a suitable potential host fungus and a small sample of material, for example soil, suspected to contain the mycoparasite is added. Precolonization of the plate with the potential host results in the utilization of most of the readily available nutrients and the production of compounds by the host fungus which are inhibitory to the growth of many other fungi. Thus the only fungi capable of growing out of the substratum are those with the ability to utilize the mycelium of the precolonist as a nutrient source, namely the mycoparasites. This technique has been used successfully to selectively isolate various species of *Pythium* using *Phialophora radicicola* as a host (Deacon, 1976; Deacon and Henry, 1978), and for *Piptocephalis xenophila* using *Penicillium* hosts. The technique was also used to survey the occurrence of mycoparasitic *Pythium* species in a variety of soils (Foley and Deacon, 1985). Samples from horticultural, grassland, arable, woodland, moorland, freshwater and coastal sites were assayed for the presence of mycoparasites by incubation on agar precolonized by *Phialophora* species. Of the total 164 samples, 84% contained one or more of the mycoparasites, *Pythium oligandrum*, *P. mycoparasiticum*, *Trichoderma viride* and *Gliocladium roseum*. *Pythium oligandrum* was found in 29% of all samples, 45% of samples from 'disturbed' sites (gardens, arable lands, managed grasslands), but only 11% of samples from 'natural' sites subject to minimal disturbance (woodlands, moorlands, permanent pastures, coastal sites and fresh-water sediments). It was commonest at pH 5.5–5.6. *Pythium mycoparasiticum* was found in 17% of all samples but was probably underestimated because of competition from *P. oligandrum* on isolation plates. Its distribution was similar to that of *P. oligandrum*, but it was much less common in horticultural and grassland sites. The mycoparasite *P. acanthicum* was found only twice, and *P. periplocum* was not found, although both could grow well on *Phialophora*-precolonized agar. Overall, mycoparasitic pythia were found in 38% of samples – at a similar frequency to *T. viride* (45%) and *G. roseum* (40%). The growth of fungi across pre-colonized plates indicates differences between antagonistic fungi. For example, *Trichoderma harzianum* can grow well across several bait fungi unsuitable for *P. oligandrum* (Laing and Deacon, 1990) and not across others that *P. oligandrum* could colonize. *Gliocladium roseum*, on the other hand, grows across most of the fungi used as baits. *Pythium acanthophoron*, originally described in 1932 (Sideris, 1932) has recently been re-isolated from soils in India and Britain (Lodha and Webster, 1990).

Zygomycota are especially susceptible to parasitism by *P. acanthophoron* although *Fusarium solani* and *Pythium myriotylum*, both pathogens of the rhizomes of ginger (*Zingiber officinale*), are also parasitized. The precolonized plate technique has been further developed as a method to determine the mycoparasitic spectrum of a soil (Mulligan and Deacon, 1992). Using a range of host fungi, the presumptive mycoparasites *Pythium oligandrum*, *Gliocladium roseum* (and related species), *Trichoderma* spp. and *Papulaspora* sp. were detected in, respectively, 18, 28, 24 and 21 of a total of 28 British soils. Most of the soils contained three or more mycoparasites, but the frequency of detection in experimental replicates suggested that only *P. oligandrum* and *G. roseum* were abundant in all the soils in which they occurred. The type of host fungus markedly influenced the efficiency of detection of different mycoparasites: *Fusarium culmorum* was most efficient for *P. oligandrum*, *Rhizoctonia solani* for *Trichoderma*, *Botrytis cinerea* for *Papulaspora* and *Trichoderma aureoviride*, *R. solani* and *B. cinerea* were equally efficient for detection of *G. roseum*. No single host was suitable for consistent detection of any single mycoparasite and the authors suggested that several hosts may therefore be needed to determine the mycoparasite spectrum of a soil. An alternative method for the isolation of *P. oligandrum* has recently been described (White *et al.*, 1992) involving the use of air-dried soil placed on 1.5% water agar containing 0.1% glucose.

Specialized techniques can also be used for the isolation of those mycoparasites that grow on soil-borne resting structures of their hosts, such as sclerotia. Sclerotia can be extracted from soil by a combination of wet and dry sieving and are then incubated in damp chambers to allow the development of any associated mycoparasites. Alternatively, laboratory-grown sclerotia can be buried in soils as bait for later recovery and examination in the laboratory (see also Chapter 7) in the hope of obtaining natural antagonistic parasites likely to be of use in the biocontrol of sclerotium-forming plant parasites (Adams and Ayers, 1985). The sclerotia of *Sclerotinia sclerotiorum* isolated from infected sunflower stems have been used in this way (Zazzerini and Tosi, 1985). The fungus can be grown in culture on discs of carrot where it produces numerous sclerotia. The sclerotia to be used as bait are buried in different types of soil for about a month, then recovered. They are washed, surface sterilized, plated onto potato dextrose agar then incubated at room temperature for ten days. Fungi which grow from the sclerotia are isolated and may be tested for their ability to antagonize the mycelium of *S. sclerotiorum* in pure dual culture. *Fusarium solani*, shown to be a vigorous antagonist of *S. sclerotiorum*, was isolated in this way. If antagonistic fungi are to be used as biocontrol agents, they have to be carefully selected for their ability to act as antagonists in the soil without themselves parasitizing the crop plant being protected. Isolates of *F. solani* are not of interest in this respect as their value as antagonists is almost certainly outweighed by their activity as plant pathogens. A sclerotial baiting technique using sclerotia of *S. sclerotiorum* has been used successfully to isolate mycoparasitic *Pythium* species from 92% of the soils tested (Ribeiro and Butler, 1992). *Pythium acanthicum*, *P. oligandrum* and *P. periplocum* were present in 80%, 27% and 13% of the soils respectively, yet no evidence for parasitism was found in the sclerotia used as baits.

A baiting approach has also been used for the study of those fungi which attack oospores. Oospores of a susceptible host fungus are produced in culture and then buried in soil or placed in a soil extract which, after a suitable incubation period, is recovered together with any attendant mycoparasites. Oospores of *Pythium* species treated in this way yield various parasitic fungi to which they appear to be susceptible in the soil. If relatively large numbers of propagules of the potential host fungi are added to the soil it is assumed that the probability of attracting mycoparasites is enhanced thus facilitating the detection of mycoparasitism. Oospores of *Pythium myriotylum* added to soil in large numbers at the rate of 500,000 g^{-1} have been used to detect *Hyphochytrium catenoides* which infects and destroys them (Ayers and Lumsden, 1977). As *H. catenoides*, which is capable of attacking various oosporic fungi, has been detected in several soils and it is able to reproduce rapidly in the parasitic phase it has been suggested that this mycoparasite may contribute to the natural decline of oosporic fungi in soils with time.

Oospores of *Phytophthora erythroseptica* have also been used as a bait. They have been applied to membrane filters, placed in nylon mesh bags then buried in soil for 21 days. When the oospores are recovered and transferred to agar in Petri dishes, those that have become infected by the mycelium of parasitic fungi can be observed microscopically and transferred to fresh agar medium. If the parasites are facultative, they can be cultured on laboratory media (Epton and Wynn, 1984). The survival of oospores of *Phytophthora syringae* formed in apple leaves has also been studied in a natural ecosystem. Spores were stored on or beneath soil in an apple orchard from March 1980 to November 1982. A large proportion (80%) of the buried oospores were destroyed in the first seven months and most of these were colonized by *Microdochium fusarioides* (Harris, 1985b). This fungus, however, caused only minor losses of oospores in surface-stored leaves. Several mycoparasites attack the spores of arbuscular mycorrhizal fungi (see section 6.3). However, when spores of these fungi have been used as bait in soils the recovery of mycoparasites has generally been low, ranging from 1 to 3% although higher rates of recovery, up to 16%, have been recorded (Paulitz and Menge, 1986).

6.2 Natural Interfungal Relationships

Owing to the difficulties associated with the direct observation of living fungi in the soil most of the information concerning mycoparasitism has been obtained from *in vitro* dual culture experiments and by inference from field observations. The direct examination of a thin layer of soil spread onto a microscope slide can be instructive. Normally, the fungal structures have to be heat-fixed and their cytoplasm stained with erythrosin, for example, in order to be able to detect and observe them clearly. Relatively large host structures such as the oospores of *Pythium* species (Ayers and Lumsden, 1977) seen in soil smears or the spores of arbuscular mycorrhizal fungi taken from the soil frequently appear to be infected by the very fine hyphae of mycoparasitic fungi. Incubation of oospores of *Phytophthora sojae* on membrane filters buried in soil has allowed natural infection by mycoparasites to be observed (Sneh, 1977). Chytridiaceous fungi, such as *Hyphochytrium catenoides*, frequently invaded the spores and produced their own spores in them within 3–4 days.

Some of the most obvious dual fungal associations in the field involve toadstools where the supposed parasite growing on the cap can readily be seen either with the naked eye or with the aid of a hand lens. These relationships have been discussed in more detail in section 2.1.2. Other obvious examples of mycoparasitism in the natural environment are the penetration of spores of arbuscular mycorrhizal fungi, the colonization of sclerotia, and the growth of fungicolous associates within the pustules of rust fungi. Each of these is discussed in more detail in subsequent sections of this chapter.

Mycoparasitic fungi can sometimes be seen in the laboratory if samples from the field are collected and observed daily. If freshly dropped balls of horse dung are incubated in the light in a relatively humid atmosphere, a succession of sporulation structures of mucoralean fungi will invariably develop within a few days. These often include species of *Mucor*, *Thamnidium* and *Pilobolus* all of which may be attacked by various mycoparasitic fungi such as *Chaetocladium*, *Piptocephalis* and *Syncephalis*. The succession may be readily observed over a period of about 12 days after which the mycoparasites of the Mucorales themselves are supplanted by their successors, usually Ascomycota, whose perithecia and other sporulation structures then become evident in the surface of the dung. Whether the mycoparasites are responsible for the eventual demise of their hosts is a matter of conjecture as there is insufficient evidence to demonstrate that they actually harm them. The greater probability seems to be that the physiological condition of the substratum changes through microbiological activity in favour of those fungi which develop later in the succession, or that a longer time period is required to produce the sporulating structures of the higher fungi.

The unequivocal demonstration of a mycoparasitic association under laboratory conditions cannot be taken as proof that a similar parasitic relationship exists in the field. *In vitro* tests of the host range merely provide a strong indication of susceptible and immune fungi in the absence of natural competitors (e.g. Tables 4.2 and 4.3). Under field conditions involving a community of microorganisms, however, the situation could be different. Environmental factors such as the competition for nutrients, ability to withstand antibiotics produced by other competing microorganisms, and the availability of carbon and nitrogen might affect the intensity of attack and the susceptibility of the potential hosts. Interactions of mycoparasites and their hosts *in vitro* is known to vary depending on the nutrient conditions or environmental conditions. For example, Whipps (1987a) investigated the interactions *in vitro* of six plant pathogens with four antagonistic species on a nutrient-poor substrate, tapwater agar (TWA), on soil-extract agar (SEA) with no added nutrients, and on the relatively nutrient-rich potato dextrose agar (PDA). Both the pathogens and antagonists used could be split into groups where known ecological behaviour fitted with observations obtained on the media *in vitro*. For instance *Rhizoctonia solani*, *Fusarium oxysporum*, *Pyrenochaeta lycopersici* and *Phomopsis sclerotioides* are soil-borne pathogens which survive in soil as sclerotia, resistant spores or hyphae, or within plant debris, and which can actively grow through soil to infect roots or necrotic leaves touching the soil surface. In the tests, these pathogens all had the greatest growth rate on SEA but differences between this and rates on PDA and TWA were relatively small, indicating an adaptation to growth under a range of nutrient conditions likely to be found in soils and plants. *Sclerotinia sclerotiorum* and *Botrytis cinerea*, however, grew relatively poorly on TWA and SEA compared with PDA and this again probably reflects natural ecology. These pathogens survive on plant residues, or as sclerotia in the soil, and infect aerial parts of plants where normal growth occurs under nutrient-rich conditions. In general, they show little or no ability to grow as mycelium through the soil, and have limited competitive saprotrophic ability. Similar ecological groupings can be seen with the antagonists. *Trichoderma* species and *Coniothyrium minitans* grow poorly in the low-nutrient media but at least double their growth rates on PDA. The former are well known for their ability to grow and sporulate profusely where large amounts of nutrients are made available in the soil, overcoming soil **fungistasis** (Papavizas, 1985). The latter survives as spores in the soil until a suitable sclerotium is encountered and then germination, penetration and rapid growth occurs within the protected, nutrient-rich conditions of the sclerotium. *Gliocladium roseum* has similar growth rates on all media and is a ubiquitous soil organism able to grow and survive in soil (Papavizas, 1985). *Pythium oligandrum* similarly has been isolated from soil (Foley and Deacon, 1985) and in common with other *Pythium* spp. survives as thick-walled resting spores and is capable of rapid growth when nutrients become available (Deacon, 1976). However, *P. oligandrum* had very fast growth rates on TWA and SEA but a comparatively slow growth rate on PDA, probably explained by the fact that commercial preparations of PDA contain inhibitors of mycoparasitic *Pythium* spp.

Little is known of the natural occurrence of biotrophic mycoparasitism. The highly specialized nature of the biotrophic haustorial mycoparasites coupled with the difficulty encountered in inducing them to grow in the laboratory in the absence of the host leaves little doubt that they normally grow as parasites in nature. Where the parasite can be observed growing together with the host on the natural substratum, for example species of *Piptocephalis* and *Dimargaris* on living mucoralean hosts on the dung of herbivorous animals, the natural parasitic relationship seems clear. *Piptocephalis* species are nevertheless extremely common in soil or on dung samples incubated in damp chambers. Richardson and Leadbeater (1972) and Bawcutt (1983) reported that both genera could be frequently isolated from litter and A horizon soil samples, especially from woodland and pasture habitats. It has been suggested that the occurrence of *Piptocephalis* on dung samples is not the result of the suitability of dung as a habitat for the parasite, but a combination of the ability of dung to support a rich mucorine flora and the invasion of already-established *Piptocephalis* from the underlying soil and litter (Richardson and Leadbeater, 1972). Sometimes, natural mycoparasitic relationships can be inferred as mycoparasites can cause quite distinctive growth abnormalities in the hyphae of the host, especially when such morphological alterations have also been noted in samples obtained directly from the field. Hyphal swellings in *Rhizopus oryzae* have been used in this way to infer mycoparasitic activity of *Syncephalis californica* in both naturally and artificially-infested agricultural soils. Such cases are exceptional, however, and most mycoparasitic associations have only been demonstrated in the laboratory, usually after isolation of the host and parasite from the substratum on which they were growing. For this reason, the observations made on *S. californica* provide some of the most convincing evidence for the natural occurrence of mycoparasitism.

Syncephalis californica induces giant hyphal swellings in species of Mucorales including *Mucor* and *Rhizopus* (Hunter and Butler, 1975). Of 116 fungi tested as potential hosts, selected from all of the main groups, only Mucorales were parasitized. Of these, *Rhizopus oryzae* was found to be the most susceptible to attack in pure dual culture and it is normally the young, actively growing aerial hyphae which are affected. Spores of *S. californica* germinate on the surface of potato dextrose agar and produce germ-tubes which grow towards the hyphae of the potential host by chance and not, apparently, by means of directed growth. A germ-tube which contacts the hypha of *R. oryzae* develops an appressorium which consists of either one or several lobes. An infection hypha arises from the appressorium which penetrates the hyphal wall and grows within the hypha of the host. The visible effect of the hyphae of *S. californica* growing through those of the host is to induce swellings of the latter which become irregular, multilobed and massive in comparison with the width of the original hypha. Hyphal swellings are packed with dense cytoplasm for several days which then degenerates. At this stage, hyphae of the parasite can be detected within the swellings. Infection appears to be restricted to the aerial hyphae of the vegetative mycelium as the swellings have not been observed in the sporangiophores, rhizoidal or submerged hyphae of *R. oryzae*. Macroscopically, the hyphal swellings appear as whitish patches on the growing colony. Apparently the hyphal swellings serve as reservoirs of protoplasm which fill with a concentration of hyphae of the parasite. Secondly, the swellings act as zones of high inoculum density from which the parasitic hyphae emerge ready to infect further hyphae of the host. The influence of environment on parasitism of *Rhizopus oryzae* by *S. californica* was studied *in vitro* in non-sterilized field soil (Hunter *et al.*, 1977). *Syncephalis californica* was observed frequently parasitizing *R. oryzae* in nature on decaying apricot fruit on soil. Host and parasite populations in soil were measured at depths of 3 and 20 cm during May to November 1973; propagules per gram of dry soil were in the range 23–1847 for *Rhizopus* spp. and from 40–233 for *S. californica*. There were no consistent relationships between the populations of the hosts and parasite in the field. Parasitism occurred *in vitro* at temperatures of 15–36°C, at pH values of 4.0–7.5, and at four different carbon:nitrogen ratios. Parasitism was observed *in vitro* and in non-sterilized orchard soil at oxygen concentrations of 1–21% and at carbon dioxide concentrations of 0.03–5%. Soil moisture levels between saturation and field capacity, but not including saturation, were suitable for parasitism. Infection by *S. californica* suppressed sporulation of *R. oryzae* under most conditions that were suitable for growth of the host. Thus *S. californica* is an aggressive mycoparasite capable of attacking *R. oryzae* under a wide set of soil conditions.

In some circumstances, especially among necrotrophic mycoparasites which show a marked saprotrophic ability in culture, it may be that the parasitic mode of growth plays only a secondary role in nature. In such instances, the mycoparasitic nature is usually deduced as a result of laboratory trials rather than from field observations. It is speculative to consider whether the necrotrophic ability of a mycoparasite confers any advantage in a natural environment. This is especially difficult to prove without positive evidence of significant nutrient transfer between host and antagonist, or without a demonstration that the saprotrophic activity of the host is drastically curtailed by the competing antagonist. Antagonistic interactions of mycelia can be extremely important in the colonization of dead organic substrates. For example, many *Trichoderma* species, besides being mycoparasitic on other genera, are also antagonistic towards each other and this has been demonstrated using spruce litter. Leaves colonized initially by one species are subjected to a high dosage of the conidia of a second species, the challenge species, and incubated for several weeks. Samples taken at intervals throughout the incubation period at 25°C reveal that *T. viride* is easily displaced by *T. hamatum* and *T. koningii* which gain control of the substrate. The ability of *Trichoderma* species to colonize the leaves, however, appears also to be temperature dependent, *T. hamatum* and *T. koningii* competing better at higher temperatures (15–25°C), *T. viride* competing better at lower temperatures (10–15°C). Possibly the competitive ability of *T. viride* enables this fungus to colonize substrates in the soil during the cooler winter weather being displaced by *T. hamatum* and *T. koningii* during the warmer summer months. *Trichoderma hamatum* and *T. harzianum* are commonly isolated from hardwood bark composted in bins. The compost added to soil in containers used for growing radish seedlings suppresses the growth of *Rhizoctonia solani*, and *T. hamatum* and *T. harzianum* are believed to be the most efficient antagonists in this respect (Nelson *et al.*, 1988). The role of mycoparasitism in secondary resource capture has also been discussed in section 1.2 in relation to other wood decay fungi.

In order to determine the role played by a mycoparasitic fungus in the natural cycle of events a carefully planned programme of research may have to be implemented. The population dynamics of two *Pythium* species, the plant-pathogenic *P. ultimum* and the mycoparasitic *P. oligandrum* have been investigated in soils either conducive or suppressive to pre-emergence damping-off of cotton caused by *P. ultimum* (Martin and Hancock, 1986). In suppressive soils, *P. oligandrum* was the most commonly isolated fungus, and tended to be found at higher propagule densities than in conducive soils. When propagule densities were increased artificially in conducive soils, colonization of plants and subsequent inoculum increases of *P. ultimum* were reduced. Soil suppressiveness was overcome by amendments with dried leaf debris, which also resulted in reductions in the frequencies of colonization of isolation plates by *P. oligandrum. Pythium oligandrum* was more tolerant of chloride additions than *P. ultimum*, and apparently soils with elevated chloride concentrations allowed *P. oligandrum* to successfully compete with *P. ultimum* and increase its population density so as to further suppress the saprotrophic activity of *P. ultimum.* A similar correlation between soil suppressiveness and levels of propagules of mycoparasitic fungi has been reported for *P. ultimum* and *Laetisaria arvalis* in pasteurized and untreated field soils planted with table beets (Martin *et al.*, 1983). Soils that were repeatedly planted with this crop developed suppressiveness characterized by reduced numbers of *P. ultimum* propagules, increased numbers of *L. arvalis* sclerotia, and lower seedling disease incidence. There are many other examples where the ecological importance of mycoparasites is inferred from their role in suppressing disease in natural soils. The same phenomenon can also occur on plant substrates, and the common presence of *Gliocladium roseum*, for example, in strawberry crowns has been cited as a reason why *Verticillium dahliae* was not readily isolated from these regions, yet occurred frequently on roots and petioles (Gourley and MacNab, 1964). Other similar examples will be discussed in Chapter 7 where a role in biocontrol of plant disease is important.

A problem which requires investigation is the relation between *Verticillium psalliotae* and *Rhopalomyces elegans. Verticillium psalliotae* is a mycoparasite of the soil-inhabiting mucoralean fungus *R. elegans*, which is itself capable of parasitizing the eggs of nematodes belonging to the genus *Rhabditis. Verticillium psalliotae* is also a mycopathogen of the cultivated mushroom. The broad sporophores and large spores of *Rhopalomyces elegans* become filled with hyphae of the parasite. Although the parasitized spores may not be viable, it has been suggested that the thick walls could protect the parasite when the spores are dispersed. *Rhopalomyces elegans* has been isolated from rotting dung in soils, faeces of geese and from soil debris in glasshouses, but whether *R. elegans* can influence the survival of the nematode population in the soil is not known. Experiments would have to be devised to determine whether this fungus is capable of significantly affecting their survival in the field. Invasion of the sporophores and spores of *R. elegans* by *V. psalliotae* may mean that the natural host of the mycoparasite acts both as a reservoir and vector of infective units of the latter.

6.3 Parasitism of Spores of Arbuscular Mycorrhizal Fungi

From an ecological point of view, in order to control many plant-pathogenic fungi, various mycoparasites are employed (see sections 7.2 and 7.3). The reverse situation occurs, however, when the mycoparasite mitigates the effect of a useful fungus. This phenomenon is noted when the population dynamics of mycorrhizal fungi are affected by mycoparasitic activities. For example, parasitism by Chytridiales is probably a widespread phenomenon and may limit the populations of mycorrhizal fungi in wet soils (Sylvia and Schenck, 1983). Mycoparasites isolated from the spores of arbuscular mycorrhizal fungi (AMF) appear to be facultative with some degree of saprotrophic ability and, therefore not entirely dependent on the presence of the spores for survival (Paulitz and Menge, 1986). The colonization of the roots of plants by AMF represents one of the most interesting obligate associations of fungi with higher plants. This dual association of plant roots and certain soil fungi is ubiquitous in natural ecosystems and is responsible for the nutritional well-being of many plants. The fungi involved are related to the zygomycetes, and include the genera *Acaulospora*, *Glomus*, *Gigaspora*, *Scutellospora* and *Sclerocystis*. The vegetative hyphae of AMF resemble the broad, mainly non-septate hyphae of Mucorales. Their obligate mycorrhizal association, however, essentially sets them apart from the Mucorales as none of

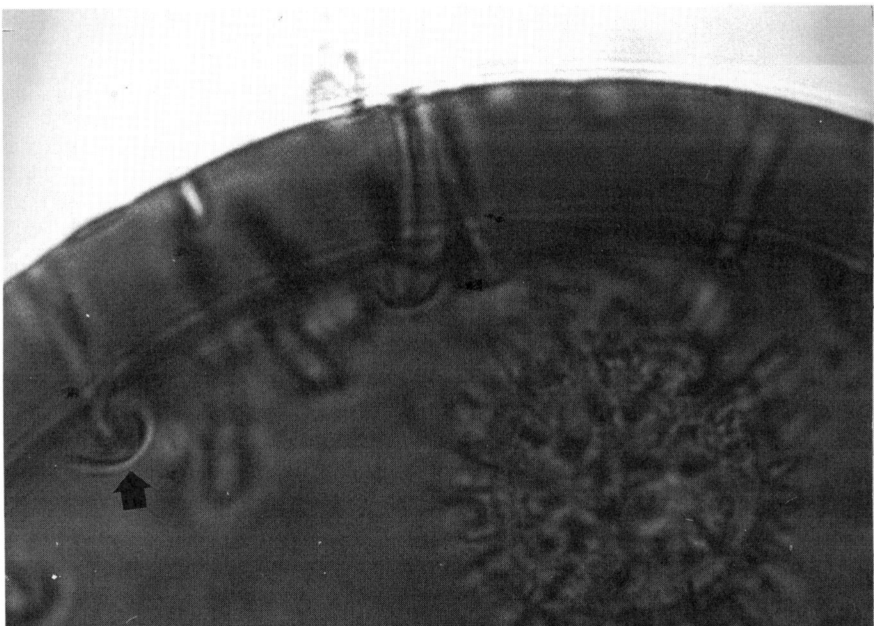

Fig. 6.1. The wall of a spore of *Glomus* which has been penetrated by other fungi. A papilla is arrowed and the spore contains a chytridiaceous sporangium (LM: × 2200).

the latter are obligately mycorrhizal. These fungi are grouped in the Glomales, on the basis of vegetative and reproductive morphology, although this order is not universally accepted (Pegler *et al.*, 1993). Much remains to be learnt about the life cycles of AMF, however, and Glomales may constitute an order of taxonomically heterogeneous fungi. Extensive evidence from pot trials and field work shows that AMF are beneficial to the development of the plant principally by improving phosphorus uptake in nutrient-depleted soils. The fungus benefits from such an association by obtaining organic carbon substrates from the plant, and as most angiosperm plants have arbuscular mycorrhizal associations, the implications for agriculture and forestry are enormous (Jeffries, 1987). The spores of some AMF are the largest known within the fungal kingdom, e.g. *Gigaspora* (Taber, 1982). They can be extracted from field soils by a simple wet-sieving procedure. In any sample of extracted spores there is usually a proportion, sometimes the majority, which appear to be parasitized. The most frequently observed parasites of spores of AMF are members of the Oomycota, including *Spizellomyces* and *Pythium*-like fungi (Boyetchko and Tewari, 1991).

Examination of individual spores shows that the wall may be perforated by many fine radial canals (Fig. 6.1), which have been well-illustrated in *Glomus microcarpum* (Malençon, 1930). The canals are usually said to arise through penetration of the wall by the hyphae of mycoparasitic fungi and, indeed, hyphae are occasionally seen to penetrate the canals. In some cases, however, it has been suggested that amoeba-like organisms could also be responsible (Boyetchko and Tewari, 1991), especially when no bacteria or fungi can be seen inside the perforated spores. In other examples, however, the hyphae of mycoparasitic fungi such as *Humicola fuscoatra* and *Anguillospora pseudolongissima* have been detected within and isolated from the spores of *Glomus fasciculatum* and *G. versiforme* (= *G. epigeum*) grown in glasshouse pot culture (Daniels and Menge, 1980). A high percentage of these spores are infected when they are inoculated on to tapwater agar or become so when placed in sterile sand containing these mycoparasites. Radial canals in the wall of the spores of *Glomus constrictum* and *G. macrocarpum* are often associated with ingrowths on the inner surface of the spore wall which are reminiscent of the papillae induced by the penetration pegs of *Piptocephalis* species on mucoralean hosts. The ingrowths are synthesized as an invading hypha penetrates the spore wall, indicating that the spore must have been alive at the time of attack. Some spores may contain many such papillae and this feature has been mentioned in some of the original descriptions of these fungi, especially *Glomus* species. Several zoosporic fungi are able to attack the spores of members of the Glomales, and either sporulate within the spores themselves, or on the outer surface. The colonization of these spores in this manner may enhance the survival and dispersal of the mycoparasites in a way analogous to that suggested for *Verticillium psalliotae* infecting *Rhopalomyces elegans*. Some of the observations of spores described as being parasitized do not indicate whether the spores were viable or dead when attacked. In the absence of papillae, this is not always clear. *Spizellomyces punctatum* has been shown to invade non-viable spores of *Gigaspora margarita* (Paulitz and Menge, 1984), and it was suggested that this fungus is primarily a saprotroph that attacks dead spores. This evidence suggests that some previous reports of *Phlyctochytrium* species as mycoparasites of spores may need reinterpretation. For example, treatment of agricultural soil used for growing peanuts and soybeans with the fumigant methyl bromide resulted in the enhanced development of *G. macrocarpum* in the roots and the soil during the growing season (Ross and Ruttencutter, 1977). During the following season, however, a decline in the population of *G. macrocarpum* chlamydospores appeared to be correlated with an increase in mycoparasitism by *Phlyctochytrium.* It is not clear whether this reflects an increase in necromass of the AMF that provided a suitable substrate for *Phlyctochytrium*, on whether the chytrid was directly responsible for the decline of the *Glomus.* Another chytridiaceous fungus occasionally observed associated with the chlamydospores of *Glomus* is *Rhizidiomycopsis stomatosa* (Hyphochytriales) reported as a parasite of *G. versiforme. Glomus versiforme* is an important AMF used experimentally in mycorrhizal studies of pot cultures of crop plants as it can be readily maintained and any potential reduction in growth through mycoparasitism is clearly of agricultural interest.

More detailed information is available for the antagonism of spores of *Gigaspora margarita* by *Stachybotrys chartarum* (Siqueira *et al.*, 1984). This fungus was found as a frequent contaminant of pot-cultures of inoculum of this mycorrhizal fungus. *Stachybotrys chartarum* colonizes the spores and produces conidiophores over the surface. Sometimes, hyphae were seen to penetrate the host spore, but it is not known if only dead spores were penetrated. The population dynamics of an arbuscular mycorrhizal fungus in a natural soil has been studied by Paulitz and Menge (1986), who investigated the effects of *Anguillospora pseudolongissima* parasitizing spores of *Glomus deserticola. Glomus deserticola* associates with the roots of onion plants and is responsible for improved growth of the crop especially in soil in which the availability of phosphorus is low. In order to kill the hyphae and leave the spores as the main source of the inoculum, a preparation of soil, pieces of root and spores was chopped and air-dried for seven days prior to incubation with the mycoparasite. Dilutions of the inoculum were mixed with various dilutions of the mycoparasite in sandy loam planted with onion seeds and the plants were harvested 40–100 days later. The results showed that the primary effect of *A. pseudolongissima* was to reduce the number of effective propagules of *Glomus* resulting in a delay and reduced incidence of root colonization. At low propagule densities, *A. pseudolongissima* was responsible for reducing (by up to 50%) the effective propagule density to the point at which it no longer colonized the host plant. The mycoparasitic activity indirectly affects the crop yield as the timing of colonization of the roots by the fungus is delayed, the rate of colonization is reduced and fewer plants are colonized which results in an overall reduction in the dry weight of the plants. These effects are most marked when the effective propagule density of the mycorrhizal symbiont in soil is low, for example under conditions of anaerobiosis or after a fumigation

Table 6.1. Spore populations of AMF in soils in northern Greece. (From Jeffries *et al.*, 1988)

Crop	% fractional infection of plant roots	Viable spores 100 g^{-1} soil	Ghosts (dead spores) 100 g^{-1} soil
Alfalfa	100	12	12–60
Cotton	100	15	15–75
Grapevine	95	3	3–15
Maize	65	2	2–10
Sunflower	5	20	20–100
Tobacco	60	1	1–5
Wheat	95	85	85–425

treatment, when the inoculum density of the mycopathogen is high, and when the soil is deficient in available phosphorus. Phosphorus-deficient soils are likely to occur where significant leaching or binding of this element takes place as in the acid-infertile soils of the tropics.

An assessment of the arbuscular mycorrhizal status of various crops in different agricultural soils of Northern Greece showed that AMF were associated to some extent with every crop. Selected sampling sites are illustrated in Table 6.1. Spore types, number and distribution were also determined and it was found that a consistent feature of all soils was the presence of large numbers of empty spores (ghosts) which usually outnumbered intact spores by a factor of 1–5. The ghosts were always perforate, and this was taken to indicate attack by other soil fungi. Many of the ghosts had developed papillae at the infection sites indicating that they were alive at the time of the mycoparasitic attack.

It has been suggested that the degree of melanization of the wall of the spore can influence its resistance to mycoparasitism (Daniels and Menge, 1980). *Glomus macrocarpum*, for example, was observed to be more susceptible to parasitism than the darker-coloured *Gigaspora gigantea* (Ross and Ruttencutter, 1977). In contrast, a white-spored species of *Gigaspora*, *G. candida*, appears to be very susceptible to parasitism (Bhattacharjee *et al.*, 1982). Some mycoparasitic fungi, however, do not affect the natural mycorrhizal population. For example, a lack of antagonism has been reported between *Gliocladium virens* and AMF (Paulitz and Linderman, 1991). This is particularly significant since *G. virens* is a potential biocontrol agent that has already received approval for use in commercial agriculture (see section 7.2.2).

6.4 Destruction of Sclerotia

The colonization of sclerotia by antagonistic fungi offers a further opportunity to assess the behaviour of these fungi in nature. There are several methods available for the extraction of fungal sclerotia from natural soils and the extracted sclerotia can then be examined for the presence of associated fungi. For example, the distribution in soils from samples obtained throughout the world of three important mycoparasites of plant-pathogenic sclerotium-forming fungi, mainly *Sclerotinia* species, has been assessed (Adams and Ayers, 1985). The mycoparasites are *Sporidesmium sclerotivorum*, *Teratosperma oligocladum* and *Laterispora brevirama.* To samples of soil taken from areas with a known history of diseases caused by *Sclerotinia* species were added the sclerotial bait of *Sclerotinia minor.* Subsequent colonization by mycoparasitic fungi was then assessed. *Sporidesmium sclerotivorum*, for example, was found to occur in soil samples from Australia, Canada, Finland, Japan and Norway but not in those from Bermuda, Brazil, China, Egypt, France, Germany, Pakistan and the Netherlands. Although *S. sclerotivorum* was not detected in all of the soils, it should be borne in mind that the initial samples used were small and may possibly not have been representative of the areas from which they were obtained. It is clear, however, that *S. sclerotivorum* is widely distributed on a global basis and could be an important natural biocontrol agent of sclerotial plant-pathogenic fungi. Australian soils screened for mycoparasites of the sclerotia of *Sclerotinia sclerotiorum* have revealed the presence of the important mycopathogens *Coniothyrium minitans* and *Dictyosporium elegans.* Both of these necrotrophs are capable of destroying the sclerotia in non-sterile soil (McCredie and Sivasithamparam, 1985). Populations of fungi associated with experimentally buried sclerotia of *Phymatotrichum omnivorum* (= *Phymatotrichopsis omnivora*), the causal agent of Texas root rot of cotton plants, include known mycoparasites such as *Gliocladium* and *Trichoderma* species which, it is hoped, may be of use in controlling the disease (Kenerley and Stack, 1987). The sclerotia of *Corticium rolfsii* are parasitized by the mycelium of *Aspergillus terreus*, a ubiquitous saprotroph found growing in soils throughout the world. A scanning electron microscope study has shown that mycelium of the parasite entirely overgrows and penetrates deeply into the sclerotial matrix (Shigemitsu *et al.*, 1978). Sometimes, a new species of fungus is found to be mycoparasitic through the use of a baiting technique. *Cylindrobasidium parasiticum*, described recently and known to be associated with sclerotia of *Typhula incarnata*, which causes snow rot disease in winter barley, has been isolated in this way. This mycoparasite, through its association with samples of sclerotia from different locations is believed to be widely distributed throughout the United States. The frequency of occurrence of mycoparasitic fungi on sclerotia can be quite variable. For example, in the course of investigations of neck rot disease of onions caused by *Botrytis allii*, *Gliocladium roseum* was found with almost all sclerotia (Walker and Maude, 1975). Generally, though, these associations are less frequent. The relationship between *Sclerotinia sclerotiorum*, the causal agent of wilt or basal stem rot of sunflower, and the mycoparasite *Coniothyrium minitans* was investigated during 1975 in a sunflower field naturally infested with the two organisms (Huang, 1977). Sclerotia of *S. sclerotiorum* were collected biweekly from roots and basal stems of wilted plants and analysed for infection by the mycoparasite. The results showed that *C. minitans* parasitized and killed the sclerotia produced on the root surface. This mycoparasite continued to parasitize the pathogen inside the root and upwards into the base of the stem, then infecting the sclerotia produced at these sites. By the end of the growing season, 59%, 76%, and 29% of sclerotia on the root surface, inside the root, and inside the stem had been killed by the mycoparasite, whereas 4%, 9%, and 68% of the sclerotia at these locations were healthy. Death of the rest of the sclerotia was due to organisms other than *C. minitans*. The data also indicated that *C. minitans* was more effective in parasitizing sclerotia produced on or inside the root than those produced in the basal stem. A second survey in 1978 showed that *C. minitans* occurred in 27% of 122 sunflower fields sampled in Manitoba. This mycoparasite was found throughout the entire sunflower production area except the east and north-east regions, but its population density in each field generally was low.

Gliocladium virens is an important mycoparasite isolated from the sclerotia of *Sclerotinia sclerotiorum* in the field (Phillips, 1986). Besides infecting sunflowers, *S. sclerotiorum* is also a serious pathogen of soybean and it persists in soil through the longevity of the sclerotia which are known to survive there for up to five years. Owing to the resistant sclerotia, the pathogen is difficult to control as they provide a source of inoculum from year to year. It seems, however, that the sclerotia are naturally infected by mycoparasites which eventually reduces the numbers of viable propagules in soil (Adams and Ayers, 1979). Sclerotia of *S. sclerotiorum* infected with *G. virens* in culture initially exude yellow-coloured droplets. Mycelium of the mycoparasite envelops the sclerotium, which turns green as the parasite produces conidia, and then penetrates between the rind cells growing intercellularly in the cortex and medulla. Sclerotia of *S. sclerotiorum* are also parasitized in the field by *Sporidesmium sclerotivorum* which attacks them in a similar manner (Adams and Ayers, 1983). Macroconidia of the mycoparasite applied to the surface of a sclerotium germinate within 3–5 days. It has been calculated that a minimum of approximately 5 macroconidia g^{-1} soil are needed to successfully infect a sclerotium (Adams *et al.*,1984). The germ-tubes establish the infection and the mycelium which proliferates within the sclerotium grows intercellularly. The rate at which *S. sclerotivorum* invades the sclerotia in soil, causing their decay, also depends on the initial inoculum density of the host and biocontrol of *Sclerotinia minor*, the causal agent of lettuce drop, has been obtained with 10 macroconidia g^{-1} soil at an initial sclerotium density of 16–24 100 g^{-1} soil. It has been suggested (Adams and Ayers, 1979) that the prolific development of the mycoparasite in soil containing sclerotia of susceptible species and its ability to spread through soil by growth from one sclerotium to another indicates that *S. sclerotivorum* is an important contributor to the natural destruction of sclerotia in soil.

The mechanism of parasitism of the sclerotia of *S. sclerotiorum* by *S. sclerotivorum* appears to be related to the conversion of glucans to glucose by glucanase activity of the host. The mycoparasite, which lacks glucanase and is therefore unable to digest glucans, is believed to utilize the glucose and possibly other monosaccharide sugars produced through enzyme activity of the host. An extracellular matrix associated with hyphal elements of the cortex and medulla is composed mainly of β-1,3-glucans. It is suggested that after penetrating the sclerotium, mycelium of the parasite absorbs any nutrients available from the extracellular matrix and in the intercellular spaces. As the supply of nutrients in the intercellular spaces becomes depleted through the activity of the parasite, the sclerotium responds by the conversion of stored glucans into glucose which restores the equilibrium. Thus a continuous supply of energy-containing carbon compounds is supplied which enables the mycoparasite to grow and sporulate. It has recently been demonstrated that the cortex and medullary elements of the sclerotia of *Sclerotinia minor* infected by *Sporidesmium sclerotivorum* are penetrated by hyphae which produce branched haustoria (Bullock *et al.*, 1986) and the implication is that this mycoparasite is essentially biotrophic as the haustoria are likely to absorb nutrients from the cytoplasm with which they are in close contact. A single infected sclerotium can provide sufficient energy to support the production, on average, of 15,000 spores of the mycoparasite thus providing a prodigious increase in inoculum potential. The practical effect of this form of parasitism is to reduce the food reserves of the sclerotia to a level below which they are not viable. Simultaneously, the dramatic increase in inoculum potential of the mycoparasite enables the organism to persist in the soil and continue to act as a natural biocontrol agent.

The surfaces of newly produced potato tubers can be heavily contaminated by sclerotia (black scurf) produced by the plant-pathogen *Rhizoctonia solani.* Black scurf incidence is often non-uniformly distributed over a field, and patches giving serious black scurf disease in one year do not necessarily do the same in a following potato crop. *Verticillium biguttatum* is able to infect and destroy sclerotia and hyphae of *R. solani* (Jager *et al.*, 1979; Velvis and Jager, 1983; Boogert and Jager, 1984). Inoculation of the soil of potato fields with mycelium of *R. solani* stimulates the development of soil-borne *V. biguttatum* (Boogert and Jager, 1983; Velvis *et al.*, 1989). In a survey of sclerotia of *Rhizoctonia solani* collected from potato tubers from different countries, *Verticillium biguttatum* was by far the most frequently occurring mycoparasite (Boogert and Saat, 1991). *Gliocladium roseum*, *G. solani*, other *Gliocladium* spp., *Penicillium* spp. *Trichoderma* spp., *Volutella ciliata* and *Chaetomium* spp. were occasionally observed. *Verticillium biguttatum* has now also been isolated from sclerotia of *R. solani* in the United Kingdom (Morris *et al.*, 1992). Thus the distribution of *V. biguttatum* appears to be worldwide, and does not seem to be restricted to a particular soil type or soil pH. Apparently the fungus is able to survive in different soils, varying from almost purely mineral soils to peat soils, at pH values between 4.8 and 7.9. Mycelium of *R. solani* is a prerequisite for development of *V. biguttatum* in the soil. Evidence has been provided for the potential regulation of *Rhizoctonia solani* by *V. biguttatum* on potato (Boogert and Velvis, 1992). There is a close relationship between the population dynamics of the mycoparasite and that of its host fungus. Host mycelium development on the below-ground plant material appeared to be a prerequisite for *V. biguttatum* development. An increase in *R. solani* was followed by an increase in *V. biguttatum*; the increase depended on the initial soil population densities, temperature and soil type. A positive correlation was observed between host fungus population density and the parasitized fraction suggesting that *V. biguttatum* may cause population density-dependent mortality. This may explain the decline in *Rhizoctonia* observed after growing several consecutive potato crops on the same field. In the absence of the host plant, mycoparasitism played a minor role in regulating *R. solani* as contact between host and mycoparasite could not be established. *Verticillium biguttatum* was able to survive in plant-free soil for at least 4 years. Within a range of increasing dependence on mycoparasitism in nature, *V. biguttatum* is regarded as taking an intermediate position between the fast-growing and antibiotic-producing mycoparasites (*Gliocladium* spp. and *Trichoderma* spp.) and the slow-growing antibiotic-negative *Sporidesmium sclerotivorum*. As compared with *Verticillium biguttatum*, *S. sclerotivorum* grows relatively poorly on various agar media (Barnett and Ayers, 1981; Ayers and Adams, 1983), but under natural conditions the populations of these two fungi are built up on host fungi, namely *R. solani* or *Sclerotinia* and *Sclerotium* species (Ayers and Adams, 1978). *Verticillium biguttatum* can best be described as an ecologically obligate mycoparasite, indicating the strictly parasitic habit under natural plant-soil conditions. This concept does not exclude the fact that the fungus can grow saprotrophically under axenic conditions, or conditions of low microbial activity.

Aspergillus flavus and *A. parasiticus*, important aflatoxin-generating fungi, develop sclerotia capable of overwintering in soil. The sclerotia are of economic importance because they provide a source of inoculum through which maize and peanut crops can become infested in the United States of America causing the kernels to become mouldy (Wicklow and Wilson, 1990). Viability tests of sclerotia buried in soil samples in Georgia for up to 18 months showed that *Paecilomyces lilacinus* (= *Penicillium lilacinum*) could be isolated from about 18% of the sclerotia of *A. flavus* and 22% of the *A. parasiticus* sclerotia. In a similar experiment carried out in Illinois, however, recovery of *P. lilacinus* from the sclerotia of *A. flavus* never exceeded 3%. Experimental inoculation of sclerotia of *A. flavus* with conidia of *P. lilacinus* indicated that they rot through colonization and presumably infection by this mycoparasite. *Paecilomyces lilacinus* has also been reported to parasitize the sclerotia of *Sclerotinia sclerotiorum* (Karhuvaara, 1960), *S. trifoliorum*, *S. borealis* and *Claviceps purpurea* (Makkonen and Pohjakallio, 1960). It may be a potential biocontrol agent of these fungi.

6.5 Associations with Rust and Smut Fungi

A large number of fungi have been reported from nature growing on leaves within the erumpment pustules of the rust fungi (Uredinales). Some of these have been described as fungicolous (see section 2.1.2), particularly if the nature of the relationship is unknown. Others known to penetrate directly and kill the spores have been described as invasive necrotrophs and are mentioned in section 3.2.3. In a study of the mycobiota of rust-infected and healthy plant material during decay, a number of mycoparasitic fungi were found associated with the rust pustules which were absent from the healthy leaves (McKenzie and Hudson, 1976). These included *Gonatobotrys simplex*, *Verticillium lecanii* and a species of *Ramularia*. The ability to parasitize other fungi was confirmed for the two former fungi by the observation that they overgrew colonies of *Alternaria* when they appeared in cultures from other field-collected material. *Cladosporium* spp. and *Alternaria* spp. were also observed sporulating profusely on rust pustules, whereas *Epicoccum nigrum* and *Pithomyces chartarum* were observed only on interpustule areas of leaves or stems (McKenzie and Hudson, 1976). Colonization of the aecia of *Puccinia lagenophorae* on groundsel (*Senecio vulgaris*) by secondary plant-pathogenic fungi has been shown to lead to death of the host whereas the host usually survives in a debilitated condition when infected by the rust alone. Invasion of the pycnia and aecia of *Puccinia poarum* found growing on coltsfoot (*Tussilago farfara*) by a range of secondary pathogens has also recently been reported (de Nooij and Paul, 1992). The rust pustules constitute a ready means of ingress into the leaf of the other foliicolous plant-pathogenic fungi and it has been suggested that such dual combinations of pathogenic fungi might be used as **mycoherbicides** in the biocontrol of weeds (Hallet and Ayres, 1992). Many other examples could be given, but it is not our purpose to provide an exhaustive list. Instead several representative species are quoted in Table 6.2.

A typical example is *Scytalidium uredinicola* on the western gall rust of *Pinus* species, *Endocronartium harknessii*. The hyphae of this fungus apparently grow into the sori of the rust fungus causing disintegration of the spores (Tsuneda *et al.*, 1980). They can also spread into the woody tissue of the galls and destroy rust hyphae to a depth of 300 μm below the sori. *Scytalidium* hyphae do not seem to penetrate the structures of the rust fungus, but may cause disintegration through the production of lytic enzymes (Tsuneda *et al.*, 1980). Colonization of the gall by the *Scytalidium* is relatively slow and complete inactivation of the gall is not achieved in the first year of invasion. An antifungal agent called maltol (3-hydroxy-2-methyl-4H-pyrone) has been isolated from cultures of *S. uredinicola* (Cunningham and Pickard, 1985), and shown to inhibit spore germination of *E. harknessii* as well as inhibiting the growth of several wood-inhabiting fungi.

Some rusts are also infected by *Cladosporium aecidiicola*, for example old aecia of *Puccinia phragmitis* and *Uromyces limonii*. *Cladosporium uredinicola*, on the other hand, infects the uredinia and telia of *Phragmidium mucronatum*, *Phragmidium bulbosum* (= *P. rubi* = *Puccinia rubi*) and *P. violaceum* (= *Puccinia violacea*). Species of *Tuberculina* are also reported to be parasitic on rusts, most of them being exclusively uredinicolous (Hawksworth, 1981). *Tuberculina maxima* infects the pycnia and aecia of *Cronartium flaccidum* and *C. ribicola* where it is responsible for ‘purple mould’. The parasitic nature of this relationship has been challenged (Wicker, 1981) and it is suggested that *T. maxima* does not attack the rust directly

Table 6.2. Examples of fungi reported as mycoparasites of rust spores.

Rust fungus	Mycoparasite	Reference
Aecidium clematidis	*Tuberculina persicina*	Buller (1924)
Atelocauda koae (= *Uromyces koae*)	*Tuberculina maxima*	Gardner *et al.* (1979)
Cronartium flaccidum	*Tuberculina maxima*	Ellis and Ellis (1985)
Cronartium quercuum f.sp. *fusiforme* (= *C. fusiforme*)	*Scytalidium uredinicola*	Kuhlman *et al.* (1980)
Cronartium quercuum f.sp. *fusiforme*	*Tuberculina maxima*	Gardner *et al.* (1979); Wicker (1981)
Cronartium ribicola	*Tuberculina maxima*	Ellis and Ellis (1985)
Cronartium strobilinum	*Eudarluca caricis*	Kuhlman and Matthews (1976)
Endocronartium harknessii	*Cladosporium gallicola*	Tsuneda and Hiratsuka (1979)
Endocronartium harknessii	*Scopinella gallicola*	Tsuneda and Hiratsuka (1981)
Gymnosporangium cornutum (= *G. juniperum*)	*Tuberculina persicina*	Buller (1924)
Melampsora allii-populina	*Ramularia uredinis*	Magnani (1970)
Melampsora larici-populina	*Cladosporium tenuissimum*	Sharma and Heather (1983)
Phakopsora pachyrhizi	*Verticillium psalliotae*	Saksirirat and Hoppe (1990)
Phragmidium mucronatum	*Hainesia rubi* *Micropodia oedema*	Ellis and Ellis (1985)
Phragmidium bulbosum (= *P. rubi*)	*Hainesia rubi*	Ellis and Ellis (1985)
Phragmidium violaceum	*Hainesia rubi*	Ellis and Ellis (1985)
Puccinia allii	*Cladosporium uredinicola* *Ramichloridium schulzeri* *Trichothecium roseum* *Verticillium lecanii*	Uma and Taylor (1987)
Puccinia arachidis	*Acremonium persicinum* *Eudarluca caricis* *Penicillium islandicum* *Tuberculina costaricana* *Verticillium lecanii*	Gwande (1990)
Puccinia coronata	*Eudarluca caricis*	Buller (1924)
Puccinia graminis	*Aphanocladium album*	Koc *et al.* (1981)
Puccinia phragmitis	*Cladosporium aecidiicola*	Ellis and Ellis (1985)
Puccinia poarum	*Tuberculina persicina*	Buller (1924)
Puccinia recondita	*Cladosporium uredinicola*	Ellis and Ellis (1985)
Puccinia recondita	*Eudarluca caricis*	Buller (1924); Stahle and Kranz (1984); Casulli (1990)
Puccinia vincae	*Tuberculina schrozzii*	Ellis and Ellis (1985)

Table 6.2. continued

Rust fungus	Mycoparasite	Reference
Triphragmium ulmariae	*Cladosporium uredinicola*	Ellis and Ellis (1985)
Uromyces appendiculatus (= *U. phaseoli*)	*Verticillium lecanii*	Mendgen and Casper (1980)
Uromyces cestri	*Cercospora uromycestri*	Pollack (1971)
Uromyces cytisi	*Eudarluca caricis*	Buller (1924)
Uromyces dianthi	*Verticillium lecanii*	Spencer (1980)
Uromyces limonii	*Cladosporium aecidiicola*	Ellis and Ellis (1985)

but affects the rust-infected pine cells instead. Thus, *T. maxima* suppresses the rust by enzymatically destroying the nutrient source vital to the survival of the obligate pathogen. There is no evidence that *T. maxima* derives any nutritional benefit from the rust itself but, it is presumed, merely exploits rust galls as a convenient ecological niche. *Tuberculina persicina*, the most frequently occurring species, is found in the aecia of many rusts and may prevent the development of *Gymnosporangium sabine* (= *G. fuscum*) on previously infected leaves (Mijuskovic and Vucinic, 1974). Although *T. maxima* can be grown in axenic culture the rate of development is slow (Hubert, 1935) both in culture and on the potential host (Hiratsuka and Powell, 1976) and it seems unlikely to have potential as a biocontrol agent. *Tuberculina schrozzii* infects the sori of *Puccinia vincae* (Ellis and Ellis, 1985).

Eudarluca caricis is the most widespread uredinicolous mycoparasite being known to occur on 20 genera and at least 226 species of rusts (see also section 3.2.3) which grow mainly on Cyperaceae and Gramineae (Hawksworth, 1983). There is clear evidence for the destructive invasion of rust spores. Many other genera and species of rust-infecting fungi are known (e.g. Hawksworth, 1981) which presumably have an influence in the control of the natural population of Uredinales. *Eudarluca caricis* is of particular interest in this respect as it can be grown in axenic culture (Calpouzos *et al.*, 1957) and appears to have potential as a biocontrol agent of some rust fungi (see Chapter 7).

Verticillium lecanii is one of the most studied mycoparasites of rust fungi. In addition to attacking insects, strains of this fungus can also attack and penetrate the walls of uredospores of *Uromyces*, *Puccinia* and *Hemileia*, either directly or via the germ-pore, and it has been suggested that the isolates found in rust pustules may represent a different species, *V. hemiliae*. There is no evidence for penetration of immature spores or mycelium, but it has been suggested that *V. lecanii* might produce an antagonistic metabolite which inhibits development of the rust mycelium (Spencer and Atkey, 1981). An immunofluorescence technique has been used to specifically label hyphae of *V. lecanii* within invaded rust pustules (Mendgen and Casper, 1980). This method was successfully used to show that the mycoparasite was restricted to the sporulating layer of the pustule, where it penetrated the uredospores, but it did not spread into sporogenous tissue of the rust fungus or into the tissues of the host leaf. *Verticillium psalliotae*, a fungus more usually associated with the growth of mushrooms (see section 6.6), has also been recorded as a mycoparasite of rust fungi. It occurs, for example, on the uredinia of the soybean rust (*Phakopsora pachyrhizi*) in Thailand and Taiwan, and has been previously isolated from *Puccinia coronata* and *Hemileia vastatrix* (Saksirirat and Hoppe, 1990). In laboratory tests, four Asian isolates of *V. psalliotae* were tested on soybean rust and bean rust (*Uromyces appendiculatus*) uredia. The strains of *V. psalliotae* originally isolated from the soybean rust grew rapidly on both substrates, but a European isolate of *V. lecanii* developed rapidly on bean rust uredia but only slowly on the soybean rust (Saksirirat and Hoppe, 1990). Few fungi seem able to attack the smut fungi (Ustilaginales), which are another group of important plant pathogens (Hawksworth, 1981). *Eudarluca caricis* has been reported to attack *Ustilago* species. *Ustilago bullata*, the loose smut of brome grass (*Bromus mollis*), is parasitized by *Pythium vexans* in culture and probably also in nature (Robertson *et al.*, 1990). Teliospores of the smut are attacked by germ-tubes of *P. vexans* which attach to them by means of appressoria from which the infectious pegs penetrate the thick walls of the teliospores. Cytoplasm of the teliospores appears to be rapidly destroyed by the infection and neighbouring teliospores are infected by emergent hyphae from the empty spore cases.

6.6 Antagonistic Fungi Associated with Mushroom Cultivation

6.6.1 Mushroom culture

The cultivation of edible fungi provides an artificial ecosystem in which other fungi may grow and cause problems through antagonism of mushroom mycelium. The commonest edible fungus in the Western hemisphere is the cultivated mushroom *Agaricus bisporus*, a species which is also grown and extensively distributed commercially throughout the world. The total annual crop, well in excess of 1,000,000 tonnes, accounts for about 75% of the mushrooms cultivated for sale. On a commercial basis, therefore, the cultivated mushroom serves as a tasty and nutritious food to the consumer and is an important source of income for mushroom farmers. The continuous growth of the mushroom crop on a mushroom farm is a form of monoculture. In general, mushrooms are grown on a solid substrate, rich in nutrients and organic material, which has been composted under controlled conditions and then pasteurized. The compost is inoculated with mycelium of *Agaricus* which quickly establishes and colonizes the substrate. Once colonization is complete, the mushrooms are induced to form by addition of a 'casing' layer over the compost surface. Antagonistic fungi can thus cause problems at two different stages in the process: (i) as weed moulds colonizing the compost and inhibiting the growth of *Agaricus*; (ii) as mycoparasites attacking *Agaricus* directly. These two aspects will be dealt with separately, but there can be overlap between the categories, for example when *Trichoderma* species are the antagonistic fungi. The loss of income to mushroom growers throughout the world on an annual basis through the activity of weed moulds and mycoparasites may amount to several million pounds. The compost on which mushrooms are normally grown is a mixed fermentation based on stable manure, hay, or other complex mixes which exclude horse manure. Chicken manure may be added as a nitrogen source or substituted for horse manure. Microorganisms break down sugars and starch before cellulose and lignin in the compost and the decomposition releases carbon dioxide, water, ammonia and heat. It has been shown that the most efficient decomposition is achieved by thermophilic aerobic microorganisms and the process may be monitored in order to maintain the optimum temperatures required for growth in phases I and II of the composting process. The mushroom compost is not sterile and does not therefore exclude the growth of other fungi. As the temperature of the compost rises through microbial activity, a succession of microorganisms develops. The environmental conditions under which the mycelium of the mushroom is grown are manipulated in order to favour the growth of the mushroom to the disadvantage of weed moulds (Sinden, 1971). Compost in phase I of decomposition is turned in revolving drums or by other mechanical means for several days at a temperature of 70–80°C. Phase II of composting is carried out for several days at 50–60°C and requires a plentiful supply of air which reduces the growth of weed moulds and mycoparasites, allows the decomposition to continue, liberates ammonia and provides the organic nitrogen through microbial metabolism necessary for growth of the mushroom mycelium.

At the end of phase II composting, the compost is 'spawned' with a culture of *Agaricus bisporus* grown on sterilized cereal grains. The spawn is uniformly mixed with the compost and the mycelium is allowed to colonize. For commercial purposes, strains of *A. bisporus* are selected for their productivity and appearance when cropped. Spawning takes place at a much lower temperature (about 25–28°C) than composting. Viral and fungal contamination is reduced by spawning in sheds with a filtered air flow. The process of cultivation of *A. bisporus* is designed to make growth conditions favourable to the mushroom mycelium and enable compost to be colonized within 14 days. The mycelium of *A. bisporus* produces antifungal antibiotics which enable it to dominate the compost and prevent competitors from gaining a hold. Some weed moulds, however, appear to be unaffected by the antibiotics and attempts have been made to reduce contamination of the compost with their spores by filtering the air used to ventilate the compost.

After the mycelium of *A. bisporus* has fully colonized the compost, it is 'cased' with a layer of peat/soil. The function of the casing layer is unclear, but it provides a bacterial and algal community the growth of which stimulates mushroom formation. Biotic factors and anchorage and water holding capacity are all important factors supplied by casing which enables the mushroom primordia to develop. The casing soil may be pasteurized with steam or disinfected in order to kill mycoparasitic fungi and weed moulds, but the soil must not be completely sterilized as the process would also remove the useful microorganisms. The casing soil provides an appropriate nutrient medium for the growth of competitive microorganisms. In 'healthy' compost mushrooms appear at weekly intervals called 'flushes' over about 7 weeks until nutrients are depleted in the compost.

6.6.2 Weed moulds

Many different weed moulds have been recognized and they grow in the compost competing with the mycelium of the mushroom for nutrients, water and space and may themselves be unaffected by antibiotics produced by the mushroom mycelium. Their effects on the growth of the mycelium of the mushroom are variable. Many of them have been given common names by mushroom growers which refer to the colour of spores or form of the mycelium or the reproductive structures. Thus green and brown moulds comprise an assemblage of different species of fungi. Weed moulds can cause severe reductions in yield if the composting procedure is inefficient (e.g. if too low or too high a temperature is reached, or if the material is too wet) or if soil infested with the spores of weed moulds is used in the preparation of compost and casing. In fact, the presence of certain weed moulds in mushroom beds can be used to indicate faults in the composting procedure. Such weeds are referred to as indicator moulds. For example, if the aeration is insufficient during phase I composting *Chrysosporium* (= *Myceliophthora*) *luteum*, responsible for the mat yellow mould disease of mushroom, may develop at this stage. If the compost is overmoist and conditions become less aerobic it may be colonized by *Sordaria fimicola*, an ascomycete associated with the dung of herbivorous animals used in the production of compost. If the quantity of ammonia is too high at the end of phase II composting the weed mould *Coprinus fimetarius* (= *C. radiatus*) may develop. This fungus is normally associated with the ammoniacal conditions encountered in the dung of herbivorous animals. Some weed moulds stimulate the growth of the mycelium of *A. bisporus*. *Scytalidium thermophilum*, for example, almost always develops during phase II of composting (Straatsma *et al.*, 1989). It is inactivated by the growth of *A. bisporus* which grows less vigrously when *S. thermophilum* is absent.

Other weed mould fungi can also gain the ascendency over the *Agaricus* mycelium if conditions are appropriate. For example, *Chaetomium globosum*, called the olive-green mould owing to the olive-green perithecia produced in profusion on the straw in compost, is responsible for colonizing compost and making it unsuitable for *Agaricus*, thus reducing the yield of mushrooms. This fungus grows well in ammoniacal conditions at reduced oxygen tension and its presence in the compost indicates inadequate aeration during phase II of the composting procedure.

Colonies of *Chrysosporium luteum* grow in the compost or under the casing soil usually forming a yellow–brown mat of mycelium at the interface between the casing soil and compost. This mat strongly inhibits growth of the mycelium of the mushroom. Conversely a brown mould, *Peziza ostracoderma*, a discomycete which can grow vigorously in the peat-based casing soil, has only a slightly inhibitory effect on growth of the mushroom mycelium. *Peziza ostracoderma* competes with the mushroom for nutrients and it may take longer for the mushroom basidiomes to develop. The dark-brown apothecia of the mould develop on the surface of the compost between the mushrooms.

Species of *Trichoderma*, usually ascribed to *T. viride* and *T. harzianum*, have recently become troublesome on mushroom farms in the UK (Seaby, 1987). There are many reports of species of *Trichoderma* causing problems but there is a need to be cautious over the use of species names as specific identification is difficult. We have used the terminology given in the original papers. When *Trichoderma* occurs as a weed mould it is referred to as the green mould or mildew owing to the prolific production of conidia which form a dense green cover on the casing soil of bags, blocks and trays in which mushrooms are grown. From 10 to 80% of the bags in a mushroom house may be infested which results in considerable loss of income for the grower. The mould is difficult to eradicate because the vast numbers of viable conidia produced can be spread both within and between mushroom farms in several ways. Mycophilic mites, such as *Pygmephorus mesembrinae*, which feed on the mycelium of *Trichoderma* and disperse the spores throughout the growing houses are important vectors in the spread of the mould. Mushroom flies may also transport large numbers of conidia. Mice in search of the grain spawn used as food, may also transport conidia on the hair on their bodies. If infested spawn is stored at less than 10°C, growth of the mycelium is slow and green patches indicating the presence of *Trichoderma* may not be evident. Contaminated spawn provides a source of inoculum of the green weed mould. If the casing soil is wet and not properly pasteurized, species of *Trichoderma* may develop, producing green contaminating colonies. After cutting the mushrooms, the cut stipes which remain in the casing soil provide an appropriate food base for the continued growth of *Trichoderma* which leads to further crop losses.

An interesting physiological relationship between the mycelia of *A. bisporus* and *T. harzianum* has been noted. Fresh peak-heated compost does not become infested with *Trichoderma* unless the mushroom mycelium is also present. It seems that bacteria in the compost inhibit the growth of *Trichoderma*. Mycelium of the mushroom, however, is responsible for lysis of the *Trichoderma*-inhibiting bacteria thus enabling the weed mould to become established in the compost and compete with that of the mushroom. Secondly, the competitive balance between *Trichoderma* and *Agaricus* is affected by temperature. In experiments using growth temperatures in the range 20–30°C it was found that the deleterious effects of *Trichoderma* on the growth of *Agaricus* increased with increase in temperature. This observation led to the first recommended control measure not to allow the temperature to exceed 25°C during mushroom growth (Seaby, 1987).

There is variation in the type of mushroom infestation due to species of *Trichoderma* (Sinden and Hauser, 1954). *Trichoderma viride* which causes a spot or blotch of the mushroom is associated with decaying organic matter in the casing soil. *Trichoderma viride* is responsible for action in advance of the hyphae as its mycelium grows on the killed hyphae of the mushroom. On the other hand, *T. koningii* also parasitizes the living hyphae of the vegetative mycelium and basidiomes of the mushroom. The 'soft wet rot' disease initially becomes evident on dead mushrooms in the beds then spreads to the living mushrooms as the mycelium grows over the casing soil. The caps of infected mushrooms turn purple–brown and become cracked and the mycelium softens and rots. *Trichoderma koningii* sporulates weakly in comparison with *T. viride* and the light-green sporulating areas of the mildew caused by *T. koningii* are restricted to the basidiomes. *Trichoderma* species are soil-borne and strict hygiene measures are required to control them as, once established, they tend to persist. Such measures include providing employees with overalls that have been sterilized to kill the spores, using clean disposable plastic gloves when handling boxes of spawn and the provision of antiseptic foot dips for people entering the growing rooms (Seaby, 1987).

Pythium oligandrum has also been reported as a problem in mushroom cultivation. It has been consistently isolated from black patches of mushroom compost that were not colonized by *Agaricus* (Fletcher *et al.*, 1990). The morphological relationship between the two fungi has not been determined. Weed moulds can also cause problems during cultivation of other edible fungi. *Corticium rolfsii*, for example, is a common contaminant of the beds of the mushroom *Pleurotus flabellatus* (Rajarathnam *et al.*, 1979), and several saprotrophs have been shown to inhibit growth of *Pleurotus sajorcaju* (Doshi and Singh, 1985). *Podospora faurelii*, on the other hand, grows on the spawned beds of the paddy straw mushroom, *Volvariella volvacea*, and the fruiting is strongly inhibited (Bahl and Chowdhry, 1981).

Trichoderma hamatum and *T. harzianum* can also affect the growth of the oyster mushroom *Pleurotus ostreatus* through the production of a diffusible, non-volatile toxin which kills the mycelium of the mushroom (Reper and Penninckx, 1987). The mushroom is cultivated on pasteurized straw which may also be colonized by the cellulose-competent antagonist *T. hamatum*. Where the mycelia of the two fungi contact in dual culture experiments over a range of temperatures and pH, the growth of *P. ostreatus* ceases whereas that of *T. hamatum* is unaffected. A technique used to demonstrate the production of a toxic substance by *T. hamatum* involves growing *T. hamatum* on a cellophane disc on the surface of nutrient agar for 48 h at 25°C then removing the disc bearing the mycelium. Next, the centre of the plate of agar is inoculated with mycelium of *P. ostreatus* and incubated as before for up to 10 days. The mycelium of *P. ostreatus* does not resume growth even when placed on fresh nutrient medium which indicates that it is killed by a toxic principle from the growing mycelium of *T. hamatum*.

6.6.3 Mushroom pathogens

The three most important mycopathogens of *A. bisporus* are *Verticillium fungicola* which causes 'dry bubble' disease, *Mycogone perniciosa* causing 'wet bubble' disease, and *Hypomyces rosellus*, responsible for 'cobweb disease'. Mushrooms

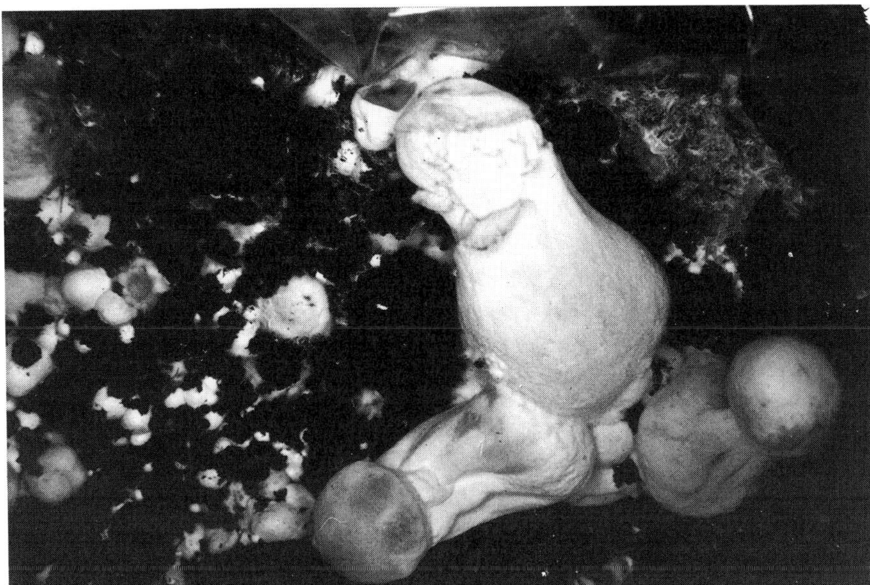

Fig. 6.2. An imperfectly formed mushroom resulting from attack by *Verticillium fungicola*. (Courtesy of Dr J.T. Fletcher.)

affected by dry or wet bubble are malformed and unsaleable. Variation in the necrotrophic reaction to these pathogens of the cultivated mushroom has been observed. With *Hypomyces rosellus*, a destructive parasite of mushrooms, and *Verticillium fungicola* var. *fungicola* the mycelium of the host is overgrown and severe necrosis occurs. The reverse reaction occurs, however, with *Mycogone perniciosa* as the vegetative mycelium of the mushroom overgrows that of the mycoparasite. Mature mushrooms may also be attacked. It has been suggested that an inability to parasitize the actively growing vegetative mycelium of the potential host could confer an adaptive advantage on the parasite as the likelihood that the mushrooms will form is increased, thus maintaining the supply of susceptible material (Gray and Morgan-Jones, 1981).

Dry bubble disease caused by *Verticillium fungicola* var. *fungicola* is manifested by swelling of the stipe, and the pileus may develop blue–grey spots from which a grey-coloured mycelium grows as a halo at the periphery of the spots and the entire mushroom tends to develop a lop-sided appearance. Symptoms vary according to the stage of primordial or mushroom development at which infection occurs. If the mushroom is infected in the button stage, it becomes a mass of undifferentiated tissue (Fig. 6.2). Young infected mushrooms develop dry non-necrotic bubble-like spheres over the surface which later becomes covered with a dusty-grey layer of conidia. *Verticillium fungicola* produces thin-walled unicellular conidia embedded in mucilage which enable them to adhere to the bodies of mushroom flies, mites, the hands of farm workers and farm implements. In an experiment using the mushroom fly *Megaselia halterata*, it was shown that flies inoculated with conidia of *V. fungicola* could effectively spread the disease (White, 1981). A low density of flies (75 m^{-2}) was responsible for the primary infection of the crop, although later spread was due primarily to watering, and the intensity of infection was related to the initial density of the flies. Thus the adult insect is an important vector in the transmission of dry bubble disease. The conidia are also dispersed by splash droplets and the movement of water in the growing rooms. The most common source of the inoculum, however, is contaminated casing. Although no resting spores are known to be produced, the mycelium from the dry bubbles is believed to be persistent (Sinden, 1971).

Other varieties of *Verticillium* are also known to parasitize the cultivated species of *Agaricus* (van Zaayen and Gams, 1982). For example, *V. fungicola* var. *aleophilum* causes a brown discoloration of the mushroom caps in *A. bitorquis* and to a lesser extent in *A. bisporus* and reduces the quality of the crop. This is perhaps explained by the fact that this mycoparasite grows better at 27°C which is the temperature at which *A. bitorquis* is grown, whereas *Verticillium fungicola* var. *fungicola* grows better at 24°C, the cultivation temperature for *A. bisporus*. The oyster mushroom *Pleurotus ostreatus*, which is widely cultivated throughout the globe on a commercial basis, including China, Europe, India, Japan and the USA, is also attacked by *Verticillium fungicola* (Marlowe and Romaine, 1982).

An unusual feature of the control of fungal parasites of the cultivated mushroom is that, as both the host and the pathogen are fungi, it is difficult to control the mycopathogen with fungicidal chemicals unless they are selective for the mycopathogen and non-toxic to the mushroom. Biocontrol methods may overcome this difficulty. Although various fungicides are used to control dry bubble disease, tolerant strains of the mycopathogen eventually emerge and a new chemical control treatment has to be devised. It has recently been demonstrated that manipulation of the environmental conditions under which the mushroom crop is grown can significantly reduce the incidence of the disease. The conditions under which mushrooms are grown are carefully controlled and it has been shown that reducing the air temperature in which the crop is grown from 20 to 14°C and the relative humidity from 90 to 80% reduces the number of mushrooms infected with dry bubble. This technique has been referred to as 'ecological control' and is important when epiphytotic outbreaks of dry bubble occur. Apparently the mushroom survives better than the mycopathogen at the lower temperature and, as all of the commercially available strains of cultivated mushrooms are susceptible, this would appear to be a promising means of controlling the disease. Since water is known to assist dispersal of the spores of *V. fungicola*, the lower relative humidity may also reduce the condensation of water on the mushrooms and limit dispersal. Alteration of the environmental conditions in this way neither reduces the quantity of mushrooms produced nor impairs their flavour.

Biocontrol of *V. fungicola* has been achieved by spraying the casing soil with propagules of *Trichoderma viride* (de Trogoff and Ricard, 1976). The mushroom

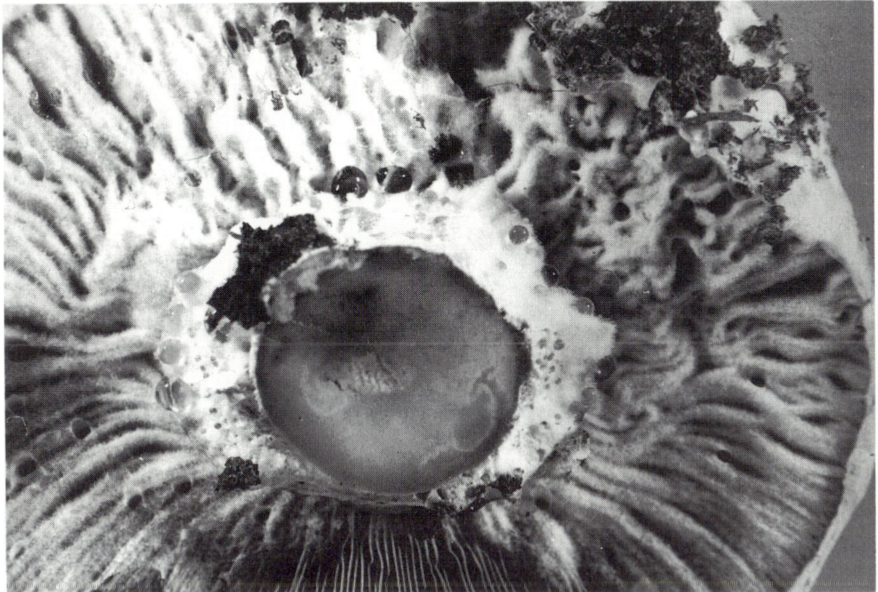

Fig. 6.3. A basidiome showing symptoms of partial colonization by *Mycogone perniciosa*. The pathogen has colonized one side of the stipe and radiated onto the gills affecting a sector which shows amber droplets of liquid typical of this disease. (Courtesy of Dr J.T. Fletcher.)

mycelium also grows in the casing soil, but it is assumed that the mycelium of *Agaricus* is not attacked by the growth of *Trichoderma* in this method of treatment. The mechanism of control is not known but it is possible that parasitism of the mycelium of *V. fungicola* is involved. The *Trichoderma* spray has been developed commercially and it has been shown that if the surface of casing soil is sprayed at the rate of 1 l m^{-2} with 1×10^8 propagules l^{-1} the disease can be controlled.

Wet bubble disease (Figs 6.3 and 6.4) caused by *Mycogone perniciosa*, infects *Agaricus bisporus*, other cultivated mushrooms and possibly native mushrooms and toadstools which grow in the wild. Outbreaks of the disease sometimes occur on isolated mushroom farms and it has generally been assumed that wild agarics are the most likely source of infection. Fletcher and Ganney (1968) were unable to demonstrate that propagules of *M. perniciosa* are air-borne, and in this respect the colonization of mycelium and spores of *Rhopalomyces elegans* by *M. perniciosa* is of interest (Barron and Fletcher, 1972). *Rhopalomyces elegans* is a parasite of nematode eggs, and is commonly associated with concentrated populations of nematodes such as might occur in mushroom composts. *Verticillium psalliotae* also parasitizes *R. elegans* and it has also been suggested that the latter fungus can act as a reservoir of the mushroom pathogens during the composting phase (Dayal and Barron, 1970). In wet bubble disease, under relatively humid conditions of growth,

Fig. 6.4. *Mycogone perniciosa* causing a large 'cauliflower-like' distortion of a mushroom, the sclerodermoid mass. (Courtesy of Dr J.T. Fletcher.)

the stipes of the young mushrooms swell disproportionately and may reach the size of a tennis ball, the pilei are small and misshapen with a cauliflower-like appearance, and the gills are either rudimentary or do not develop and no basidiospores are produced (Fig. 6.4). A white mycelium develops in place of the abortive lamellae. Amber-coloured droplets of fluid exude over the surface of the distorted mushroom (Fig. 6.3) which subsequently collapses into a wet, bubble-like, vile-smelling decaying mass (Fletcher *et al.*, 1986). In less humid conditions, however, mushrooms infected with *M. perniciosa* do not rot and the symptoms resemble those of dry bubble. Infections of mature mushrooms tend to be restricted to the base of the stipe which, if left in the compost after cropping, may provide a source of inoculum of the mycopathogen. *Mycogone perniciosa* produces two types of spore both of which can germinate and infect a mushroom. The first type is a two-celled chlamydospore, with a thick-walled deeply pigmented apical cell and a thin-walled basal cell. Both cells are capable of producing germ-tubes although after drying, only the thick-walled cell is able to do so as the cytoplasm in the basal cell dies, leaving the thick-walled cell which acts as the dormant survival spore (Holland *et al.*, 1985). The chlamydospores may survive for 2–3 years in casing soil thus providing a primary source of inoculum if infested casing soil is used to cover the compost. In samples of the survival spores placed under conditions of germination in the laboratory about 20–30% germinate but, if the spores are subjected to a period of vernali-

Fig. 6.5. A mushroom becoming colonized by *Hypomyces rosellus*. (Courtesy of Dr J.T. Fletcher.)

zation at temperatures from −5 to 10°C, over 70% of the spores can be induced to germinate. The second spore type is a thin-walled single-celled conidium. They are spread by means of splash droplets and in the surplus water which runs off the mushroom beds. The thin-walled conidia do not appear to require special conditions for germination and are relatively short-lived. The chlamydospores, however, have to be activated by one or more ethanol-soluble chemicals produced by the mycelium and the basidiome (Holland and Cooke, 1990). Extracts of the mycelium of several other Agaricales have been shown to stimulate chlamydospore germination which raises the possibility that wild mushrooms may act as a reservoir of infection. The mycelium of *M. perniciosa* does not grow through the casing soil and the spores will only germinate when close to the basidiome initials. Reliance of the mycopathogen on the proximity of a potential host for the germination of its spores appears to be an adaptation to the mycoparasitic mode of life. Fungicides containing benzimidazole have been used to control the spread of both wet and dry bubble disease, but resistant strains of the mycoparasites have now been found.

Cobweb disease of mushrooms (Figs 6.5 and 6.6) is due to overgrowth by the mycelium of *Hypomyces rosellus* (anamorph: *Cladobotryum dendroides* = *Dactylium dendroides*) which colonizes the mushrooms and also grows through and over the surface of the casing soil. *Hypomyces rosellus* is another fungus often found growing on the caps of wild mushrooms and toadstools. Infected basidiomes become engulfed by the rapid advance of the grey–white cobweb-like mycelium, the gills may fail to develop and the basidiomes appear distorted. The parasitized mushrooms turn brown as they develop a wet rot. The cobweb fungus produces perithecia which appear as dark-coloured specks over the surface of the parasitized basidiome. Initially the mycelium appears cobweb-like but later thickens to become mat-like and turns a pink–red colour from which the specific name of the fungus is derived. The mushroom caps also develop pink–brown spots. *Hypomyces rosellus* is dispersed by means of multicellular conidia carried in splash droplets and water films and by viable fragments of mycelium carried in infested casing soil.

Shaggy stipe disease of the cultivated mushroom is caused by growth of the grey–white mycelium of *Mortierella bainieri* through the casing soil and over the mushrooms. The effect of *M. bainieri* is to cause the material of the stipe to break up and peel into several layers. The pileus becomes discoloured to dark brown. *Mortierella bainieri* is a member of the Mortierellaceae, fungi which inhabit soils throughout the world, often associated with the rhizosphere. Most species of *Mor-*

Fig. 6.6. Extensive colonization of mushroom basidiomes by *Hypomyces rosellus* producing characteristic cobweb symptoms. (Courtesy of Dr J.T. Fletcher.)

tierella are characterized morphologically by the production of relatively small acolumellate asexual sporangia which contain sporangiospores usually capable of rapid germination under the appropriate conditions of temperature and nutrients. Over 70 species are known (Gams, 1977). As a natural group, they probably play an important part in the recycling of carbon and nitrogen through their saprotrophic activity in soil, thus the association of *M. bainieri* with the cultivated mushroom is especially interesting. The probability is that the relationship of *M. bainieri* to *A. bisporus* is similar to that of other Mucorales such as *Spinellus* and *Syzygites* which grow on the caps of toadstools in the field. These mycoparasitic Mucorales also grow well as saprotrophs in culture on nutrient media. *Syzygites megalocarpus* also attacks the shiitake mushroom, *Lentinula edodes* (= *Lentinus edodes*), which is cultivated widely throughout tropical Asia.

Diehliomyces microsporus which causes false truffle disease of mushroom beds, occurs in association with cultivated mushrooms throughout the world, including India, Italy, France, The Netherlands, South America, Taiwan, the USA and the UK where considerable crop losses have occurred. This fungus does not attack the basidiomes but prevents their formation. It produces cream to reddish-brown coloured ascomata 3–40 mm in diameter in the casing soil and compost which have an irregularly lobed surface and are said to resemble a Perigord truffle although, unlike the Perigord truffle, they are inedible. A single sporome may contain thousands of globose ascospores about 5–6 μm in diameter which appear smooth-walled in the scanning electron microscope. *Diehliomyces microsporus* is a persistent nuisance in mushroom beds as it appears to compete with the mycelium of the mushroom for nutrients in the compost and, in addition, is capable of antagonizing and killing the mycelium. Growth of the mycelium of the false truffle in the soil both displaces that of the mushroom and prevents it from colonizing the straw. The net result is that no mushrooms develop where the mycelium of the mycopathogen grows thus reducing the productivity of the crop by as much as 75%. Characteristically, the compost also appears black and soggy and has the pungent odour of chlorine. Freshly prepared mushroom compost is alkaline but as the mycelium of the mushroom grows it becomes acidic. Acidification of the growth medium apparently favours the growth of *D. microsporus.*

Persistence of the false truffle is believed to be due to the sporomes which survive in the soil after the mycelium has disintegrated and possibly also to the impregnation of wooden boards used in the beds with the mycelium. Breakdown of the sporomes releases the abundant ascospores into the soil and their germination is stimulated by the proximity of the mycelium of the mushroom and the high temperatures which obtain during composting. Even an inoculum potential as low as 10 spores m^{-2} is capable of producing false truffle disease (van Zaayen and van der Pol-Luiten, 1977). The mycopathogen is dispersed through infested compost and the disposal or sterilization of infested soil is required to ensure that fresh compost does not become contaminated with ascospores. Growth of the mycopathogen is said to be associated with warm conditions ($>$20°C). Strains of *Agaricus bitorquis*, selected for their resistance to viral infection, require higher growth temperatures than *A. bisporus* which, however, appear to favour the growth of *D. microsporus*. Reduction of the temperature to less than 20°C during spawning and cropping reduces the incidence of the disease but this is not a cost effective control measure. False truffle has only ever been isolated from mushroom crops. The widespread occurrence of false truffle disease throughout the world wherever *Agaricus* species are cultivated indicates that the mycopathogen is successfully adapted physiologically to the conditions associated with growth of the cultivated mushroom. In this context, the mycelium is thermophilic in growth behaviour, the optimum temperatures for growth in the range 26–32°C (van Zaayen and van der Pol-Luiten, 1977). The ascospores have been reported to require the relatively high optimum temperature of 30°C for germination to occur (Fletcher *et al.*, 1986) which correlates with the temperatures reached in the compost. Whether the ascospores also show some degree of resistance to heating is controversial, however, and requires further investigation. Various biocides have been tested for their effects on ascospore viability and it has been shown that exposure for up to 6 h to, for instance, copper sulphate (2%) or phenol (2%) may fail to kill them (Kligman, 1944). Resistance of the ascospores to such fungicidal chemicals at concentrations which could be used in practice and susceptibility of the mushroom mycelium itself to them makes it difficult to control false truffle disease by chemical means.

Other minor mycopathogens of *Agaricus* include *Aphanocladium album*, which causes brown spots on the cap, and *Hormiactis alba*, also causing cap spotting (Fletcher *et al.*, 1986). Stunting of the gills associated with a growth of white mycelium on their surfaces has been attributed to *Acremonium* (= *Cephalosporium*) spp., but there is still doubt as to the true relationship (Fletcher *et al.*, 1986). *Gliocladium virens* has been reported from India causing brown necrotic lesions on young basidiomes of *A. bisporus* (Dadwal and Jamaluddin, 1984). The mycelium and basidiomes of *A. bisporus* are also attacked in culture by a range of hyphomycetes (Gray and Morgan-Jones, 1981). Whether the vegetative mycelium or the basidiomes are parasitized, however, appears to be related to the individual mycoparasite.

7 BIOCONTROL OF PLANT PATHOGENS

7.1 Biocontrol of Plant Diseases

Fungi pathogenic to plants affect crop and timber production and are responsible for the loss of and damage to plants in horticultural practice. At least 20% of potential agricultural production in the field is lost annually through fungal diseases, representing millions of pounds worth of produce. Strategies for disease control have evolved over a long period of time, either fortuitously through good cultural practices, or deliberately once it was realized that most diseases were a result of microbial infection. The term biocontrol in relation to plant-pathogenic fungi implies the control of disease through a biological agency other than any natural defence mechanism of the host plant (Garrett, 1956). Alternatively it can be defined as the reduction of inoculum density or disease-producing activities of a pathogen in its active or dormant state, by one or more organisms, accomplished naturally or through manipulation of the environment, host or antagonist, or by mass introduction of one or more antagonists (Baker and Cook, 1974). Most plant diseases are caused by fungi, thus the use of fungi as agents of their biocontrol means that this must be accomplished through interactions of fungi. Biocontrol of plant diseases can also be achieved using other microorganisms, but this is not the concern of this book. Instead, this chapter will concentrate on the potential of mycoparasites for the biocontrol of plant diseases. One further approach to biocontrol is the use of non-pathogenic strains of fungi related to the plant pathogen in question. For example, biocontrol of *Fusarium* wilt diseases has been accomplished by introducing other non-pathogenic species of *Fusarium* into soil or into the infection court (Mandeel and Baker, 1991). These approaches seem to work through a variety of mechanisms including competition for nutrients, competition for infection sites, or through elicitation of host resistance responses. We will not consider this approach in any detail here, except in the brief discussion of the use of transmissible hypovirulence in the control of tree diseases (see also section 7.4.2). This has been used most notably for the control of chestnut blight caused by the lethal canker-causing organism *Cryphonectria parasita* (= *Endothia parasitica*). In summary the methodology involves inoculation of healthy trees with naturally occurring hypovirulent strains of the pathogen. The factor which causes hypovirulence is transmissible such that when a pathogenic strain with normal levels of virulence encounters a hypovirulent strain, hyphal fusion occurs and the pathogenic strain is converted into hypovirulence by transmissible cytoplasmic factors. Hypovirulence has also been observed in other plant pathogens such as *Gaeumannomyces graminis*, *Rhizoctonia solani* and *Bipolaris victoriae* (= *Helminthosporium victoriae*), but as it is not a form of mycoparasitism in the strictest sense it will not be discussed further here and the reader is referred to van Alfen (1982) for a detailed discussion of the biology and potential for disease control.

Control mechanisms which result in a reduction in activity and survival of the pathogen and hence the disease, may occur naturally or be contrived artificially. Plant pathologists involved in the development of methods of biocontrol are concerned with the introduction and augmentation in numbers of one or more species of biocontrol organisms and with alterations in the environmental conditions designed to favour the multiplication and activity of such organisms. Combinations of both procedures have led to several examples of the biocontrol of plant diseases by mycoparasitic fungi. Where cultural procedures, chemicals or physical agents are used to supplement the effects of a biocontrol fungus, this approach has been referred to as integrated control. The two basic methods of biocontrol of fungal pathogens are the 'classical' and the 'inundative'. In the former an exotic organism, the biocontrol agent, is introduced into an area where a non-indigenous plant pathogen has become established in the absence of natural antagonists that existed in its original ecosystem. The biocontrol agent is introduced systematically into the population of the pathogen and in successful cases may have the ability to progress to epidemic proportions thus controlling the pathogenic agent. Inundative methods, on the other hand, involve single or multiple applications of a biocontrol agent, which can be indigenous or introduced, in much higher concentrations of propagules than normally encountered by the plant-pathogen. The inundative use of indigenous species promises to be the most useful method of control using fungi (Burge, 1988).

Attempts to control the diseases produced by plant-pathogenic fungi have for many years centred around the use of chemical fungicides and the development of resistant strains of crop plants coupled with good agricultural management of farms and forests. Plant breeding programmes often involve the development of disease resistant strains but experience shows that sooner or later the pathogenic fungus is able to overcome the resistance bred into the host. Similarly, with the persistent application of fungicidal chemicals to crop plants the genetic variability of the fungi is such that resistant strains eventually emerge and further chemicals have to be screened and developed for their effectiveness against the pathogen. The costs involved in the screening of chemical compounds suitable for their application as fungicides are high but so are the costs relating to the development of mycoparasites as biocontrol agents. For example, Campbell (1986) has estimated that the primary biological screen may have to find one organism in 1000 or even 10,000. It must fail at least 99.9% of organisms and so must handle a large number to have a chance of success. Some of the numerous other problems involved in registering a fungus for commercial use as an agent of biocontrol have been described by Templeton *et al.* (1980) using a mycoherbicide as an example. Why therefore should the development of biocontrol agents involving mycoparasites be pursued? In the late 1980s and early 1990s there has been an increased global public awareness of the problems associated with the accumulation in the environment of persistent agrochemicals. Recalcitrant organic and inorganic compounds often constitute the active components of fungicides and, in a similar way to the so-called disease resistant strains of crop plants, fungicide-resistant strains of the plant pathogens develop and the chemical methods of treating plant diseases caused by fungi have to be re-evaluated. The development of methods for the biocontrol of plant-pathogenic fungi by means of mycoparasitism appears to be promising although it is clear that considerable resources will have to be devoted to this area of research if viable methods of control are to be developed.

One of the first suggestions that a mycoparasite could be used to control a plant disease was by Weindling (1932) that *Trichoderma* could be used to control certain pathogenic fungi by augmenting the soil with an abundance of this mycoparasite. Since this early observation, the assumption that mycoparasitism is a natural phenomenon has been used as the basis of measures adopted in the biocontrol of certain fungal pathogens of higher plants. Consequently, antagonistic fungal interactions have been exploited either directly by adding specific antagonists to infested soils, or indirectly, through the development of agricultural methods that enhance existing mycoparasitic relationships. An outstanding advantage of the use of a biological means to counter the effects of a fungal pathogen of crop plants in comparison with control using fungicides is the curtailment or even elimination of a possible long-term pollutant effect on the environment. Commercial pressures, however, have dictated that fungicides have become the method of choice for the farmer in controlling plant diseases when resistant cultivars of plants are not available. As a consequence there has been a lack of adoption of biological strategies of disease control in the agricultural industry, except where chemical means of control are inefficient or ill-advised. For example, the problems inherent in the introduction of fungicides into the soil has meant that soil-borne diseases are difficult to control. Biological methods to overcome this problem have thus featured strongly in the literature (e.g. Hornby, 1990). Similarly in the case of postharvest foliar diseases there are several other reasons why biocontrol has advantages over chemical control (Jeffries and Jeger, 1990):

1. Exact storage conditions can be established and maintained to suit the operation of a biocontrol agent.
2. Biomass of the harvested produce is less than that of the standing crop, thus only a limited surface area needs treatment.
3. The high net cost of harvested produce justifies a more stringent approach to disease control.
4. The limited time between treatment and consumption means that pesticide residues must be avoided.

Progress in developing fungal agents for biocontrol of both soil-borne and foliar diseases is discussed in sections 7.2 and 7.3. A comprehensive review of the use of fungi in biocontrol of plant diseases is not given here as several excellent and recent treatises are available (e.g. Burge, 1988; Whipps and Lumsden, 1989; Hornby, 1990). The present chapter concentrates on the use of selected mycoparasites in the control of soil-borne and postharvest foliar diseases, and discusses some of the considerations that need to be taken into account in the commercialization of a fungus for use in the control of plant disease (see section 7.4). It is necessary to stress, however, that biocontrol is a natural phenomenon, and that many plant diseases in the field are probably already kept under control by the types of fungal interactions discussed in the preceding chapters. When this natural control breaks down, a disease arises and artificial enhancement via the activities of a mycoparasitic fungus is then sought to restore the natural balance.

Sometimes the natural effect of a biocontrol agent can be augmented by the application of toxic chemicals to soil. For example, the fumigation of soil in citrus orchards in California with carbon disulphide led to a reduced incidence of root infection due to *Armillaria mellea* owing, apparently, to the antagonistic effect of *Trichoderma viride*, a rapidly growing, strongly saprotrophic competitor in soil which colonized the infected roots and displaced the pathogen. Similarly, the *in situ* fumigation of pine stumps with various fumigants, including carbon disulphide and methyl bromide, eradicated *A. mellea*, corresponding with a significant increase in the incidence of *Trichoderma*. Probably several factors such as the competition for nutrients, relative growth rates, enzyme- and possibly antibiotic-producing abilities play their part in the antagonism of *A. mellea* by *Trichoderma*, as well as direct mycoparasitism. A related species, *Trichoderma harzianum*, has also been used to control a range of important crop pathogens including the fungus causing rice blast *Pyricularia oryzae* (Sy *et al.*, 1990). Such observations indicate the potential for the selective manipulation of particular groups of soil fungi such that biocontrol is achieved through the activity of indigenous organisms. Another example is given by Chinn (1971) who found that spores of *Bipolaris sorokiniana* (= *Helminthosporium sativum*) were not affected by 5 ppm methyl mercury dicyandiamide, but the treatment stimulated an increase in the soil of certain microorganisms (especially *Penicillium* species) which were thought to be responsible for the disease control observed.

Some cases of antagonism of fungi in soil may be due to a background level of soil fungistasis. Fungistasis is a phenomenon associated with many soil types whereby the spores of fungi are held in the dormant state but not killed. When the appropriate conditions for germination arise in the soil, however, the dormant spores may be stimulated to germinate and the mycostasis is overcome. Why fungistasis occurs in soil is a contentious issue although it seems that the production of fungistatic metabolites by a range of microorganisms which live in the soil could provide one plausible explanation of the process. The range of fungistatic compounds known to be produced by microorganisms is wide and includes volatile materials such as carbon dioxide, acetaldehyde and ethanol (Upadhyay and Mukhopadhyay, 1983) and more complex non-volatile antibiotics. Volatile and non-volatile inhibitors of spore germination produced by *Trichoderma* and *Gliocladium* probably augment the more overtly mycoparasitic behaviour of these fungi. Conidia of the mycoparasitic fungi *Trichoderma* and *Gliocladium* are also known to be sensitive to fungistatic agents in soils which inhibit their germination (Beagle-Ristaino and Papavizas, 1985). On the other hand, the chlamydospores of *Trichoderma* and *Gliocladium* seem better able to withstand fungistatic factors in soil than their conidia and the germination ability is enhanced by the inclusion of a food base such as molasses or bran (Lewis and Papavizas, 1984) to the inoculum of chlamydospores and mycelial fragments added to soil. Once established, the chlamydospore-producing mycelium is able to proliferate and develop into a reservoir of chlamydosporic propagules believed to be responsible for the long-term survival of these fungi in soil. A major problem in determination of the precise mechanism involved in the biocontrol of plant-pathogenic fungi by mycoparasites lies in the distinction to be made between cause and effect. Whether antibiosis leads to or is an adjunct to the success of a fungus as a mycoparasite is difficult to determine. Although a fungus may behave as a mycoparasite in culture, for example through hyphal coiling, the penetration of hyphae and the exertion of deleterious effects on the cytoplasm of the host, it may not behave in this way in soil. Similarly, the demonstration of the production of an antibiotic in culture where the conditions of growth may be deliberately optimized for the synthesis and liberation of an antibiotic, does not necessarily mean that the antibiotic will be produced by the same fungus in the soil where many other environmental factors may also operate simultaneously. The physiological behaviour of a mycoparasite is likely to be affected by the soil temperature, pH, aeration, water content and relative humidity, nutrient status, the density of colony-forming units, presence of other antagonistic microorganisms, soil type and the physiological state of the potential host. This list of environmental factors, although not exhaustive, serves to illustrate the varied range of conditions acting simultaneously with which the biocontrol agent has to contend.

A case where a potential biocontrol agent failed in the field has been described by Boudreau and Andrews (1987). Earlier work (Cullen and Andrews, 1984; Cullen *et al.*, 1984) had shown that *Chaetomium globosum* was antagonistic towards the apple scab pathogen *Venturia inaequalis* under field conditions, and that an epidithiadiketopiperazine antibiotic, chetomin, was an important component for the inhibition of growth of *V. inaequalis*. However, larger-scale field trials resulted in only a 0–25% reduction in disease levels when ascospores of *C. globosum* were sprayed onto apple leaves. It was found that the antibiotic from the ascospores could diffuse passively onto the phylloplane and inhibit the growth of the apple scab pathogen; active growth of the *Chaetomium* was not necessary in this process. The loss of control arose due to abiotic degradation of the antibiotic which meant that inoculum of *Venturia* which arrived some time after the ascospore application was no longer subject to antibiosis. Antibiotics produced by *Chaetomium globosum* have also been implicated in reports of biocontrol of damping-off diseases caused by *Pythium ultimum* and *Rhizoctonia solani* (Walther and Gindrat, 1988; Di Pietro *et al.*, 1992).

7.2 Soil-borne Diseases

7.2.1 General considerations

Soil-borne diseases are a particular target for biocontrol as conventional control by agrochemicals is problematic. There are many reports in the literature of soil-borne fungi which are susceptible to antagonism or mycoparasitism by other soil fungi. A well-known example is the form genus *Rhizoctonia*, the non-sporulating anamorph of the mycelium of numerous isolates from soil and infected host plants. Many pathogenic strains have been assigned to *R. solani* in which lateral branches of the septate mycelium typically arise close to 90° to the main hyphae. A basidiospore-producing mycelium, the teleomorph belonging to *Corticium*, has been observed in some isolates of *Rhizoctonia*. The mycelium of *Rhizoctonia* is capable of saprotrophic growth through the soil in the absence of host plants. The pathogen appears to be widely distributed in soils throughout the world where it colonizes cellulosic substrates and may survive indefinitely. Although *R. solani* may not compete strongly with saprotrophs in the soil, it is capable of colonizing living cells in roots and stems which appears to provide the pathogen with a competitive advantage over them in the colonization of such substrates (Garrett, 1970). *Rhizoctonia solani* attacks roots of the seedlings of a range of crops including cereals and the potato plant and is responsible for damping-off and pre-emergence blight. Strains which attack the above ground parts of host plants apparently tend to be able to produce a resistant sclerotial stage whereas those which normally attack roots tend to survive unfavourable growth conditions by means of thick-walled, darkly pigmented hyphae in the soil. The pathogen infects the root tip region of young roots and can continue to infect newly-produced root tips as the host plant ages. Plants not killed by the pathogen may successfully outgrow the disease as *Rhizoctonia* is essentially a parasite of young tissues which have not developed resistance to the attack. *R. solani* is

Table 7.1. Some parasites of *Rhizoctonia solani.*

Mycoparasite	Reference
Arthrobotrys oligospora	Persson (1991)
Botryotrichum piluliferum	Turhan (1990)
Coniothyrium sporulosum	
Dicyma olivacea	
Gliocladium catenulatum	Jager *et al.* (1979)
G. roseum	
G. virens	
Hormiactis fimicola	
Laetisaria arvalis	Burdsall *et al.* (1980)
Papulaspora stoveri	Warren (1948)
Penicillium vermiculatum	Boosalis (1956)
Pythium oligandrum	
Stachybotrys chartarum	Turhan (1990)
S. elegans	
Stachylidium bicolor	
Talaromyces flavus	McLaren *et al.* (1986)
Trichoderma hamatum	Chet *et al.* (1981); Chet and Baker (1981); Chu and Wu (1981); Elad *et al.* (1983b)
T. harzianum	Roy and Sayre (1984)
T. longibrachiatum	Chu and Wu (1981)
T. pseudokoningii	
T. viride	Weindling (1932)
Trichothecium roseum	Turhan (1990)
Verticillium biguttatum	Boogert (1989a,b)
V. chlamydosporium	Turhan (1990)
V. lamellicola	Kuter (1984)
V. lecanii	
V. nigrescens	
V. psalliotae	
V. tenerum	Turhan (1990)
Volutella ciliata	Jager *et al.* (1979)

susceptible to attack by mycoparasitic fungi and it can also behave as a mycoparasite itself. Some fungal parasites of *R. solani* are summarized in Table 7.1.

In glasshouse studies with pure dual cultures of *Gliocladium virens* and *Rhizoctonia solani*, responsible for the development of root rot in white beans, severity of the disease is significantly curtailed. Some strains of *R. solani* causing black-root disease of sugar beet also appear to be susceptible to infection by *Papulaspora stoveri* in the soil which results in reduced incidence of the disease. It is worth noting that most applications of biocontrol through mycoparasitism involve plant pathogens which form long-lived resistant propagules within the

Table 7.2. Examples of fungal antagonists reported to control diseases caused by species of *Pythium*. (Modified from Whipps and Lumsden, 1991.)

Antagonist	Pathogen	Host	Application method
Acrophialospora levis	*P. ultimum*	Pepper	Incorporation
Chaetomium sp.	*P. debaryanum*	Squash	Seed
	P. ultimum		Seed
Gliocladium catenulatum	*P. ultimum*	Pea	Incorporation
G. virens	*P. ultimum*	Cabbage	Incorporation
		Cotton	Incorporation
			Seed
		Pea	Incorporation
		Zinnia	Incorporation
Glomus fasciculatum	*P. ultimum*	Poinsettia	Incorporation
Laetisaria arvalis	*P. ultimum*	Sugar beet	Incorporation
Myrothecium verrucaria	*P. ultimum*	Pea	Incorporation
Neocosmospora vasinfecta	*P. ultimum*	Pepper	Incorporation
Penicillium glabrum	*Pythium* species	Sugar beet	Seed
P. jensenii	*Pythium* species	White mustard	Seed
P. janzewskii	*Pythium* species	White mustard	Seed
P. oxalicum	*P. ultimum*	Chickpea	Seed
	Pythium species	Chickpea	Seed
Pythium nunn	*P. ultimum*	Cucumber	Incorporation
P. oligandrum	*P. ultimum*	Cress	Seed
		Cucumber	Seed
		Sugar beet	Seed/incorporation
	Pythium species	Chickpea	Seed
Trichoderma hamatum	*P. ultimum*	Pea	Incorporation
	Pythium species	Pea	Seed
		Radish	Seed
T. harzianum	*P. aphanidermatum*	Cucumber	Incorporation
		Gypsophila	Incorporation
		Pea	Incorporation
		Pepper	Incorporation
		Tobacco	Incorporation
		Tomato	Incorporation
	P. ultimum	Cotton	Seed
		Cucumber	Seed
		Lettuce	Incorporation
		Maize	Seed
		Pea	Seed
		Snapbean	Seed

Table 7.2. continued

Antagonist	Pathogen	Host	Application method
		Wheat	Seed
T. koningii	*Pythium* species	Pea	Seed
	Pythium species	Pea	Seed
T. viride	*P. aphanidermatum*	Poinsettia	Incorporation
	P. debaryanum	Alfalfa	Seed
	P. splendens	Geranium	Incorporation
	P. ultimum	Pea	Seed
	Pythium species	Sugar beet	Seed
		White mustard	Seed

soil and they are usually responsible for soil-borne diseases and, as mentioned earlier, this type of inoculum is very difficult to destroy using conventional fungicidal treatments. Diseases caused by *Pythium* provide a further example where many fungal antagonists have been suggested to be potential agents of control (Table 7.2). The biocontrol agent is normally either applied to the seeds of the crop plant to be protected or is incorporated into the soil via a carrier material.

Living organisms respond to alterations in environmental conditions and are prone to changing their behaviour under differing circumstances. Investigations into these problems have been carried out on the interaction of *R. solani* and the antagonist *Verticillium biguttatum.* The viability of sclerotia and hyphae of the potato plant pathogen *R. solani* is significantly reduced in soil through parasitism by *V. biguttatum* (Velvis *et al.*, 1989). Although *V. biguttatum* is capable of saprotrophic growth in pure culture, it behaves as an ecologically obligate parasite of *R. solani* in the soil as the hyphae and spores only become active in the presence of the potential host. The inoculum of *V. biguttatum* increases in the soil solely through the mycoparasitic activity of the fungus (Boogert, 1989a,b) and, in fact, the experimental addition of living mycelium of the host fungus *R. solani* to the soil stimulates the development of the mycoparasitic *V. biguttatum* and *Gliocladium roseum* and reduces the infection of sprouts of infected seed potatoes (Boogert and Jager, 1983). The conidia of *V. biguttatum* inoculated onto seed potatoes are believed to be transported passively on the surface of potato plant stolons, and to a lesser extent the roots, as they grow through the soil. The conidial cover enables *Verticillium* to colonize the hyphae of *Rhizoctonia* growing in the vicinity of the potato plants and so produce a mycelium and conidiophores, thus replenishing the inoculum in the soil. There is a marked reduction in the number of sclerotia of *R. solani* which form on the newly formed potato tubers and the mycoparasite also kills the sclerotia. To several experimental soils in the laboratory varying from sandy, sandy clay, loam clay to clay, of known pH and maintained at 50% moisture holding capacity, standard numbers of sclerotia of *R. solani* were added. It was found that at 10°C all sclerotia germinated whereas at 20°C the sclerotium germination was severely curtailed within two weeks in those soils containing a relatively high inoculum density of *V. biguttatum.* In experimental field soils, however, the proportion of sclerotia infected never exceeded 6%, sclerotium viability decreased gradually to virtually zero after two years in sandy soils but this decline proceeded more slowly in clay soils. It seems from this example that the type of soil, soil temperature and inoculum density of the parasite are related to the intensity of the infection but it also underlines the view that the results from field trials may be very different from those obtained from the more precisely controlled experimental conditions of the laboratory. The loss of viability of test sclerotia in the field could not be attributed to the parasite and it has been suggested that the energy reserves of the sclerotia become so depleted over a period of two years in the soil that they are no longer able to germinate.

The successful demonstration of the control of the growth of a plant-pathogenic fungus such as *R. solani* in field, pot and laboratory trials does not necessarily mean that the mycopathogen can be used commercially to control the disease on a long-term basis, as other factors may come into play. If the mycopathogen is unable to grow in the soil situation over the same temperature range as the plant pathogen there could be times during the growing season of the host plant when the plant is not protected by the mycoparasite and becomes vulnerable to infection by the pathogen. In theory, the problem can be overcome by the application of two or more mycoparasitic fungi simultaneously to seed potatoes in order to provide a protective cover over the entire temperature range in which the pathogen can grow. Combinations of *V. biguttatum* with other mycoparasites have resulted in a statistically significant reduction of the infection of potato plants by *R. solani* although it seems to be difficult to sustain the protective effects of these antagonistic combinations in soils (Boogert and Jager, 1984).

Other factors which have to be taken into consideration in the survival of mycoparasitic fungal additions to soils in order to curb the growth of *R. solani* are the age and density of the inoculum, time of year at which the inoculum is added, infective behaviour of the strain of the mycoparasite used and susceptibility of the pathogen, and the physical factors of soil type, pH, and moisture content. Although factors such as these, which undoubtedly influence the efficacy of mycoparasitic fungi, can be singled out in controlled laboratory experiments the situation in field soils is more complex owing to their interaction. Three-day-old cultures of a strain *Trichoderma hamatum* comprising the actively growing hyphae of germinated spores significantly arrest the development of *R. solani* on sugar beet and other crop plants whereas 40-day-old cultures containing conidia and chlamydospores but lacking actively growing hyphae do not (Lewis and Papavizas, 1987). The reduction in survival of the pathogen is greatest when the germinating spores are added to soil either at the same time or within seven days of planting *Rhizoctonia*-infested seed, an observation also recorded in the biocontrol of *Pythium ultimum* by the destructive mycoparasite *Gliocladium virens* (Lumsden and Locke, 1989). The density of the inoculum is equally important in controlling the pathogen as the addition of 0.1% inoculum is as effective as the addition of 1% inoculum. The interaction of soil pH, moisture and temperature also affects the survival of *R. solani*, the pathogen being most effectively controlled at low to moderate temperature (15–23°C) in acidic to almost neutral soil (pH 5.0–6.7). As these conditions are known to occur in agricultural soils during spring planting it has been suggested that *T. hamatum* may be a potentially useful agent for the biocontrol of *R. solani*.

From a practical point of view, experimental work involving determination of the effects of mycoparasites on their hosts is very time consuming as field trials may require several years to complete. The investigation of natural ecological systems also demands painstaking efforts on the part of the investigators as an apparent correlation between the incidence of mycoparasitism and reduction of disease may be due to the interaction of several factors. One of the main objectives of developing methods of biocontrol using mycoparasites is to reduce the environmental impact of the use of fungicides. Fungicidal compounds may also be toxic to humans, animals and plants and therefore have to be handled with caution and applied carefully. Organomercury and copper compounds used in many fungicides can also pollute the environment and the pollutant elements may persist in soils for many years. Plant-pathogenic fungi may also exhibit resistance to fungicides and the processes involved in their development, production and application may be expensive. The development of methods of biocontrol by means of mycoparasites represents an alternative means of controlling plant-pathogenic fungi whilst minimizing the damage caused to environment by the application of control measures. In order that a method of biocontrol should be commercially viable, however, the measure of control obtained has to be equivalent to that achieved by more conventional means. Frequently, the effect of the biocontrol fungus is enhanced if the soil is first treated with a fumigant such as methyl bromide, or pasteurized by steam or heated by solarization.

A problem associated with the establishment of fungi known to behave as potentially beneficial mycoparasites is to maintain the levels of inoculum which have to be added to soil in order to initiate the action of the parasite. The ecological balance has to be altered so as to favour the mycoparasite to the detriment of the pathogen. The desired imbalance may be achieved in various ways but, eventually, if the ecological system is not manipulated properly it is likely to revert in favour of the pathogen. For example, species of *Pythium* are cosmopolitan fungi in soils on a worldwide basis and are responsible for rotting seeds, pre-emergence blight and damping-off disease resulting in the collapse of the stems of seedlings at soil level. The pathogens are responsible for considerable crop, horticultural and silvicultural losses from seed beds and nurseries every year. *Trichoderma* species are known to be active in controlling *Pythium* infections but they grow more readily in non-rhizosphere soil than in the zone of soil which is associated with developing roots. Coating seeds, with *Trichoderma* before planting, however, leads to a significant increase in the population of the mycoparasite (Nelson *et al.*, 1988), presumably because the inoculum concentration of the mycoparasite is sufficiently high initially to enable it to gain a stronghold from which it repels the pathogen. The incorporation of food bases such as carboxylic acids or polysaccharides (Nelson *et al.*, 1988) into the seed treatment further enhances the establishment of the mycoparasite probably because spore germination is stimulated although another possibility is that the increase in acidity due to carboxylic acids may provide an environment favourable to growth of *Trichoderma* and less favourable to the growth of *Pythium*. Damping-off of cucumber seedlings caused by *Pythium aphanidermatum* has been controlled in laboratory trials by *Trichoderma harzianum* and *Talaromyces stipitatus* to an extent equivalent to that obtained by the application of a fungicide (Sharif *et al.*, 1988). Pre-emergence blight appears to be more readily controllable by these fungi than post-emergence damping-off, possibly because by the time the seedlings have reached this stage of development the inoculum potential of the control agents has declined to a level which is no longer effective. Both of these antagonistic fungi produce compounds toxic to *Pythium* but the ability to do this apparently declines with time and the explanation favoured for their ability to control the disease is through their activity as mycoparasites. A contrary view, however, is derived from a study of the processes apparently involved in the suppression of infection of pea seeds by *Pythium* spp. over a period of 50 h (Baker, 1989). Control of the pathogen when *Trichoderma* is applied to the testas of pea seeds has been attributed to the action of an antibiotic. When the hypha of *Trichoderma* approaches that of the pathogen, the cytoplasm in the hyphal tip of *Pythium* rapidly retracts and growth slows. The antibiotic which alters the course of events has been termed the 'routing factor' and this mechanism is operative before mycoparasitism of the pathogen occurs. To illustrate these general principles it is worth considering some specific examples.

7.2.2 *Trichoderma, Gliocladium* and *Talaromyces*

Species of *Trichoderma*, *Gliocladium* and *Talaromyces* have been extensively tested and used both as potential and actual biocontrol agents of a wide range of plant pathogens. Their strongly antagonistic behaviour towards other fungi and the ease with which they may be cultured in the laboratory has resulted in them being among the most favoured fungi for studies of the biocontrol of plant pathogens. Much of the relevant information has been summarized by Papavizas (1985). *Trichoderma* is commonly found in soils throughout the world. *Trichoderma harzianum*, in particular, has shown much promise in biological control. This species can act by mycoparasitism, together with the production of volatile pyrone antibiotics and cell wall-degrading enzymes. It also produces a powerful cellulase and is a dominant colonist of straw, which is a major substrate for microorganisms in soils used for arable cultivation. In comparison with a large range of fungi reputed to have good biocontrol potential, an isolate of *T. harzianum* (IMI 275950) exhibited the greatest activity *in vitro* (Lynch, 1987). This fungus has also shown potential in the field; for example, control of bottom rot of lettuce caused by *Rhizoctonia solani* can be achieved in mature lettuce plants grown commercially in polythene tunnels (Coley-Smith *et al.*, 1991). Although the level of control was not always as good as that achieved with a standard fungicide (tolclofos-methyl), the enhanced growth of lettuce plants following treatment with *Trichoderma* could produce marketable yields of lettuce similar to those obtained using the fungicide alone. In other trials in the USA, an application of 140 kg ha^{-1} of inoculum of *T. harzianum* was used successfully to control damage to peanuts caused by *Corticium rolfsii* over a three-year period. Control by this method was equivalent to that obtained using conventional fungicides (Backman and Rodriguez-Kabana, 1975). *In vitro* tests of the antagonistic activity of *Trichoderma* species show them to be active against members of most of the Oomycota, Ascomycota, Basidiomycota and deuteromycetes, which indicates the broad spectrum of potential hosts. *Trichoderma hamatum*, for example, interacts with *Pythium* species and *Rhizoctonia solani*, responsible for damping-off diseases in the seedlings of plants, and also *C. rolfsii*, which causes root rot. The primary antagonistic relationship between *T. hamatum* and these pathogens involves hyphal coiling and penetration, and appears to be destructively mycoparasitic. Thus *Trichoderma* species have been used effectively in controlling the plant diseases caused by these pathogens. *Trichoderma harzianum*, for example, is a promising biocontrol agent specifically targeted to application to seeds for protection against pre-emergence damping-off caused by *Pythium* species. The fungus has been formulated in a preparation with the trade name F-Stop®, with US Environmental Protection Agency registration pending (Whipps and Lumsden, 1991). *Trichoderma* species have also been shown to stimulate the growth of plants even in the absence of pathogenic fungi, and this has also prompted the use of this fungus as a growth-promoting supplement to a wide range of horticultural crop species (Baker, 1988, 1989; Lynch *et al.*, 1991). In a comparison of the use of autoclaved or non-autoclaved inoculum of *Trichoderma harzianum* for the biocontrol of *Rhizoctonia solani* on lettuce, Maplestone *et al.* (1991) found that both preparations decreased disease levels and increased the yield of plants. The incorporation of ethyl acetate extracted autoclaved inoculum of *T. harzianum* resulted in similar levels of biocontrol and improved plant growth as did incorporation of non-autoclaved and autoclaved inoculum. It was thus apparent that *T. harzianum* produces some form of biologically active, heat-stable metabolite. Ethyl acetate extraction should remove the alkyl pyrones thought to be involved in biocontrol (Claydon *et al.*, 1987) so it is likely that other chemicals are responsible for the decrease in disease incidence, or a growth-stimulating effect.

Although species of *Trichoderma* have been extensively used as potential agents of the control of a range of plant-parasitic fungi both in laboratory and field trials, the precise mechanism involved in the control of the growth of one fungus by another is difficult to determine. It cannot be tacitly assumed that the growth behaviour of the biocontrol fungus in experiments performed under controlled environmental conditions in the laboratory will necessarily mirror that in the field as it is more difficult to control the experimental conditions in field trials and hence more difficult to monitor and assess the results. Although a fungus may be shown to act as a mycoparasite in culture, evidence that the same fungus behaves in this way in the soil is difficult to obtain. Mycoparasitism, however, is often assumed to play a part in the biocontrol of plant-pathogenic fungi in the soil. The vigorous competitive saprotrophic behaviour of species of *Trichoderma* in non-rhizosphere soil (Papavizas, 1981) appears essentially to be related to such factors as the rapid growth of the mycelium, an ability to recover rapidly after soil fumigation and the ability of growth propagules to withstand such treatment, prolific spore production and rapid spore germination, antibiotic production, the production of volatile inhibitors of growth and the ability to parasitize a potential host with which it may also be in competition in the soil. The mycelium also secretes a range of extracellular hydrolytic enzymes which enable the fungus to colonize a food base in the soil and may also bring about lysis of the cell walls and cytoplasm of competitors (Cherif and Benhamou, 1990). The broad enzymic competence of *Trichoderma* enables the fungus to colonize cellulosic substrates in soil, especially straw, which provide a substantial energy-rich food base from which the mycelium radiates and are a major substrate found in arable soils. In addition, the extracellular wall-degrading enzymes cellulase and β-1-3-glucanase (Chet and Baker, 1981; Sivan and Chet, 1989) are known to be responsible for the lysis of cellulose and glucan respectively in the hyphal and oospore walls of plant-pathogenic *Pythium* spp. Chitinase, capable of degrading chitin which contains polyglucosamine units, in combination with β-glucanase, provides an effective means of lysing the hyphal wall of a potential host such as *Rhizoctonia solani*. The mycolytic propensity of *Trichoderma* probably constitutes an important part of the invasive armoury of this destructive necrotrophic mycoparasite. The sclerotia of *Corticium rolfsii* are lysed by the enzymic activity of *Trichoderma* spp. and those of *R. solani* by *Gliocladium* spp. which may partially account for the decline in sclerotial numbers recorded in test soils. The secretion of antifungal antibiotics by *Trichoderma* is also believed to participate in the biocontrol mechanism. Indeed, antibiosis may be the major determinant although strains of some species either do not produce antibiotics or the quantities produced in culture are below the limits of detection.

An important property for the success of *Trichoderma* species in the control of soil-borne disease is the possession of 'rhizosphere-competence' (Ahmad and Baker, 1987a). This term expresses the ability of the fungus to proliferate and establish along the developing rhizosphere after it has been applied to seeds of appropriate crop plants. Seed treatments with spores of beneficial fungi are an attractive way to apply agents of biocontrol. **Rhizosphere competence** can be measured experimentally and some fungi possess a natural ability to colonize the soil adjacent to developing root systems. In other cases, however, rhizosphere competence is not always expressed. For example, when seeds of several crops were coated with conidia of *T. harzianum* and germinated in soil, the fungus was not detected in the rhizosphere of roots at 1–8 cm depth after 8 days. Rhizosphere competence can, however, be induced in wild strains by mutagenesis or

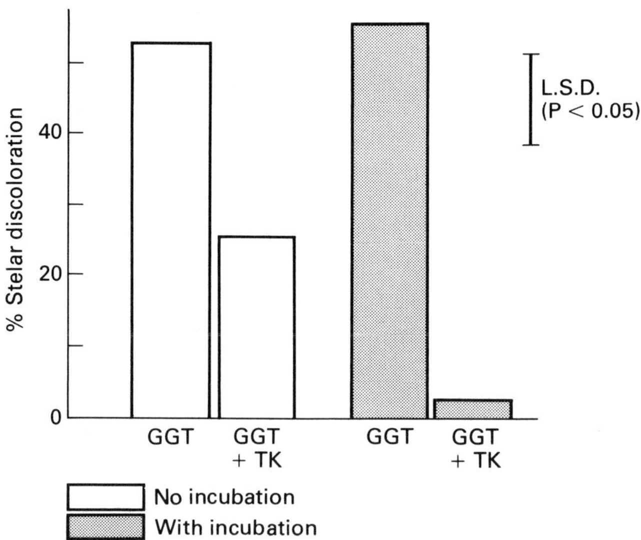

Fig. 7.1. Effect of inoculation of soil with *Gaeumannomyces graminis* var. *tritici* (GGT) and *Trichoderma koningii* (TK) 2 weeks before sowing on severity of take-all of 4-week-old wheat seedlings measured as length of seminal roots with discoloured stele. No incubation: pathogen and antagonist added to soil at time of sowing with wheat seeds. With incubation: pathogen and antagonist added to soil and incubated at 25°C for two weeks before sowing. (Redrawn from Simon, 1989.)

genetic manipulation (Sivan and Harman, 1991). Ahmad and Baker (1987a) induced rhizosphere competent strains of *T. harzianum* during experiments to mutate for tolerance to benomyl. In these studies it was not clear whether the attribute of benomyl tolerance contributed to the mechanism(s) involved in induction of rhizosphere competence; however, the mutants produced significantly higher quantities of cellulase and had an increased competitive saprotrophic ability on cellulose-rich substrates added to soil (Ahmad and Baker, 1987b). Thus the hypothesis was advanced that the mutants were rhizosphere competent because they utilized the mucigel, rich in cellulose covering the roots. Further evidence for this hypothesis was provided by the discovery that benomyl tolerance did not seem to be a necessary attribute for rhizosphere competence (Ahmad and Baker, 1988). More recently (Sivan and Harman, 1991) improvement in the rhizosphere competence of *T. harzianum* has been obtained as a result of a protoplast fusion technique. Integrated approaches to control have also been demonstrated for damping-off of radish plants caused by *Rhizoctonia solani*. Strains of *T. harzianum* tolerant to pentachloronitrobenzene (PCNB) have been used successfully with this agrochemical to control damping-off, and in a similar way other isolates of *Trichoderma* have been used in association with various fungicides to control *Corticium rolfsii* (Montealegre and Henriquez, 1990).

There are many reports of the suppression of soil-borne pathogens in the saprotrophic phase of growth or saprotrophic phase of survival as resistant propagules in soil (Simon and Sivasithamparam, 1989) by components of the soil itself, considered to be the normal microbial flora. For example, it appears that *Trichoderma koningii* acts as a suppressive agent to the harmful pathogen of wheat *Gaeumannomyces graminis* var. *tritici* (Fig. 7.1). If a small amount (1% w/v) of soil known to be suppressive is added to a non-suppressive (conducive) soil and the saprotrophic growth or survival of the pathogenic fungus is suppressed, it implies that owing to the small volume of suppressive soil added, the suppressive effect is more likely to be due to the activity of microorganisms than to an alteration in the nutritional status of the conducive soil. Suppression of the saprotrophic growth of *G. graminis* var. *tritici* in soil in which wheat is continuously grown is enhanced by a yearly application of ammonium sulphate whereas the addition of lime to ammonium sulphate-treated soil reduces suppression of the pathogen. The activity of *T. koningii* is affected by soil pH, being greater in acidified soil. The addition of *T. koningii* to acidified soil also results in enhanced suppression of the pathogen whereas no enhancement occurs when the biocontrol agent is added to limed soil. The artificial addition of *T. koningii* to pathogen-infested soil increases its inoculum potential, reduces the inoculum potential of the pathogen and reduces the incidence of disease. A problem associated with the biocontrol of take-all of wheat, however, is the need to demonstrate that the methods evolved are significant and consistent under field conditions. The disease occurs irregularly throughout infested fields and its intensity is affected by a range of environmental factors including climatic conditions, soil type, pH, moisture content, and cultural practices such as the application of fertilizers and previous cropping history. Thus the control methods have to be carefully monitored. A strain of *T. koningii* isolated from an Australian soil suppressive to take-all disease has been shown to inhibit the growth of *G. graminis* var. *tritici* and to control the incidence of take-all, under controlled environmental conditions (Simon, 1989). The field situation is much more complex however, but it is clear that biological suppression of the saprotrophic phase of this soil-borne plant-pathogen is an important component in disease reduction, a concept discussed in detail by Simon and Sivasithamparam (1989). A measure of interest in the development of new agents of biocontrol for *G. graminis* var. *tritici* is apparent by the recent assessment of *in vitro* screening systems for potential biocontrol agents (Renwick *et al.*, 1991), and the use of immunological methods to determine the amount of *G. graminis* var. *tritici* on wheat roots within 6 days of seeding (El-Nashaar *et al.*, 1986). This latter technique could also show promise in screening potential antagonists to *G. graminis* var. *tritici in vivo* within 8–10 days of planting host material.

Other examples of the suppression of plant pathogens in soil by the addition of propagules of *Trichoderma* to the soil include *Trichoderma hamatum* which is effective against *Rhizoctonia solani* (Chet and Baker, 1981) and *T. harzianum*, active against *Corticium rolfsii* infecting sugar beet (Upadhayay and Mukhopadhyay, 1983), *Pythium aphanidermatum* which infects tobacco plants (Mukhopadhay *et al.*, 1986), *Fusarium oxysporum* infecting tomato plants (Sivan *et al.*, 1987) and *Rosellinia necatrix* responsible for the white root rot disease of fruit trees such as apple, pear and avocado (Sztejnberg *et al.*, 1987). Application of *T. harzianum* to soybean residues in the field decreased the incidence of *Fusarium graminearum*, *Glomerella glycines*, and *Macrophomina phaseolina* (Fernandez, 1992a), whereas inoculation of wheat straw reduced the incidence of *Cochliobolus sativus* (Fernandez, 1992b).

Gliocladium species have also often been suggested as potential agents of biocontrol (e.g. against *Verticillium dahliae*, Keinath *et al.*, 1991; against *C. rolfsii*, Papavizas and Collins, 1990; against *Pythium ultimum*, Howell, 1991). The production of antibiotics, especially gliotoxin, and the ability to parasitize hyphae of other fungi both play a role in the antagonistic nature of *Gliocladium* isolates (see also section 3.1.3). A selective medium has recently been developed (Park *et al.*, 1992) which can exclusively differentiate *Gliocladium* from *Trichoderma*. The development of *Gliocladium virens* (isolate GL-21) as a biocontrol agent has developed to the stage of registration by the US Environmental Protection Agency for use against damping-off caused by *Pythium* and *Rhizoctonia* (Anon., 1991). This fungus is thus the first to be cleared by the regulatory agency for future commercial use in the United States against a plant pathogen. An alginate–wheat bran formulation has been used to produce an easily applied prill which resembles a granular fertilizer (see also section 7.4.1). Mutants of *G. virens* tolerant to benomyl at a concentration of 10 μg ml^{-1} have been isolated from aqueous suspensions of conidia tested with both ultraviolet irradiation and ethyl methanesulphonate (Papavizas *et al.*, 1990). Unfortunately, the acquisition of benomyl tolerance coincided with a loss of ability to control diseases caused by *Corticium rolfsii*, the test organism used for these experiments. More recently, benomyl resistance has been incorporated into *G. virens* isolate GL-21 by regenerating protoplasts and transforming them with a *Neurospora crassa* β-tubulin gene (Ossanna and Mischke, 1990). This was the first report of the addition by transformation of a beneficial trait allowing potential use of a biocontrol fungus in an integrated pest management situation. The effect of transformation on the ability to suppress disease in the field has not yet been reported. *Gliocladium virens* has also been used in combination with soil solarization as an integrated method to control diseases caused by *C. rolfsii* in tomatoes (Ristaino *et al.*, 1991). When disease pressure was high, the combined effect of solarization and treatment with *G. virens* was greater than either treatment alone in reducing disease.

Talaromyces flavus is considered to be a potential biocontrol agent of *Verticillium dahliae* and *Sclerotinia sclerotiorum* (Kim *et al.*, 1990a,b). Under agronomic production conditions in the field it has been shown to reduce levels of *Verticillium* wilt of eggplant (*Solanum melongena*) by up to 76% (Marois *et al.*,

1982). *Talaromyces flavus* attacks sclerotia of target fungi directly, but has also been shown to produce a metabolite in culture ('talaron') that retarded radial growth of *V. dahliae in vitro* and killed microsclerotia of *V. dahliae* in soil (Fravel *et al.*, 1987). This metabolite has since been identified as the enzyme glucose oxidase, which catalyses the oxidation of glucose to gluconate and hydrogen peroxide. Hydrogen peroxide, in turn, killed the microsclerotia of *V. dahliae* (Kim *et al.*, 1990a). An additional antimicrobial agent, vermiculine, has also been isolated from culture filtrates of *T. flavus* (Fuska *et al.*, 1972) but its significance in biocontrol has not been investigated. The survival of inoculum in the soil is positively correlated with potassium and zinc concentrations, and is inversely related to sodium concentrations (Fravel and Marois, 1986), and it apparently reaches higher levels in the rhizosphere than in non-rhizosphere soils (Marois *et al.*, 1984). Mutants of *T. flavus* resistant to benomyl and dicarboximide fungicides have been developed in the laboratory (Katan *et al.*, 1984) using conventional genetic techniques. This will enable the use of this fungus in integrated systems of disease control. Inheritance of the ability of strains of *T. flavus* to secrete wall-degrading enzymes and glucose oxidase has been studied using conventional genetic techniques (Madi *et al.*, 1992). No single factor was identified as a major determinant of antagonism, but the fact that one progeny strain was able to reduce bean root rot more effectively than either parent strain demonstrated that genetic improvement of a biocontrol agent is possible by conventional breeding.

The likelihood of agricultural use of *T. flavus* as a biocontrol agent has progressed more recently with the development of formulations of pyrophyllite carriers of ascospores of the mycoparasite (Papavizas *et al.*, 1987). However, the use of these formulations in the field did not seem to significantly reduce disease development by *V. dahliae*, even though survival of *T. flavus* was markedly improved (Spink and Rowe, 1989).

7.2.3 *Pythium* as a biocontrol agent

Species of *Pythium*, responsible for pre-emergence blights and damping-off diseases, form another group of important soil-borne pathogens of seeds and seedlings. *Pythium oligandrum*, *P. acanthicum* (Deacon and Henry, 1978) and *P. nunn* (Paulitz *et al.*, 1990) are mycoparasites. It is of interest that aggressive plant-pathogenic species of *Pythium* such as *P. ultimum* and *P. debaryanum* are themselves parasitized by *Pythium oligandrum* although its host range is not restricted to species of *Pythium*. For example *P. oligandrum* identified by its spiny oospores was detected in 18 of 28 test soils using various susceptible fungi as bait. *Fusarium culmorum* was most frequently colonized by *P. oligandrum*, followed by *Botrytis cinerea*, then *Trichoderma aureoviride*. *Pythium oligandrum* also often occurred in soil with the necrotrophic mycoparasite *Gliocladium roseum* although the most efficient hosts for the detection of *G. roseum* were

Botrytis cinerea, *Rhizoctonia solani* and *T. aureoviride* (Mulligan and Deacon, 1992). *Pythium oligandrum* can behave as a weak pathogen of higher plants, for example sugar beet seedlings, although in the presence of susceptible pythia in the soil its tendency to act as a mycoparasite appears to gain the ascendancy (Vesely, 1977). Experimental inoculation of soil with *P. oligandrum* protects emerging sugar beet plants against both soil-inhabiting and seed-borne pathogens enabling the crop to produce a higher yield (Vesely, 1978) and the fungus has been used in various trials as a biocontrol agent (e.g. Al-Hamdani *et al.*, 1983; Lutchmeah and Cooke, 1984, 1985; McQuilken *et al.*, 1990a, 1992a). Investigation of the range of osmotic potential over which the oospores are capable of germinating and the mycelium can grow shows that both are affected although not to the same extent. Mycelial growth is more sensitive to a decrease in osmotic potential than oospore germination. The range of osmotic potential over which the mycoparasite can grow may be paralleled by soil-inhabiting phytopathogenic fungi and this physical parameter is therefore likely to influence the mycopathogenicity of *P. oligandrum* (McQuilken *et al.*, 1992b). Dual and triple inoculation of the soil with *P. oligandrum* and the other species of *Pythium* results in significantly increased numbers of healthy plants than in those combinations lacking *P. oligandrum* and the effect has been attributed to its ability to act as a mycoparasite.

Pythium nunn and *P. oligandrum* have both been developed as biocontrol agents for use against the plant-pathogenic *P. ultimum*. *Pythium nunn*, for example, has been shown to reduce the population density of *P. ultimum* and the subsequent degree of damping-off and root rot of cucumber (Lifshitz *et al.*, 1984a; Paulitz and Baker, 1987a). Although *P. nunn* has mycoparasitic activity, evidence indicates that *P. nunn* requires organic substrates rather than living fungal hosts for antagonism and disease suppression to operate (Paulitz and Baker, 1987a) and that it competes with *P. ultimum* for these substrates because the two fungi occupy overlapping environmental niches (Paulitz and Baker, 1987b). It has recently also been applied to cucumber plants together with *Trichoderma harzianum* to demonstrate that two biocontrol agents can be combined to give additional control of *Pythium ultimum* (Paulitz *et al.*, 1990). This dual combination of biocontrol agents appears to be compatible and a possibility is that mycoparasitism of the plant pathogen *P. ultimum* is involved as mycoparasitism of *P. ultimum* by *P. nunn in vitro*, at least, has been demonstrated (Lifshitz *et al.*, 1984a,b). *Pythium oligandrum* has been most successful when used as a seed coating for control of *P. ultimum* on sugar beet (Lutchmeah and Cooke, 1985; Martin and Hancock, 1987; McQuilken *et al.*, 1990a) or cress (Al-Hamdani *et al.*, 1983; Lutchmeah and Cooke, 1985; McQuilken *et al.*, 1990a), and these formulations are discussed in more detail in section 7.4.1. A commercial preparation of *P. oligandrum*, Polygandron, has been produced in Czechoslovakia and has undergone several trials on sugar beet and other crops in Europe but with variable results (Vesely, 1989).

7.2.4 Sclerotial mycoparasites

Sclerotium production by plant-pathogenic fungi constitutes a means of survival in the soil in the period between cropping and the planting of new susceptible crop plants. The taxonomy and germination of the major sclerotium-producing fungal genera have been reviewed (Schoen, 1983; Phillips, 1987) and it is significant that the sclerotial fungi are often important plant pathogens. Mycoparasites may exert a strong influence on populations of potential host fungi by reducing the longevity of these survival structures (Sneh, 1977). A large number of sclerotial mycoparasites have been reported to have potential as biocontrol agents of sclerotial plant-pathogens, including several of those discussed in the preceding paragraphs. Some attributes that would be required by a mycoparasite for biocontrol of plant diseases caused by sclerotium-forming fungi, where the strategy is to reduce the number of sclerotia in the soil, are an ability adversely to affect: (i) propagule germination and/or viability; (ii) growth potential through soil; (iii) activity over a range of environmental conditions; and (iv) survival and reproductive potential (Kenerley and Stack, 1987). Various screening methods have been devised to isolate mycoparasites with these attributes from soil (e.g. Mueller *et al.*, 1985; Kenerley and Stack, 1987; Whipps, 1987b; Whipps and Budge, 1990; Gerlagh and Vos, 1991; Jackson *et al.*, 1991b). In many cases a small number of species are regularly identified in this way, particularly *Coniothyrium minitans*, *Sporidesmium sclerotivorum*, *Verticillium biguttatum* and, inevitably, various species of *Gliocladium* and *Trichoderma*. Most of these screens have been used to isolate potential biocontrol agents for species of *Sclerotinia* and *Sclerotium.*

Sclerotinia sclerotiorum, the causal agent of white mould of a variety of crops, is a plant pathogen known to be susceptible to attack by several species of mycoparasitic fungi (e.g. Trutmann and Keane, 1990). These include *Coniothyrium minitans*, *Gliocladium catenulatum*, *G. virens*, *Sporidesmium sclerotivorum*, *Teratosperma oligocladum* and *Trichoderma koningii* (Trutmann and Keane, 1990). In dual cultures, *G. virens* effectively inhibits sclerotium formation by *S. sclerotiorum* and also parasitizes them thus preventing the production of ascospores which constitute the primary infective inoculum of the crop plant. Strains of *Trichoderma* are also among the most important antagonists of sclerotia and attempts to determine the mechanism of antagonism have shown that mycoparasitism of sclerotia could well play an important part in the control of diseases caused by the pathogens *Rhizoctonia solani* and *Corticium rolfsii* (Henis, 1984). Fresh sclerotia of *C. rolfsii* in natural soil do not germinate and are not invaded by other microorganisms. Sclerotial germination of *C. rolfsii* is also suppressed in experimental compost (Hadar and Gorodecki, 1991). The diseases caused by *C. rolfsii* and *Rhizoctonia solani* are also suppressed by composts used in experimental container media although the mechanism of suppression has not been determined (Gorodecki and Hadar, 1990). Whether the pathogen is suppressed, however, may depend upon physical factors, for example temperature. Suppression of diseases caused by *Rhizoctonia* with *Trichoderma hamatum* or *T. harzianum* in compost may be interfered with, albeit temporarily, by thermophilic fungi such as *Humicola* spp. which are natural inhabitants of the compost (Chung and Hoitink, 1990). If the sclerotia become dry, however, this appears to stimulate germination when they also become susceptible to attack by *Trichoderma harzianum* in the soil. Exposure of the sclerotia in soil to metham sodium (sodium *N*-metho-di-thiocarbonate) for 1 h or to a heat shock of about 100°C for 15 s also increases their germinability and susceptibility to degradation by *T. harzianum*. It seems that nutrients are released from sclerotia which are subjected to physical or chemical trauma. The nutrients are probably competed for strongly by *Trichoderma* which is then able to colonize the sclerotia. *Trichoderma* isolates apparently differ in their physiological capacity to penetrate the sclerotia of *C. rolfsii* and experiments have shown that *Trichoderma* is best able to colonize living host mycelium weakened by contact with killed mycelium rather than actively-growing mycelium. *Trichoderma* behaves in this instance as an opportunistic necrotrophic parasite which requires a food base from which to operate and it is likely that dead or weakened mycelium of the host supports that requirement.

The sclerotia of *Sclerotinia sclerotiorum*, *S. minor*, *Corticium rolfsii* and *Botrytis cinerea* are also parasitized and decayed by *Gliocladium virens* when grown on nutrient media in the laboratory. Screening tests for sclerotial mycoparasites of *S. sclerotiorum* have shown that, of ten fungi known to be mycoparasites, *Gliocladium virens* and *Coniothyrium minitans* are consistently the most effective in infecting the sclerotia and reducing their viability (Whipps and Budge, 1990). Both of these test mycoparasites were isolated from sclerotia originally and are known to be persistently associated with them in field conditions. In order to determine whether sclerotia of *S. sclerotiorum* are susceptible to infection by a test mycoparasite the sclerotia are immersed in a concentrated suspension of spores then incubated in sand or soil for 4 weeks at 18°C. After incubation, growth and sporulation of the mycoparasite on the surface of the sclerotium may be taken to indicate, but does not prove, that mycoparasitism has occurred because the mycoparasite may survive saprotrophically on nutrient sclerotial exudates. Thus association of the mycoparasite with the surface of the sclerotium is not necessarily related either to infection ability or a reduction in the number of sclerotia in soil. The next stage in the screening procedure involves surface sterilization of the colonized sclerotia which are then cut into halves. The half sclerotia are incubated for ten days on potato dextrose agar containing antibiotics to prevent the growth of bacteria then examined for the growth of mycelium into the agar and the presence of the mycoparasite. If no mycelium of *S. sclerotiorum* grows the sclerotium is not viable and, it is assumed, has been killed by the mycoparasite.

Survival of the sclerotia of *S. sclerotiorum* in the soil is reduced by the activity of *G. virens* (Phillips, 1986), and the amount of infection increases with increase in concentration of the conidial inoculum to the soil up to 10^{10} kg^{-1} soil.

Sclerotia are parasitized over a wide range (10–80%) of soil moisture content at pH 5–8. *Sclerotinia sclerotiorum* is difficult to control by biological means owing to the longevity of the sclerotia in soil which may survive there for up five years. In culture, however, at 25°C the sclerotia can be decayed within 10 days of inoculation with *G. virens* although in the soil, parasitism is markedly reduced at 15°C and probably little parasitism occurs below 15°C. The marked difference between survival of artificially inoculated sclerotia in laboratory experiments and that in the field is related to the interaction of environmental factors one or more of which may limit the germination of conidia and growth of the mycoparasite thus reducing its efficiency in the infection and destruction of the sclerotia of the pathogen. Macroconidia of *Sporidesmium sclerotivorum* which germinate in soil in the vicinity of susceptible sclerotia produce hyphae which infect them. Each colonized sclerotium constitutes a food base by providing a nutrient supply for the mycoparasite. Mycelium of the mycoparasite is thus able to grow through the soil whence further sclerotial infections may be effected. Abundant sporulation of the mycoparasite on the surface of the infected sclerotium also provides a reservoir of propagules which act as a further potential source of infection. For a mycoparasite to remain effective after addition to a soil, and for the process to be cost effective, it is important that growth and sporulation or the production of other types of survival propagule should occur. The results of experimental infections of the plant pathogen with the mycopathogen appear promising and *S. sclerotivorum* has been patented as a biocontrol organism (Deacon, 1988a).

Coniothyrium minitans, which grows well as a saprotroph in culture, may be a promising biocontrol agent as it is known to infect the sclerotia of various plant-pathogenic fungi (Whipps and Budge, 1990; Budge and Whipps, 1991). *Coniothyrium minitans* parasitizes the sclerotia of *Sclerotinia sclerotiorum*, *S. trifoliorum*, *Sclerotium cepivorum* (the fungus causing white rot of onions), and *Botrytis allii*, also a pathogen of the onion plant and *B. cinerea* (Searle and Tribe, 1984). Research with a pycnial dust of this mycoparasite applied to a field plot after sclerotia of *S. trifoliorum* were laid on the soil surface indicated that the degree of infection of sclerotia was inconsistent (Turner and Tribe, 1975). When *C. minitans* was applied directly to the seed furrow during planting, however, 42–78% disease control of *Sclerotinia* wilt of sunflowers was reported (Huang, 1980). The inoculum used in the latter case was a mixture of cereal grains in which the *Coniothyrium* had been grown. Unfortunately, this meant that as much as 1 kg of inoculum was needed per 6 m of furrow (equivalent to 6000 kg ha^{-1}), an amount which would be impractical in a commercial situation. More recently, a foliar spray of fresh spore suspensions of the fungus has been used to control the development of *S. sclerotiorum* on the aerial parts of bean plants (Trutmann *et al.*, 1982). *Coniothyrium minitans* has also been shown to persist in soil in glasshouses after trial applications in order to control disease of celery and lettuce plants caused by *Sclerotinia sclerotiorum* (Whipps *et al.*, 1989). The conditions of growth in glasshouses can be carefully controlled as watering, the supply of nutrients, the relative humidity and temperature can be automated whereas in the field the soil moisture content and temperature, for example, are subject to vagaries of the prevailing climatic conditions. Grain-based inoculum has also been used to control sclerotinia diseases of lettuce and celery in the glasshouse (Budge and Whipps, 1991), but it was suggested that the fungus may be effective as a biocontrol agent only under low disease pressure or immediately following application of inoculum. Further refinements in application techniques for this fungus are still in progress, and the biology and potential for use in biocontrol has recently been reviewed (Whipps and Gerlagh, 1992).

The biotroph *Sporidesmium sclerotivorum*, like *Coniothyrium minitans*, parasitizes the sclerotia of *Sclerotium cepivorum*. It also parasitizes those of *Sclerotinia minor*, a serious pathogen of lettuce plants, and those of *Botrytis cinerea* (Bullock *et al.*, 1986; Adams, 1987; Litkei, 1989; Adams and Fravel, 1990). *Sporidesmium sclerotivorum* has been reported to successfully control sclerotinial diseases in the field, especially lettuce drop caused by *Sclerotinia minor* (Adams, 1982; Adams and Fravel, 1990). The efficiency of *Sporidesmium sclerotivorum* as a mycoparasite of *Sclerotinia minor* has been determined by comparing the energy content of the sclerotia with that of the mycelium and macroconidia of the parasite in relation to the quantity of energy utilized in the substrate (Adams *et al.*, 1985). The result is expressed as an energy coefficient as follows:

$$\text{Energy Coefficient} = \frac{\text{Energy of mycelium or spores} \times 100}{\text{Energy of substrate utilized}}$$

The formula can be used to determine energy coefficients of host–parasite associations both in defined laboratory conditions and in the soil. It has been estimated that one sclerotium of *Sporidesmium sclerotivorum* produces an average of 15,000 macroconidia in soil (Adams *et al.*, 1984). Calculation of the conversion of energy from that in sclerotia to macroconidia of the parasite indicates that mycoparasitism by *Sporidesmium sclerotivorum* under natural conditions in the soil is 2–9 times more efficient than that calculated for growth of the parasite *in vitro*. *Sclerotinia minor* can produce over 12,000 sclerotia on a single lettuce plant (Adams, 1987) although in the soil the average inoculum density is 5–6 sclerotia g^{-1} soil. The potential for using *S. sclerotivorum* as a biocontrol agent has been well summarized by Adams (1990) who considers it to be a 'near-perfect biocontrol agent'. This view is supported by successful trials (Adams and Fravel, 1990) in which economical control of lettuce drop was obtained under field conditions using a homogenate of cultures of the fungus grown in vermiculite soaked in nutrient medium. From 1987 to 1989, five successive lettuce crops were grown with no additional application of the biocontrol agent other than an initial treatment of 0.2, 2 or 20 kg ha^{-1} of inoculum. By the third crop, all three application rates had resulted in significant decreases in disease incidence (53%, 68% and 72% disease control successively). Estimated costs of the three rates of application were \$2, \$20 and \$200 ha^{-1} respectively based on the production of 10^6 macroconidia g^{-1} of fresh inoculum (Adams and Fravel, 1990). Since a 10% disease loss would represent about \$1100 ha^{-1}, a 50% improvement in disease control would save \$550 ha^{-1}, well above the economic cost of the application of the biocontrol agent.

Another example of a potential biocontrol agent is *Verticillium biguttatum* which is being investigated for commercial use for the control of black scurf on potatoes. The causal agent, *Rhizoctonia solani*, is an important source of damage in potato production and until recently 5-year rotations of crops were the only practical way to limit infection from the soil (Boogert *et al.*, 1990). As discussed earlier (sections 6.4 and 7.1), *V. biguttatum* is completely dependent on *R. solani*, and field results have been reported in which applications of *V. biguttatum* have reduced disease levels of black scurf below the threshold levels set by the Dutch Inspection Service for seed certification of seed potatoes. *Verticillium biguttatum* also appears to be unaffected by the fungicides currently used to control potato diseases (Boogert *et al.*, 1990) and it seems likely that an integrated pest management system will be developed in Holland in the near future. Several other fungi have also been reported as potential biocontrol agents of *R. solani* on sugar beet and beans including *Laetisaria arvalis* and other *Rhizoctonia* species (Hoch and Abawi, 1979b; Herr, 1988). Bran formulations of *L. arvalis*, moistened and incubated for 5–15 days before use, have been used at a rate of 0.5% (w/v) to reduce seedling diseases in natural soils (Lewis and Papavizas, 1992).

7.3 Foliar Diseases

7.3.1 General considerations

Although most of the work involving the control of plant pathogens by mycoparasitic fungi has involved root-infecting fungi there are examples of the control of leaf and stem pathogens. Many of these have been highlighted in the review by Rishbeth (1988). A noteworthy example of antagonism, originally thought to involve mycoparasitism is the tripartite interaction of the rust fungus *Cronartium ribicola*, *Pinus* spp. the plant host and *Tuberculina maxima*, the purple mould or lilac fungus, the antagonist of the obligate parasite. *Cronartium ribicola* is an obligate parasite of *Pinus* spp. The infected pine seedlings become chlorotic, develop cankerous lesions of the bark and stem and the cortex cells undergo gradual necrosis. The mycelium of the rust appears to be associated with degradation of the middle lamellae of the walls of cortex cells but the tissues are not macerated. *Tuberculina maxima* appears to infect the aecial and pycnial lesions of the rust. It suppresses the production of aecia and pycnia, inhibits growth and development of the cankers, and either partially or sometimes entirely inactivates them thus reducing the inoculum potential of the rust. It seems, however, that only those tissues of the host plant which are rust-infected are attacked and *T. maxima* appears to be unable to infect non-rusted tissues. The effect of the rust on the host plant apparently predisposes the tissue of the host associated with the rust cankers to colonization by *T. maxima* which displaces the rust fungus (Wicker and Woo, 1973). From histological studies of pine tissues subjected to single and dual infections, it appears that cells invaded by *T. maxima* are destroyed and this is concomitant with the destruction of haustoria, hyphae and spore-producing elements of the rust. There is no direct evidence of the parasitism of *Cronartium* by *T. maxima* and an interesting implication of this work is that other cases of the apparent infection of the fungal pathogens of leaves and stems by mycoparasites might require re-evaluation.

The powdery mildews *Erysiphe cichoracearum* and *Sphaerotheca fuliginea* infect the leaves of cucumber plants grown in glasshouses and are responsible for loss of the cucumber crop. In an experimental situation, some measure of control of the disease has been obtained by spraying cucumber plants with the conidia of either of the mycoparasitic fungi *Ampelomyces quisqualis* or *Verticillium lecanii* (Sundheim, 1982, 1986). Conidium and cleistothecium production of the powdery mildews is inhibited by the mycoparasites. In order to improve the effectiveness of treatment of infected cucumber plants with the mycoparasite, the plants were sprayed weekly with spores of the mycoparasite and at fortnightly intervals with the fungicide triforine which did not appear to affect *A. quisqualis*. The application of spores of the mycoparasite alone provided only limited protection against the pathogens, the most efficient control being effected by augmentation of the mycoparasite with the application of a fungicide. A problem associated with the use of *A. quisqualis* to control powdery mildew of cucumber, however, is that the leaves and fruits of the cucumber plant appear themselves be susceptible to parasitism by *A. quisqualis* (Jarvis and Slingsby, 1977). Lesions caused by *A. quisqualis* develop on the cotyledons and leaves. Lesions on the cucumbers reduce their marketable value. In addition to infecting powdery mildews, *A. quisqualis* also parasitizes other fungi including *Botrytis cinerea*, the grey mould of many soft fruits, and *Alternaria solani*, a pathogen of potato plants. Thus it is possible that this mycoparasite could be a potential biocontrol agent of other plant-pathogenic fungi. An isolate of *A. quisqualis* obtained from an *Oidium* species has been isolated from Israel and shown to be effective in field trials in the biocontrol of powdery mildews of cucumber, carrot and mango (Sztejnberg *et al.*, 1989). This isolate was tolerant to many fungicides currently used to control powdery mildews and successful experiments were carried out in which the use of *A. quisqualis* in combination with the fungicide pyrazophos gave greater levels of control than when the mycoparasite was used alone. Fungicide tolerance in this fungus had been previously reported (Sundheim and Amundsen, 1982) and there is potential for the future development of this fungus as a biocontrol agent for use in integrated control of powdery mildews. Appropriate conditions for the mass production of infective spores of *A. quisqualis* have now been determined (Sztejnberg *et al.*, 1990), and a formulation has been devised using paraffins which can reduce the humidity demands of the mycoparasite and enhance the efficiency of control (Philipp *et al.*, 1990). Significant control of cucumber powdery mildew has also been obtained using *Tilletiopsis minor*. In glasshouse conditions, low relative humidities reduced the efficacy of the biocontrol agent but the incorporation of lipophilic substances into the spray formulation restored some of the lost activity (Hijwegen, 1992).

Trichoderma pseudokoningii produces non-volatile inhibitors of fungal growth and also parasitizes *Botrytis cinerea*, the causal agent of dry eye rot of apple trees, by coiling around and penetrating its hyphae. Although the development of dry eye rot can be suppressed under controlled conditions by inoculating apple flowers with the conidia of *T. pseudokoningii*, comparable trials in orchards have not successfully reduced the incidence of disease below the level normally expected in the apple tree population (Tronsmo and Raa, 1977). On the other hand, the preharvest rot of strawberries caused by *B. cinerea* and *Mucor mucedo* has been reduced to levels similar to those achieved by the application of dichlorofluanid by spraying with spore suspensions of *T. viride* and *T. harzianum* (Tronsmo and Dennis, 1977).

Verticillium species are usually thought of as responsible for vascular wilt disease in higher plants but some behave as mycoparasites. *Verticillium psalliotae*, a pathogen of the commercially grown mushroom *Agaricus bisporus*, also parasitizes uredosori of the coffee rust pathogen *Hemileia vastatrix* (Lim and Wan, 1983). Hyphae of the mycoparasite which penetrate the uredospores destroy the cytoplasm and render the uredospores non-viable. Coffee is grown as a cash crop in various parts of the world and the reduced yield due to the growth of the rust fungus is responsible for loss of income to the farming community. Destruction of the uredospore phase on the leaves of coffee plants may serve to limit natural infections of the coffee plant in plantations. A second species, *V. lecanii*, is of especial interest and has been used both experimentally and on a small commercial scale to control aphids and whiteflies on a range of plants throughout the world. The fungus attacks these insects and colonizes them before they die. A fresh batch of spores is then produced over the entire surface of the insect after emergence of the hyphae through the insect cuticle. Further, *V. lecanii* has been shown to parasitize the oidia of the powdery mildew *Uncinula necator* (Heintz and Blaich, 1990) and teliospores of various rust pathogens of flowering plants including the chrysanthemum rust *Puccinia horiana* and carnation rust *Puccinia arenariae* (= *P. dianthi*) and reduce the amount of infection produced by these pathogens (Srivastava *et al.*, 1985b). Destruction of the teliospore stage of the pathogen is significant because this is the resting stage of the fungus which can withstand adverse conditions and provides the inoculum for infection of the next generation of host plants. Similarly, eradication of the uredospores reduces the capacity of the rust to infect healthy crop plants during the growing period. The facility to control the relative humidity and ambient temperature in glasshouses probably affords the plant pathologist a better chance to control such pathogens than in the field situation. As *V. lecanii* infects both fungi and insects, there is the interesting possibility that both insect pests and plant-pathogenic fungi could be controlled simultaneously in certain situations. It is, perhaps, not surprising that a single parasitic fungus can attack both insects and other fungi as insect exoskeletons and the hyphal walls of many fungi contain significant quantities of the structural organic polymer chitin. The cells of fungi and insects may also contain chemically similar storage compounds such as glycogen and soluble carbohydrates which could provide a common nutritional base for a mycopathogen and entomopathogen such as *V. lecanii.*

Potentially useful mycoherbicidal fungi have been isolated from maize crops attacked by plant-pathogenic fungi (Vakili, 1985; Vakili and Bailey, 1989). This study considered the natural fungicolous flora associated with phytopathogenic fungi of maize and attempted to determine the frequency of association of the mycoparasites with the plant pathogens. Cornstalk rot is due to infection by several pathogenic fungi and Vakili (1985) has shown that the following are associated with mycoparasites in descending order of frequency; *Fusarium moniliforme*, *Cochliobolus carbonum*, *Colletotrichum graminicola*, *Cladosporium herbarum*, *Fusarium graminearum*, *Pyrenochaeta terrestris* and *Stenocarpella maydis.* The mycoparasitic fungi associated with the phytopathogens in descending order of frequency of isolation were *Sphaeronemella helvellae*, *Trichoderma viride*, *Gliocladium roseum*, *Exobasidiellum* spp., *Gonatobotrys simplex*, *Trichothecium roseum*, *Pythium acanthicum* and a sterile fungus. As none of the mycoparasites isolated are known to be pathogenic to maize it is possible that they might naturally affect the incidence and severity of stalk rot disease. These mycoparasites also seem promising on a commercial basis as a source of potential biocontrol agents of the causal fungi of stalk rot disease.

Although a mycoparasitic fungus may appear to be a potentially useful biocontrol agent of a plant-pathogenic fungus, the host range of the mycoparasite may be restricted to a single strain of the pathogen, or the conditions necessary for growth of the mycoparasite may also be conducive to the growth of the pathogen and the usefulness of the mycoparasite is reduced. *Claviceps purpurea*, responsible for the ergot disease of rye, is parasitized by *Fusarium heterosporum* (Hornok and Walcz, 1983). The grain of rye used to produce flour used in breadmaking can become contaminated by the sclerotia or ergots of *C. purpurea.* The ergots are known to contain powerful alkaloid poisons which if ingested in traces can make the consumer seriously ill and death may also ensue. *Claviceps purpurea* is not restricted to growth on rye as it also infects wheat and the ergots can be found on the flowering spikes of a range of grasses which enables the pathogen to proliferate in the wild. Experiments involving spraying ergots with spores of *F. heterosporum* have shown that some ergots become infected whereas others do not. Analysis of the results indicates that pathogenicity of *F. heterosporum* towards *C. purpurea* is related to the species of host plant from which the *C. purpurea* is isolated. The problem of control is further compounded because *Fusarium* requires prolonged exposure to moisture, an environmental condition also conducive to the survival of *C. purpurea.*

Dicyma pulvinata is a mycoparasite reported to parasitize hyphae and spores of *Cercospora* and *Cercosporidium* species (Mitchell *et al.*, 1987). It has been proposed as a biocontrol agent for leafspot of peanuts (causal agent *Cercosporidium personatum* = *Phaeoisariopsis personata*) in the USA and has been used successfully under field conditions (Mitchell *et al.*, 1987). Mutants have been selected for growth under low relative humidity, high temperature, and resistance to benomyl (Mitchell *et al.*, 1986) and a suitable formulation using carboxymethyl cellulose has been developed to improve performance in field use. These experiments have demonstrated the clear potential for improvements in the performance of existing biocontrol agents through genetic modification or formulation techniques. *Dicyma pulvinata* (= *Hansfordia pulvinata*) has also been reported to parasitize *Fulvia fulva* on tomato (Peresse and Le Picard, 1980) and *Asperisporium caricae* on papaya (Hepperly, 1986). It rapidly colonizes the leaf-spot lesions of its host and necrotrophically destroys the conidiophores and conidia. It has also been shown to produce a fungitoxic sesquiterpene, 1,3-desoxyphomenone, that is active against *Chaetomium* sp. (Tirilly *et al.*, 1991). The use of the fungicide Fosetyl-Al in conjunction with *D. pulvinata* has also been suggested for integrated control of *F. fulva* (Tirilly, 1990).

7.3.2 Postharvest diseases

Numerous examples are now available (Table 7.3) where bioaugmentation has shown potential in the control of postharvest diseases of fruits. It is not always clear, however, if these agents act via mycoparasitism or by direct competition for nutrients. Such agents can be applied at the preharvest stage or at harvest, and an example will be used here to illustrate the problems and progress in the development of fungal agents for biocontrol. This involves the work of Tronsmo (1986) with the preharvest use of a filamentous fungus, *Trichoderma*, to control the grey mould disease of strawberries caused by *Botrytis cinerea*. The early experiments consisted of spraying the blossoms of strawberry plants with conidial suspensions of various *Trichoderma* species. A strain of *T. pseudokoningii*, in particular, was found significantly to reduce postharvest disease levels when compared to water-sprayed controls. Dichlorofluanid treatment, however, was much more effective, and no advantage could therefore be seen in the biological approach. Subsequently, it was realized that the isolate of *T. pseudokoningii* used in these trials did not grow below 9°C and yet the critical infection temperatures for strawberries lay between 7.7 and 13.3°C. Thus the biocontrol agent did not operate across the range of conditions under which the disease agent was capable of causing infection (Tronsmo, 1986). A screening for *Trichoderma* isolates with lower temperature requirements was undertaken and an isolate was found that did grow and antagonize *Botrytis* at low temperatures. Disease control levels in field trials using the organism were much better and equivalent to the use of the conventional treatment with the fungicide dichlorofluanid (Tronsmo, 1986). This method, however, has not yet been widely adopted by the industry as appropriate chemical controls are still available. Similar successes with *Trichoderma*

Table 7.3. Some successful examples of biocontrol of postharvest diseases.

Crop	Disease	Antagonist	Reference
Apple	*Botrytis* rot	*Cryptococcus laurentii*	Roberts (1990)
Apple	*Botrytis* rot	*Trichoderma* sp.	Tronsmo and Ystaas (1980)
Apple/pear	*Botrytis* rot/ *Penicillium* rot	*Acremonium breve*	Janisiewicz (1987, 1988)
Citrus	*Penicillium* green mould	*Candida guilliermondii*	Wilson and Chalutz (1989); Droby *et al.* (1989)
Pineapple	*Penicillium* rot	Attenuated strains of *Penicillium* sp.	Lim and Rohrbach (1980)
Strawberry	*Botrytis* rot	*Trichoderma* sp.	Tronsmo and Dennis (1977)

have been achieved in France for the biocontrol of *Botrytis* diseases of grapes (Dubos, 1987).

More recent progress has been made in the development of antagonistic yeasts isolated from the fructoplane for the biocontrol of green mould of citrus caused by *Penicillium digitatum.* A strain of *Candida guilliermondii* (= *Debaryomyces hansenii*) (Wilson and Chalutz, 1989; Chalutz and Wilson, 1990), originally described as *Debaryomyces hansenii* (McLaughlin *et al.*, 1990a), has proved to be very successful in early trials, and does not seem to inhibit the *Penicillium* through the production of antibiotics. This offers an advantage for commercialization in that it should gain better public acceptance as exotic antibiotics are less likely to be introduced into the food chain if the organism is applied as a postharvest treatment of citrus fruit. The antagonist seems to act by competing with the *Penicillium* for nutrients (Droby *et al.*, 1989). Similar yeasts have also been proposed as potential agents of biocontrol for postharvest disease of apples (McLaughlin *et al.*, 1990b) and their effectiveness has been considerably improved by supplementing the treatment with solutions of calcium chloride. A basidiomycetous yeast that naturally occurs on apple leaves, buds and fruit, *Cryptococcus laurentii*, has been shown to control grey mould decay of apples (causal agent: *Botrytis cinerea*) when wounds are treated with suspensions of the antagonist (Roberts, 1990). This yeast is psycrophilic and may have a potential role in fruit stored at low temperatures.

There are some interesting examples where the foliar development of pathogens has been shown to be inhibited by the activities of phylloplane microorganisms (e.g. Fokkema, 1978; Williamson and Fokkema, 1985; Jeger and Jeffries, 1988) but few reports address the control of diseases of fruits. However, Kwee and Teik (1982) reported promising results in field trials in Malaysia for the control of anthracnose disease of mangoes.

More recently it has also been shown that treatment of mango trees with preharvest sprays of fungicide can significantly reduce the level of indigenous microorganisms, whereas a preharvest spray of nutrient-rich materials can stimulate their development (Jeffries and Koomen, 1992). In this latter case, however, significant control of mango anthracnose disease did not result from the use of nutrient sprays in the absence of fungicide. Fungicide treatment can exacerbate the occurrence of coffee berry disease, caused by *Colletotrichum kahawae* (= *C. coffeanum*), presumably through suppression of competition by the indigenous microbial flora (Firman and Waller, 1977). Preharvest treatment of trees with biocontrol agents can also reduce the amount of fruit loss through control of twig blights and blossom blights caused by severe infestations of pathogens which usually cause postharvest fruit rots. Sprays of conidia of *Penicillium glabrum* (= *P. frequentans*) applied in nutrient suspensions have been used (de Cal *et al.*, 1990) to control peach twig blight (causal agent: *Monilinia laxa*). Significant reductions in disease severity (from 38–80%) were recorded, comparable to that given by the fungicide captan. The antagonist apparently acted through the synthesis of at least two antibiotic metabolites (de Cal *et al.*, 1988).

7.4 Commercial Considerations

7.4.1 Production and formulation

Once a particular strain of a biocontrol fungus has shown potential for disease control based on laboratory, greenhouse, and field tests, production of an effective biomass becomes a major concern. Both liquid and semi-solid fermentation systems have been used for this purpose (Lumsden and Lewis, 1989). The commercial application of fungi as agents of biocontrol of plant diseases is fraught with difficulties. Formulation of the antagonist into a form which is easy to handle and deliver to the site of action can be problematical. With *Trichoderma viride* and *T. harzianum*, the population densities increase by 10^4 and 10^3 respectively in natural soil during the first three weeks of incubation when the antagonists are added on bran as mycelium from conidia allowed to germinate for 1–3 days before addition to soil (Lewis and Papavizas, 1984). The populations do not increase when ungerminated conidia on bran particles are added to soil which indicates the importance of colonization of the substrate to effective antagonism. The technique of substantially raising the number of colony-forming units of a biocontrol agent in soil forms the basis of the application of this method of the control of several plant-pathogenic fungi on a commercial basis.

Propagules of *Coniothyrium minitans*, *Gliocladium roseum* and *Trichoderma viride* incorporated in pellets of sodium alginate lose their viability over time. Experiments have shown that, over a period of three months, the number of colony-forming units per pellet when air-dried and stored at the ambient temperature may decline sharply (Magan and Whipps, 1988). The useful storage life of such pellets produced for commercial purposes is an important aspect that has to be considered in their manufacture. Further, in tests of the growth of mycelium from pellets plated onto agar the number of colony-forming units is reduced at low water potential and low nutrient levels. The formulation of a delivery system which is effective under the environmental conditions encountered in the field and in the maintenance of the biocontrol control agent has hampered commercial exploitation. The interaction of physical, chemical and biological factors in the field can be monitored but cannot be controlled as readily as the conditions in laboratory trials.

For the application of a biocontrol mycoparasitic fungus to soil to be effective requires the development of an inert carrier substance capable of retaining its integrity in the soil whilst carrying the maximum number of propagules of the antagonist on the minimum volume of the carrier. Various methods of applying biocontrol agents to the soil have been tested, including the liquids used in the production medium for fermenter biomass, as well as peat, seeds, vermiculite and various types of clay (Fravel *et al.*, 1985). Peat and bran tend to be too light for mechanical application and vermiculite is of unsuitable particle size. Liquid carriers are relatively bulky and expensive to transport. Diatomaceous earth possesses many of the important properties necessary to carry the inoculant fungus as it swells little when saturated with moisture, has an excellent absorptive capacity, withstands autoclaving and does not disintegrate when stirred in the growth medium. *Trichoderma* propagules also remain viable on the granules after air-drying for up to one month in cold storage. *Trichoderma harzianum* has been applied to peanut fields in Alabama to reduce infection of the plants by *Corticium rolfsii* (Backman and Rodriguez-Kabana, 1975). It is assumed that the pathogen is parasitized by *T. harzianum.* The antagonist is grown on sterile granules of diatomaceous earth impregnated with 10% molasses for four days which is used to provide the carrier substrate. *Trichoderma harzianum* is applied to the infested soil 70–100 days after planting peanut plants. As well as significantly reducing infection the yield of peanuts is increased. Advantages of using granules are that small volumes of the carrier substrate impregnated with large numbers of propagules can be readily applied to the soil. Furthermore, unlike chemical fungicides, the carrier substrate does not contaminate the soil with potentially toxic organic residues which may also induce the development of strains resistant to the pathogen to be controlled. *Chondrostereum purpureum* which causes the silver leaf disease of plum trees can also be controlled by the application of propagules of *T. viride* and *T. polysporum* which have been developed commercially for this purpose (Ricard, 1981). Aqueous suspensions of conidia and a variety of other formulations of *Gliocladium virens* and various

Trichoderma species have been compared for use in the glasshouse to control damping-off and blight of bean caused by *Corticium rolfsii* (Papavizas and Lewis, 1989) and of cotton damping-off caused by *Rhizoctonia solani. Gliocladium virens* was the most effective biocontrol agent and of four formulations of *G. virens* tested against germinating spores of *C. rolfsii* germlings, alginate–bran–fermenter biomass pellets and pyrax–fermenter biomass mixtures reduced disease considerably, and all three formulations were more effective than conidia in aqueous suspension. Other research (Papavizas *et al.*, 1984; Beagle-Ristaino and Papavizas, 1985) has also indicated that fermenter biomass preparations formulated as a powder with a diluent such as talc or Pyrax, or alginate pellets containing a food base such as bran, were also effective and the fungus not only proliferated greatly in soil, but also suppressed disease more effectively than naked conidia or chlamydospores of *Gliocladium* or *Trichoderma.* Pellets of fermenter biomass of these fungi also reduced the survival and growth of *Rhizoctonia solani* in soil and prevented damping-off of cotton and sugar beet in the greenhouse and fruit rot on tomato caused by this pathogen (Lewis and Papavizas, 1987). These various results have shown that preparations containing fermenter biomass are more effective than conidia in the biocontrol of *C. rolfsii.* Since industry is better equipped to produce fermenter biomass than conidia, it appears that fermenter biomass preparations will be the biotechnological approach of the future because of superior performance and practical adaptability in preparation and delivery. Another consideration for fermenter production of biocontrol agents is the cost of the process. Molasses is a relatively inexpensive substrate used in liquid fermentation biotechnological processes for the production of large biomasses of biocontrol fungi such as *Gliocladium virens* and species of *Trichoderma* (Papavizas *et al.*, 1984). In experiments simulating large-scale industrial production, the fungi are incubated in fermenter vessels and the liquid medium is aerated by stirring, shaking or bubbling air through, which enables the mycelium to grow and sporulate. The fungal biomass produced after growth in a fermenter for 15 days consists of mycelium, conidia and substantial numbers of chlamydospores. The mycelium is filtered, dried, milled and mixed with a carrier such as anhydrous aluminium silicate to increase the volume for addition to soil. *Trichoderma* and *Gliocladium* added to soil in relatively low concentrations, e.g. 2×10^3 colony-forming units g^{-1} soil rapidly proliferate producing populations of 5×10^6 organisms g^{-1} after 10–20 days. Chlamydospores rather than conidia appear to be responsible for the increased number of colony-forming units in soil. An implication from the results of such experiments is that the chlamydospore of *Trichoderma* or *Gliocladium* should be considered as the most important propagule produced in fermentation biotechnology for their development as biocontrol agents.

Another two-stage carrier–food-base experimental system involves the use of lignite granules steeped in stillage, a by-product of ethanol production. Lignite granules serve as a carrier for the nitrogen-fixing bacterium *Rhizobium* which is used to coat the seeds of leguminous crop plants prior to sowing and its use has subsequently been tested successfully in the biocontrol of *Rhizoctonia solani* by *Trichoderma harzianum* and *Gliocladium virens* (Jones *et al.*, 1984). Stillage comprises about 6% organic nutrients including nitrogenous matter, starch and sugars, sufficient to serve as the food base for the chosen biocontrol agent. The lignite is ground into granules 425–2000 μm diameter, steeped in stillage then sterilized by autoclaving. The granules are then inoculated with a conidial suspension of the selected antagonist and incubated for seven days to enable the fungus to colonize them. Next, the granules are dried in air and can then be stored at 20°C for up to four months, the inoculum retaining over 90% viability. The granules are sufficiently dense to penetrate a leaf cover when applied to the soil. When uncolonized granules are added to soil the incidence of damping-off due to *Rhizoctonia* increased thus demonstrating that the pathogen, although able to use the food base, is effectively prevented from doing so by the biocontrol fungus.

The oospores of *Pythium oligandrum*, a fungus which strongly parasitizes *P. ultimum*, have been incorporated into a seed dressing containing carboxymethyl cellulose which acts as an inert carrier material for the inoculum of the mycoparasite (Al-Hamdani *et al.*, 1983). Between 42 and 200 oospores are associated with each individual seed and, although fewer than 20% germinate, this is sufficient to ward off the pathogen. When the coated seeds are germinated in soil, the oospores also germinate and the mycelium which develops is believed to form a protective mycoparasitic sheath around them. As the oospores withstand desiccation better than the vegetative hyphae and some can probably survive in the soil for 18 months, it seems likely that they may constitute a residual source of inoculum in the soil. *Pythium oligandrum* does not itself parasitize the crop plants under protection. Several examples of crop plants are known to be protected by *P. oligandrum* from damping-off and pre-emergence blight and this fungus appears to be an important biocontrol agent. A useful method of producing large volumes of inoculum for pelleting seeds with the oospores of *P. oligandrum* has been developed (Lutchmeah and Cooke, 1985; McQuilken *et al.*, 1990b, 1992a). Oospore-producing mycelium is grown in nutrient liquid culture containing 30 g l^{-1} cane molasses, a waste product of sugar refinement. Mycelium develops during the first week of growth followed by oospore production. The energy required for oospore development is derived from the mycelium and after about three weeks the cultures, which then consist of a mass of oospores, are bulked and dried gently overnight at 18–21°C. Any mycelium is killed but the oospores survive. The oospore powder is mixed with an equal weight of dry clay then applied to the surface of the seeds. Tests have shown that the dried, pelleted oospores on seeds of cress, sugar beet and carrot may remain viable for four years when stored at laboratory temperature. The mortality of the seedlings of cress and sugar beet due to infection by *P. ultimum* is significantly reduced by the application of oospores of *P. oligandrum* to the seeds. Oospore preparations formulated as a kaolin dust, on perlite or as alginate pellets, incorporated into sand gave little or no control of disease compared to seed coatings

(McQuilken *et al.*, 1992b). Seed coating with preparations of *Gliocladium virens* has also been used as an effective means of biocontrol of diseases caused by *Pythium ultimum* (Howell, 1991) and the general considerations for the development of an effective biological seed treatment system have been discussed in detail by Harman and Taylor (1990).

In other experiments to determine the efficiency of 50 isolates of bacteria and fungi in the biocontrol of damping-off of cotton and cabbage seedlings caused by *Pythium ultimum* and *Rhizoctonia solani*, the most effective control was obtained with *Gliocladium virens.* Bedding plants grown in glasshouses in non-sterile soil-free potting media are susceptible to damping-off diseases which results in loss of the crops. The application of conidia of the mycoparasite to the soilless mix prior to planting the seeds provides a means of controlling the disease. Pellets of 2% sodium alginate with calcium chloride as the gelling agent containing large numbers of conidia provide sufficient colony-forming units to maintain the mycoparasite in the growth medium in sufficient numbers to control the growth of the pathogens during the early growth of the seedlings when they are most susceptible to attack (Lumsden and Locke, 1989). Whether control of the pathogens is due in part to the mycoparasitic activity of the antagonist is questionable. Some evidence suggests the probability that the biocontrol is largely due to the production of metabolites which are toxic to the pathogens (Howell, 1987). Strains of *Gliocladium virens* produce the antibiotics viridin, gliotoxin and gliovirin. Gliovirin inhibits the growth of *P. ultimum* and the sclerotia of *R. solani* can be killed by these antibiotics. Mutants of *G. virens* obtained by irradiation with ultraviolet light have been found to be unable to parasitize the hyphae of the pathogens although they still grow in close proximity to them. The cytoplasm of the hyphae of the pathogens growing in the vicinity of the mycoparasite often degenerates and this is believed to indicate that the mechanism of antibiosis is in operation. Further, the pathogens appear to be controlled equally well by the non-parasitic and mycoparasitic strains of *G. virens* which indicates that mycoparasitism is not necessary to the control of damping-off diseases caused by *P. ultimum* and *R. solani.* It seems, therefore, that mycoparasitism is unlikely to be the primary determinant responsible for the control of damping-off diseases caused by *G. virens* although its involvement in disease control cannot be ruled out.

The delivery of inocula containing spores and mycelium of *Trichoderma* and *Gliocladium* species to soil is a problem which has been overcome to a large extent by the development of sodium alginate gel which acts as a carrier and the incorporation of either bran or kaolin clay as the bulking agent. Bran is more effective than kaolin in maintaining the number of viable propagules in the soil as it also provides a food base for the mycoparasite. The sodium alginate probably also insulates *Gliocladium* from the effects of other the soil organisms initially, enabling the viable propagules to establish a strong hold on the food base. The soil inoculation agent is produced in the form of spongy, gelatinous pellets, 3–4 mm diam. which are dehydrated to form granules 1–2 mm diam. Each granule contains 10^5–10^7 viable colony-forming units. The active ingredient of the pellets consists of approximately 5% biomass by weight and is more effective when there is a higher proportion of chlamydospores than conidia. The number of colony-forming units in the soil after the addition of pellets increases to a maximum during the first three weeks of incubation, ranging from 10^6–10^{10} g^{-1} of soil and the population density remains high for about two months, only declining gradually during that time. This is an unusual example of the proliferation of the antagonist in soil without the need to pretreat it by means of steam sterilization or fumigants in order to eliminate competitors. Inoculum in the pellets stored at 5°C remains more viable than that stored at 25°C. Although the viability of propagules declines to less than 10% after six months in the latter, the number of colony-forming units is still comparable to that provided by freshly prepared pellets (Lewis and Papavizas, 1985). The length of storage life of the inoculum is important owing to the almost inevitable time-lag between its production, distribution and application. From a practical point of view the sodium alginate pellets are simple and cost effective to produce, are stable for several weeks at 5°C, retain their physical integrity in the soil, are probably not toxic to non-target organisms and can be applied with existing implements used for granular fertilizers and pesticides (Lewis and Papavizas, 1987). Growth of *Trichoderma* and *Gliocladium* from sodium alginate pellets is reduced *in vitro* at low water potentials and low nutrient concentrations, environmental conditions which may also affect their growth in soil (Magan and Whipps, 1988). *Trichoderma* and *Gliocladium* are also antagonists of several phytopathogens and can survive in different soil types over a range of pH which increases their effectiveness as biocontrol agents in comparison with antagonists with a smaller host range and lower tolerance of the variation in soil conditions. The application of a potential biocontrol agent to the soil on a sodium alginate pelleted carrier does not necessarily guarantee successful control of the plant pathogen. Evaluation of *Talaromyces flavus* as a biocontrol agent of the potato pathogen *Verticillium dahliae* using either wheat bran or Pyrax (pyrophyllite dust, hydrous aluminium silicate) as the bulking agent in controlled experiments has shown that development of the disease is not significantly affected by the treatment (Spink and Rowe, 1989). The greater colonization of the rhizosphere and surrounding soil with the bran-based pellets as opposed to the inert Pyrax-based pellets implies the need to include a food base in order to enable the mycopathogen to proliferate in the soil. Why the mycopathogen did not behave as a biocontrol agent in this case has not been fully explained. Perhaps the mycopathogen needs to inhabit the soil for several months before planting the potato crop in order to allow the inoculum of the potential biocontrol agent to increase to a level at which it can become effective. An alternative to the alginate pellets is the use of air-dried formulations. A mixture of vermiculite and powdered wheat bran, along with wet or dry fermenter biomass of *Trichoderma* spp. or *G. virens*, has the advantage that it can be prepared under non-sterile conditions (Lewis *et al.*, 1991). After a 2–3 day reactivation process, in which the product is moistened and incubated at room temperature, the biocontrol fungus proliferates to greater than 10^7 colony-forming units g^{-1} of soil.

7.4.2 Successful biocontrol agents

There are now several commercial examples of the use of mycoparasites in agriculture and horticulture. One of the earliest successful methods of biocontrol adopted in forest pathology involves the antagonism of *Heterobasidion annosum* by *Phanerochaete gigantea* (Rishbeth, 1963). *Heterobasidion annosum* attacks pine and other coniferous trees, especially Norway Spruce, in which it produces root and butt rot, a white heart rot of the wood. Rhizomorphs infect the living roots and the mycelium proceeds upwards through the tree. The basidiome is a bracket-like structure bearing numerous pores through which the basidiospores are liberated from tubes. The basidiospores can germinate on the stumps of felled trees and the food base so provided enables the parasite to maintain a hold in the soil from which further infections of living trees may originate. The pathogen also produces a conidial stage (*Oedocephalum*-like) although these spores appear to be less strongly infective than the basidiospores (Garrett, 1970). Living, freshly exposed pine stumps are strongly selective for the pathogen, which invades them by means of aerially dispersed basidiospores. *Phanerochaete gigantea* was chosen because it is the commonest and most effective competitor of *H. annosum* on pine stumps. It was calculated that the application of a dose of 1×10^4 oidia of *P. gigantea* to a pine stump allowed a satisfactory margin of safety for effective biocontrol of the largest spore inoculum likely to be deposited naturally on pine stumps of 16 cm wood diameter. A problem associated with the production of the fungal inoculum in bulk is storage until it can be applied in the field at the appropriate time. The oidia remain viable for up to two months when dehydrated and sachets were developed for inoculating the stumps. The viability of oidia in samples of the sachets is tested in order to ensure that *Heterobasidion annosum* can be controlled. New strains of *P. gigantea* are also selected for use which are aggressive and also serve to maintain genetic heterogeneity in the population (Parker, 1977). When required, the oidia are suspended in water and applied to the cut surfaces of the tree stumps. *Phanerochaete gigantea* outcompetes *H. annosum* for possession of the wood through hyphal interference (Section 3.1.2) so warding off the pathogen and its presence also prevents infection of the stump by *H. annosum,* from the roots of the tree in the soil. *Phanerochaete gigantea* may also parasitize *H. annosum* thus it is possible that mycoparasitism plays some part in the process of antagonism. Each sachet of *P. gigantea* contains 5×10^6 viable oidia. The contents of a sachet are mixed with 5 l of distilled water together with 5 g of a violet coloured dye Hexacol. The suspension is brushed over the tree stump to ensure an even dispersion of the spores and the dye provides a visual indication that the tree has been treated. A 5 l aliquot is sufficient to paint the surface of at least 100×20 cm diameter stumps. The procedure is no more labour intensive than that required to coat stumps with chemical fungicides and is equally effective in preventing establishment of the pathogen. A critical factor in this method of control is the need to treat the stumps of freshly felled trees with a large inoculum potential of the oidia of *P. gigantea* before the pathogen has the opportunity to become established (Parker, 1977). Decay of the wood of standing trees infected with *H. annosum* is arrested when treated with *Trichoderma* species (Lundborg and Unestam, 1980) thus also providing a useful potential means of biocontrol of the pathogen by means of this antagonist.

The history of commercialization of a *Trichoderma*-based mycofungicide has been described by Ricard (1981). The author has developed the use of *Trichoderma harzianum* and *T. polysporum* as a mixed inoculum to combat silver leaf disease of fruit trees. The efficiency of this preparation as a curative treatment for the disease has been demonstrated in the field (Corke, 1978). The first sales permit was obtained in France in 1976, the second in the United Kingdom in 1982, then five other countries followed including the United States of America. It is sold as BINAB T in the form of pellets for insertion into holes drilled into the branches of the tree, or as a BINAB T WP (wettable powder) for spray application. Over the past 15 years it has been estimated that the cumulative total of trees treated exceeds one million (Ricard, personal communication).

Biocontrol by fungi may be enhanced if the procedure also involves the use of chemical protectants. Fumigation of the soil of a wheat field with chloropicrin-methyl bromide at 440 kg ha^{-1} followed by sampling the fungi which recolonize the soil at intervals up to four months after the treatment show that *Trichoderma viride* is a strong recolonizing agent (Warcup, 1976). *Trichoderma viride* either survives the fumigation treatment or is an early colonizer and may sometimes dominate the soil microflora after treatment. In Egyptian soils treated with the fumigating agents allyl alcohol and formalin, *Trichoderma* is a dominant recolonizer (Moubasher and Mazen, 1971). *Trichoderma* has also been shown to dominate soil after fumigation with carbon disulphide for the removal of *Armillaria mellea* from the infected roots of citrus plants (Bliss, 1951). Similarly, the injection of the fumigants methyl bromide, Vorlex, chloropicrin, and carbon disulphide, into holes drilled into ponderosa pine stumps infected with *A. mellea* leads to eradication of the pathogen and a significant increase in *Trichoderma* in the previously infected stumps in comparison with untreated control stumps (Filip and Roth, 1977). Although *Trichoderma* does not necessarily tolerate carbon disulphide as well as other soil saprotrophs such as *Aspergillus niger* and *Penicillium chrysogenum* it is capable of more rapid growth which enables the fungus to colonize a larger volume of soil than the more slowly growing competitors. The fumigants appear to have a selective effect on the soil microflora thus enabling fungi such as *Trichoderma* to outcompete root-infecting pathogens. Infections of tobacco plants due to *Rhizoctonia solani* and *Fusarium solani* can be reduced by the addition of *Trichoderma harzianum* to methyl bromide-fumigated soil before the seeds are sown (Cole and Zvenyika, 1988). When the treatment is integrated with the addition of the fungicide Triadimenol, control of the disease is further improved.

Where fungicidal compounds are applied in conjunction with mycoparasites in order to control disease caused by plant-pathogenic fungi, the mycoparasite itself must be able to withstand the fungicidal treatment. For example, *Pythium oligandrum* has commercial potential as a biocontrol agent and an extensive programme of testing, especially for treatment of the seeds of susceptible plants, is in progress (see section 7.2.3). In field soils, repeated applications of the fungicides metalaxyl and mancozeb considerably reduced the population density of propagules of this fungus (White *et al.*, 1992). Selection for fungicide-tolerant strains may be necessary prior to commercial development. Promising results have been obtained following exposure during growth of *P. oligandrum* to increasing concentrations of the fungicide benomyl and the selection of a highly fungicide-tolerant strain has been made (Lewis *et al.*, 1989). Control of damping-off diseases through the combined agency of a fungicide and a fungicide-tolerant strain of *P. oligandrum* was subsequently achieved. The use of a compatible biocontrol agent in association with reduced rates of currently used fungicides is the most likely way that biocontrol agents will be used in the future. The combination of different mechanisms of control into an integrated programme, i.e. integrated pest management, is the ultimate goal of many agriculturalists. The selection or improvement of strains of biocontrol fungi with respect to tolerance to fungicides is important in improving the effectiveness of their use. An example of the successful development of a biocontrol agent with tolerance to fungicides is the selection of strains of *Epicoccum nigrum* for the biocontrol of *Sclerotinia sclerotiorum* (Zhou and Reeleder, 1990). Integrated control of *Rhizoctonia solani* has also been reported (Strashnow *et al.*, 1985) by using a combination of methyl bromide treatment and addition of *Trichoderma harzianum.* Under laboratory conditions, *T. harzianum* was found to be tolerant of up to 20,000 ppm methyl bromide, whereas *R. solani* was susceptible to a dose of less than 9000 ppm. Thus in a glasshouse, a combination of *T. harzianum* and a reduced dose of methyl bromide (equivalent to 200 kg ha^{-1}) completely controlled disease incidence of *R. solani* in bean seedlings. In addition, the biocontrol agent was also able successfully to prevent reinfestation of the fumigated soils by *R. solani*, as would normally be expected in a soil treated with methyl bromide alone.

The rapidity of growth of *Trichoderma* through the soil may be because it is capable of utilizing very low levels of nutrients which are insufficient for the sustained growth of other saprotrophic competitors. Isolates of *Trichoderma* are capable of growing in both acid and alkaline soils which doubtless also contributes to their recolonization ability. Augmentation of the effect of a biocontrol agent with a chemical treatment may result in death of the antagonist unless it can be induced to resist the chemical in some way. The induction of benomyl-resistant mutants of *Trichoderma* spp., however, by irradiation with ultraviolet light increases their biocontrol efficiency (Papavizas, 1985). Thus genetic manipulation could be of significance in this respect.

7.4.3 Biological protection of wood products

The preservation of wood products in external use is desirable to prolong their service life. The traditional approach is to infiltrate the product with creosote, pentachlorophenol, or arsenical preservatives. Because of the hazards that these compounds present to the environment, efforts are being made to find new treatment methods. One approach has been to use antagonistic fungi to inhibit the growth of the wood-rotting fungi that cause biodeterioration, thus reducing their inoculum potential and their ability to invade the wood and cause significant damage. This strategy, biological protection, has yet to be used extensively in commercial situations, but a clear potential has been demonstrated (Freitag *et al.*, 1991). Two genera, *Scytalidium* and *Trichoderma*, have emerged as the favoured potential bioprotectants or biocontrol organisms for the commercial protection of wood. 'Bioprotection' in this sense is the use of a biological agent to render wood products more capable of resisting microbial attack, whereas 'biocontrol' is the use of biological agents to restrict the growth of damaging microorganisms within the wood before significant defects arise (Freitag *et al.*, 1991). *Scytalidium* species were favoured because of their ability to produce diffusible antibiotics such as scytalidin (Strunz *et al.*, 1972), whereas *Trichoderma* species were used because of their combined competitive saprotrophic, antibiotic and mycoparasitic abilities described earlier (Chapter 3). These organisms were employed mainly because of good, short-term performance in agar plate or small wood sample laboratory tests; their overall ability to arrest decay often decreased dramatically in field tests (Freitag *et al.*, 1991). The major difficulty encountered in field trials has been the inability to obtain uniform colonization of the substrate by bioprotectants (Bruce *et al.*, 1990). The prophylactic application of *Trichoderma* species to undecayed creosoted electricity distribution poles in the UK remains the pioneering work in this area. In this particular example of the use of antagonistic fungi in disease control, advantage is taken of the knowledge that volatile fungistatic and fungicidal materials produced by species of *Trichoderma* inhibit growth and induce lysis of the mycelium of *Neolentinus lepideus*, a toadstool capable of growing on and decaying wood treated with the preservative creosote (Bruce *et al.*, 1984). Wooden poles are usually decayed at ground level, the decay commencing in sapwood which may not have been impregnated by the creosote treatment. The potential importance of *Trichoderma* species as agents of the biocontrol of *N. lepideus* is striking because creosoted wood is used in the construction of pit props, railway line sleepers and telegraph poles in countries throughout the world. This method of preventing the premature failure of telegraph poles through decay by *N. lepideus* could be valuable if it is shown to be commercially viable. For example, the electricity supply industries in the United Kingdom alone are responsible for the maintenance of an estimated 6,000,000 poles (Bruce and King, 1986a,b). In practice, the antagonist has to be capable of surviving the low temperatures associated with winter in the Northern Hemisphere in order to become established in the centre of the pole and it has been shown to do this under laboratory conditions using wood block tests. Volatile exudates were implicated as responsible for the killing effect of *Trichoderma* because the wood blocks impregnated with *Trichoderma* retained their ability to withstand colonization by *N. lepideus* after death of the antagonist and leaching of water soluble antibiotics and nutrients (Bruce and King, 1986a). The retention of protection after death of the antagonist also means that it would not be necessary for it to remain alive in the wood in order to inhibit the growth of *N. lepideus* (Bruce and King, 1986b). On a global basis, the potential for this method of biocontrol is enormous.

In field studies, creosoted distribution poles were inoculated with a commercial bioprotection product BINAB FYT™, containing a mixture of *Trichoderma polysporum*, *T. harzianum* and *Scytalidium* FY. Although the *Trichoderma* became established in the interior of undecayed poles, it failed to spread adequately throughout the ground region of some poles, even after incubation in the field for six years. Colonization failure was related to the population density of other non-decay fungi resident in poles, which had a detrimental effect on the extent of bioprotection when the poles were subsequently subjected to attack by *N. lepideus*. Wood material sampled from the areas of poles successfully colonized by *Trichoderma* was still decay-resistant in soil block tests (Bruce and Highley, 1989). These results indicated that *Trichoderma* can remain viable and continue to protect the wood for periods of up to seven years, and suggest that the general approach of bioprotection of poles against specific decay fungi is valid if improved colonization by the bioprotectant can be achieved (Bruce *et al.*, 1990).

BINAB FYT™ was also tested against several common decay fungi on utility poles in North America (Morrell and Sexton, 1990). Brown-rot fungi were inhibited in laboratory studies, but white-rot fungi appeared to be more resistant. Use of this product for bioprotection may thus result in a change in the population of decay organisms present on products rather than prevention of decay (Freitag *et al.*, 1991). Nevertheless, it is likely that the use of a mixture of organisms will be essential for successful development of wood bioprotectants so that a variety of different mechanisms can be exploited in the antagonism of wood-decay fungi. In addition, their activities might be supplemented by the use of fluoride or boron-based chemicals which could selectively inhibit decay fungi without affecting the bioprotection agents.

The remedial treatment of decaying wood is a more difficult prospect than prevention of initial colonization, and it is in this context that most experiments have been conducted with *Scytalidium* strains, especially *Scytalidium* 'FY', an undescribed species first isolated from undecayed Douglas fir poles in Oregon (Ricard *et al.*, 1969). This fungus was shown to inhibit growth and subsequent decay associated with *Antrodia* (= *Poria*) *carbonica* in small-scale field and laboratory studies. Subsequent field trials again produced mixed results, largely due to the small numbers of poles treated and further research was suspended in favour of conventional fungicides (Graham, 1973). *Scytalidium* appears to be antagonistic due to the production of diffusible, water-soluble antibiotics. It is also of interest that other workers (e.g. Mercer and Kirk, 1984a,b) have attempted to control fungi causing wood decay in wounds by living trees. One of the most successful organisms in this respect was *Trichoderma viride.*

7.5 Prospects for Use in Agriculture

Although the use of fungal antagonists to control fungal plant pathogens has not dramatically solved agricultural problems so far, it is likely that significant advances will be made over the coming decades. Adoption of fungal biocontrol agents for use against plant disease will be slow due to the complexities of the interactions involved. The use of fungi in the biocontrol of other plant pests has indicated the potential that is available. Successful examples of mycoherbicides and mycoinsecticides are now available (e.g. Burge, 1988; Whipps and Lumsden, 1989) and their use will increase dramatically in the future. It has taken several years for the use of microbial pesticides to become acceptable for agricultural use, but the major hurdles of consumer resistance are being broken down as people become more aware of the environmental consequences of overuse of agrochemicals. An increased understanding is necessary to overcome some of the resilient problems of biocontrol agents, such as their unpredictability and the difficulties in formulation. More information is needed about the basic ecology of fungal interactions in the natural environment such that the new techniques of genetic manipulation and strain improvement can be utilized to enhance the performance of the chosen agents. It is also important that the successful examples of the use of biocontrol agents are introduced into existing agricultural practices to set precedents for industry and crop producers to show that this approach is worthy of the efforts to register and commercialize such products (Lumsden and Lewis, 1989). Antagonistic fungal interactions will thus have important applications in the future development of the agricultural industry in addition to being a fascinating area of study for their own sake.

APPENDIX

Table A.1. Some fungicolous fungi and their hosts. (Compiled from Deighton, 1969; Deighton and Pirozynski, 1972.)

Fungicolous fungi	Host fungi
Acremoniula sarcinellae	*Clypeolella* sp.
	C. ricini
	Sarcinella sp.
	S. fumago
	S. oreophila
	Schniffnerula periplocae
Acrostaurus turneri	*Asterina* sp.
	Cirsosia manaosensis
Annellodochium ramulisporum	*Diatrype* sp.
Atractilina asterinae	*Asterina diplocarpa*
	A. tertia var. *africana*
A. parasitica	*Amazonia tetrorchidii*
	Asteridiella sp.
	A. anastomosans
	A. combreti var. *leonensis*
	A. deightonii
	A. naucleae var. *libericae*
	A. melastomatacearum
	A. premnae
	A. sarcocephali
	A. terminaliae

Table A.1. continued

Fungicolous fungi	Host fungi
A. parasitica continued	*A. tetracerae*
	A. thunbergiae-chrysopis
	Balladyna sp.
	B. velutina
	Irenopsis cryptocarpa
	Meliola sp.
	M. affinis
	M. artabotrydis
	M. baphiae-polygalaceae
	M. bicornis
	M. brillantaisiae
	M. capensis var. *allophyllicola*
	M. cissi
	M. clerodendri
	M. clerodendricola
	M. coffeae
	M. cookeana
	M. cookeana var. *viticis*
	M. dieffenbachiae
	M. garcinae
	M. hyptidis
	M. inermis (= *Asteridiella anastomosans* var. *macilenta*)
	M. khayae var. *minor*
	M. kibirae var. *randiae*
	M. landolphiae-floridеae
	M. landolphiicola
	M. lasiotricha
	M. maesobotryae
	M. memecyli
	M. millettiae-rhodanthae
	M. mitragynicola
	M. newbouldiae
	M. opposita var. *africana*
	M. panici
	M. psychotriae
	M. psychotriae var. *coffeae*
	M. psychotriae var. *longiseta*
	M. sempeiensis
	M. setariae
	M. simillima
	M. sorindeiae
	M. soroceae

Table A.1. continued

Fungicolous fungi	Host fungi
A. parasitica continued	*M. tecleae* var. *toddaliae-asiaticeae* *M. tenella* var. *atalantiae* *M. trichiliae* *M. woodiana*
Calcarisporium echinosporum	*Meliola millettiae-rhodanthae*
C. setiphilum	*Meliola clerodendri*
Chionomyces chorleyi	*Meliola capensis* var. *lecaniodisci* *M. capensis* var. *pancoviae* *M. capensis* var. *thomasii* *M. chaetachmes* *M. chorleyi* *M. hirsuta* *M. khavae* *M. landolphiae* *M. landolphiae-florideae*
C. meliolicola	*Amazonia tetrorchidii* *Irenopsis seyboensis* *Meliola capensis* var. *thomasii* *M. loranthi* var. *bangwensis* *M. rhois* var. *africana* *M. rigida* var. *ugandae* *M. tecleae*
C. sclerochitonis	*Puccinia holosericea*
Cladosporiella asterinae	*Asterina contigua*
C. uredinicola	*Puccinia eucomi* *Ravenelia albigiae* var. *zygiae*
C. uredinis	*Puccinia polygoni-amphibii* *P. scleriae* *P. solmsii* *P. thaliae*
Cylindrocarpon luteoviride	'*Microthelia*' sp.
C. macrosporum	*Asteridiella melastomatacearum* *A. sepulta* *Asterina aulica* *Balladyna magnifica* *Dothidella derridis* *Lembosina* sp. *Meliola* sp.

Table A.1. continued

Fungicolous fungi	Host fungi
C. macrosporum continued	*M. bicornis*
	M. brachyodonta var. *dummeri*
	M. clerodendri
	M. holarrhenicola
	M. malacotricha
	Millettia barteri
	Parasterina cynophallophorae
	Parmularia hankei
	Phyllachora sp.
	P. acaciae subsp. *leonensis*
	P. neurophila
	Rhytidenglerula homalanthi
	R. trematis
	Stigmochora albiziae
C. ugandense	*Phyllachora ficuum*
Domingoella asterinarum	*Asterina* sp.
	A. acalyphae
	A. capparidis
	A. centropogonis
	A. connectilis
	A. cubensis
	A. diplocarpa
	A. drimydicola
	A. excoecariae
	A. hendersonii
	A. homaliicola
	A. lawsonii
	A. manihotis
	A. pulchella
	A. pycnanthi
	A. scruposa var. *longipoda*
	A. spectabilis
	A. thunbergiicola
	A. trichiliae
	A. turraeae
	Asterolibertia sp.
	A. burchelliae
	A. parinarii
	Diedickea piptadeniae
	Englerula sp.
	Hysterostomella tetracerae
	Linotexis deightonii

Table A.1. continued

Fungicolous fungi	Host fungi
Elletevera parasitica	*Phyllachora* sp.
	P. graminis
Eriocercospora balladynae	*Asterina* sp.
	A. hendersonii
	Balladyna sp.
	B. tenuis
	B. velutina
	Balladynocallia glabra
	Balladynopsis entebbeensis
	Clypeolella sp.
	Sarcinella sp.
	Schiffnerula sp.
	S. hippocrateae
	S. mirabilis
	S. periplocae
	S. toddaliae
Eriomycopsis biseptata	*Irenina entebbeensis*
E. bonplandi	*Appendiculella natalensis* var. *ugandensis*
	Asteridiella hansfordii var. *densa*
	A. scabra
	A. terminaliae
	A. trematis
	A. voacangae
	Irenopsis scutiae
	Meliola sp.
	M. bicornis
	M. citricola var. *amyridis*
	M. cylindropoda
	M. ganglifera
	M. groteana
	M. landolphiae
	M. maitlandii
	M. millettiae-rhodanthae
	M. motandrae
	M. ostreoderridis
	M. scolopiae
	M. thunbergiicola
	M. ventilaginicola
	M. voacangae
	M. voacangae var. *conopharyngiae*
E. flagellata	*Asteridiella* sp.

Table A.1. continued

Fungicolous fungi	Host fungi
E. flagellata continued	*A. cyclopoda*
	A. entebbeensis
	A. hansfordii
	A. sarcocephali
	A. scabra
	A. trematis
	Asterina aburiensis
	Balladyna velutina
	Meliola sp.
	M. artabotrydis
	M. chlorophorae
	M. cissi
	M. popwiae
	M. ramicola
	M. rigida var. *ugandae*
	M. theissenii
	M. voacangae var. *carpodini*
E. meliolinae	*Meliolina* sp.
E. minima	*Asteridiella entebbeensis*
	A. hansfordii var. *densa*
	A. naucleae var. *libericae*
	A. sarcocephali
	A. terminaliae
	Irenopsis sp.
	I. colicola
	I. cryptocarpa
	I. macarangae
	I. tortuosa
	Meliola sp.
	M. albiziae
	M. albiziae var. *zygiae*
	M. anisophylleae
	M. bicornis
	M. butleri
	M. calami
	M. capensis var. *thomasii*
	M. comata
	M. crotonis-macrostachyi
	M. dioscoreicola
	M. groteana
	M. heudelotii
	M. mezoneuri

Table A.1. continued

Fungicolous fungi	Host fungi
E. minima continued	*M. paulliniae* *M. pterocarpicola* *M. rhois* var. *africana* *M. sorindeiae* *M. tenella* var. *atalantiae*
E. minuta	*Melanographium citri*
Irpicomyces schiffnerulae	*Schiffnerula solani*
Monosporiella meliolicola	*Meliola bidentata*
Paratrichoconis chinensis	*Asterina linderae*
Pseudofusidium hansfordii	*Mycovellosiella* sp.
Redbia pucciniicola	*Puccinia holosericea*
Scolecobasidium pusillium	*Exosporium stilbaceum*
Spermatoloncha maticola	*Meliola clerodendricola*
Spermosporella aggregata	*Meliola geniculata* var. *eggelingii*
S. pulvinata	*Dexteria pulchella* *Meliola deinbolliae*
Stenospora uredinicola	*Puccinia kraussiana*
Sympodiophora didyma	*Dothidella derridis*
S. meliolae	*Meliola paulliniae*
S. mycophila	*Bulgaria* sp. *Hirneola* sp. *Marasmius* sp.
S. pulchella	*Dothidella derridis*
S. stereicola	*Stereum hirsutum*
S. tenuis	*Paradioposis stevensii*
S. varanasiensis	*Cercospora vestita*
S. venezuelensis	*Colletotrichum* sp.
Trichoconis africiana	*Asterina* sp. *Irenopsis* sp. *I. aciculosa* *I. leeae* *I. njalaensis* *Meliola* sp.

Table A.1. continued

Fungicolous fungi	Host fungi
Trichoconis africiana continued	*M. baphiae-polygalaceae*
	M. bicornis
	M. capensis var. *allophyllicola*
	M. canthii var. *leonensis*
	M. chaetachmes
	M. clerodendri
	M. clerodendricola
	M. densa
	M. hirsuta
	M. homalii-dolichophylli
	M. kibirae var. *randiae*
	M. linacearum
	M. lychnodisci
	M. mitragynicola
	M. paulliniae
	M. reflexa
	M. simillima
	M. thalliformis
	M. voacangae var. *carpodini*
	M. xylopiae var. *leonensis*
T. appendiculata	*Asterina* sp.
	Asterinella tecleae
T. caudata	*Botryosphaeria rhodina*
T. englerulae	*Englerula macarangae*
T. hamata	*Meliola* sp.
T. hibernica	*Appendiculella calostroma*
T. malloti	*Clypeolella craterispermi*
	C. gymnosporiae
	C. ricini
	C. salaciae
	C. toddaliae
	C. ugandensis
	Sarcinella sp.
	Schiffnerula sp.
	S. brideliae
	S. hippocrateae
	S. mirabilis
	S. viticis
T. schiffnerulae	*Sarcinella* spp.

Table A.1. continued

Fungicolous fungi	Host fungi
T. schiffnerulae continued	*Schiffnerula allophyli* *S. cassiae* *S. etandrophragmitis* *S. hippocrateae* *S. mirabilis* *S. solani* *S. spectabilis* *S. toddaliae*
T. sigmoidea	*Meliola kawandensis*
T. trichiliae	*Asterina* sp. *Asterinella* sp.
T. viridula	*Clypeolella ricini* *Schiffnerula mirabilis* *S. solani*
Triposporina uredinicola	*Puccinia periodica*
Tuberculispora jamaicensis	*Irenopsis aciculosa* *I. cryptocarpa*
Vermispora grandispora	*Irenopsis aciculosa*

GLOSSARY

Agaric: Mushrooms and toadstools classified in the Agaricales.

Agaricoid: A sporome resembling a mushroom.

Anamorph: The asexual (imperfect) state of a fungus with a pleomorphic life cycle.

Antagonism: Either direct involving physical contact of two fungi or indirect and effected through the diffusion of antibiotics and lytic enzymes between fungi not in direct contact.

Antagonistic: An interaction in which either one or each partner in a symbiotic association adversely affects the other(s).

Antibiosis: The inhibitory effect of volatile or non-volatile inhibitory or toxic compounds produced by and active against competing fungi.

Apoplast(ic): The region of an organism which comprises the walls and extra-plasmalemmal zone through which solutes can be transported.

Appressorium: Hyphal enlargement of a parasite, adpressed to the host, from which a penetration peg develops.

Arbuscular Mycorrhizal Fungus (AMF): Produces arbuscules, fine haustorioid hyphal branches, and frequently vesicles in the root cortex cells of the plant symbiont.

Ascoma: The ascus-containing structure of the Ascomycota.

Auxotrophy: Requiring specific chemicals essential for growth, especially biotrophic mycoparasites.

Axenic Culture: Growing alone, in the absence of a host.

Basidiome: The basidiospore producing apparatus of homobasidiomycetes.

Biocontrol: Used here to refer to the utilization of a fungus, sometimes in conjunction with chemical treatment (integrated control), to control the development of plant-pathogenic fungi. In the broad sense, biocontrol refers to any organism used to control the development of organisms potentially detrimental to the activity of humans.

Biotrophic Parasite: Obtains nutrients entirely from the cytoplasm of a living host.

Commensalism: A neutral mutual relationship of dissimilar organisms which live in close association.

Competitive Interaction: Mycelium of one fungal species in competition with that of another affects and may itself also be affected by the antagonist.

Competitive Saprotrophic Ability: The summation of physiological characteristics that make for success in competitive colonization of dead organic substrates.

Conidium: An exogenously-formed asexual spore.

Contact Biotroph: See Fusion Biotroph.

Contact Necrotroph: Contacts and kills the host without penetrating it, then obtains nutrients from the dead cytoplasm.

Cytotoxin: Poisonous to protoplasm or specific organelles.

Destructive Mycoparasite: See Necrotrophic Mycoparasite.

Dimorphic: A fungus which can exist in either a filamentous or a yeast-like growth form.

Endobiotic Chytrid: The thallus grows internal to the wall of the host.

Epibiotic Chytrid: The thallus grows on the external surface of the host.

Epiphyte: An organism which lives on the surface of a plant, usually the leaves and stem.

Exotoxin: Poisonous compound which acts external to the hyphae in which it has been produced.

Fungicolous: Any interfungal association regardless of the trophic or symbiotic nature of the relationship.

Fungistasis: The inhibition of fungal growth; a phenomenon often found in soil whereby the germination of spores is inhibited (= mycostasis).

Fusion Biotroph: Produces a buffer host contact cell through which nutrients are probably transferred to the mycoparasite via direct cytoplasmic channels.

Gall Formation: The induction of swellings and abnormal branches, often complex morphologically, in the hyphae of one of the partners in a symbiotic association by another member of the association.

Haustorium: A short intracellular hyphal branch believed to mediate the transfer of nutrients from the cytoplasm of a host to that of a biotrophic parasite.

Heterotrophic: Requiring an organic source of energy.

Host–Parasite Interface: The region of apposition or contact through which materials are transferred between host and parasite.

Hyperparasitic: An association in which a parasite is itself parasitized by another parasitic organism.

Hypcrtrophy: Abnormal growth of hyphae resulting in gall formation.

Hyphal Interference: A reduction in growth rate of hyphae caused by the close proximity of hyphae of an unrelated fungus, often resulting in death.

Hypovirulence: The reduction of the phytopathogenicity through the physiological combination of a virulent strain of a pathogen with a strain of lower virulence.

Intracellular Biotroph: A biotrophic mycoparasite whose protoplast resides entirely within the hyphae of the host, as in Chytridiales.

Invasive Necrotroph: Penetrates the host, kills it and lives on the dead cytoplasm.

Lichenicolous Fungus: A fungus which inhabits a lichen.

Mutualistic: A relationship in which the partners derive mutual benefit.

Mycobiont: A fungus regularly associated with another living organism as, for example, the fungal partner in a lichen.

Mycoherbicide: A plant-pathogenic fungus used in the biocontrol of weeds.

Mycolysis: Refers to the enzymic breakdown of fungal hyphae in unsterile soil.

Mycoparasitism: An interfungal association in which the host provides the source of nutrients for the parasite which may depend on the host for its survival.

Mycopathogen: A pathogen which infects a fungus.

Mycophilic Fungus: A fungus which associates constantly with one or more other species of fungi.

Mycophthorous: See Mycophilic Fungus.

Mycotrophein: An unidentified compound from a host fungus which stimulates the growth of a mycoparasite.

Mycotrophic: An interfungal relationship in which the nutritional aspect of the association is stressed.

Myxomyceticolous: Living in association with the sporangial stage of a myxomycete.

Myxomyceticolous Lichen: Lichen in which a myxomycete is the fungal symbiont.

Neckband: An electron-dense collar around the neck of a haustorium which seals the plasma membrane to the haustorial apparatus.

Necrotrophic Mycoparasite: Utilizes nutrients from dead elements of the host killed through the activity of the parasite.

Neutralistic: A relationship which is neither advantageous nor detrimental to the members of the association.

Papilla: An apposition of wall-like material laid down around the region where a wall is damaged.

Parasitic: Living in association with, deriving nutrients from, and usually considered to harm the host.

Pathogenic: Having the ability to cause disease.

Penetration Jacket: An electron-dense sleeve formed around the infection peg of certain pathogenic fungi.

Penetration Peg: A fine parasitic hypha which penetrates the wall of a potential host.

Photobiont: The photosynthetic partner in a lichen.

Phylloplane Fungi: Fungi which live in contact with the surface of a living leaf.

Primary Resource Capture: The initial colonization of a substrate by a fungus.

Rhizosphere Competence: The ability of a fungus applied to a seed to establish in the rhizosphere once the seed germinates.

Saprotrophic: Deriving nutrients from dead organic matter.

Secondary Resource Capture: The replacement, through competition, of one species of fungus by another within a niche.

Sporome: A spore-producing structure.

Sporophagous Hymenomycetes: Parasitize the basidiospores of other hymenomycetes.

Symbiosis: An association of two or more organisms which may be advantageous, disadvantageous or neutral to any of the partners.

Teleomorph: The sexual (perfect) state.

Weed Mould: A fungus which grows within the beds of cultivated mushrooms and usually reduces the yield.

Y-Phase: Yeast-phase produced by dimorphic fungi.

References

Adams, P.B. (1982) Biological control of *Sclerotinia* lettuce drop in the field by *Sporidesmium sclerotivorum. Phytopathology* 72, 485–488.

Adams, P.B. (1987) Comparison of isolates of *Sporidesmium sclerotivorum in vitro* and in soil for potential as active agents in microbial pesticides. *Phytopathology* 77, 575–578.

Adams, P.B. (1990) The potential of mycoparasites for biological control of plant diseases. *Annual Review of Phytopathology* 28, 59–72.

Adams, P.B. and Ayers, W.A. (1979) Mycoparasitism of sclerotia of *Sclerotinia* and *Sclerotium* species by *Sporidesmium sclerotivorum. Canadian Journal of Microbiology* 25, 17–23.

Adams, P.B. and Ayers, W.A. (1983) Histological and physiological aspects of infection of sclerotia of two *Sclerotinia* species by two mycoparasites. *Phytopathology* 73, 1072–1076.

Adams, P.B. and Ayers, W.A. (1985) The world distribution of the mycoparasites *Sporidesmium sclerotivorum, Teratosperma oligocladum* and *Laterispora brevirama. Soil Biology and Biochemistry* 17, 583–584.

Adams, P.B. and Fravel, D.R. (1990) Economic biological control of *Sclerotinia* lettuce drop by *Sporidesmium sclerotivorum. Phytopathology* 80, 1120–1124.

Adams, P.B., Marois, J.J. and Ayers, W.A. (1984) Population dynamics of the mycoparasite, *Sporidesmium sclerotivorum*, and its host, *Sclerotinia minor*, in soil. *Soil Biology and Biochemistry* 16, 626–633.

Adams, P.B., Ayers, W.A. and Marois, J.J. (1985) Energy efficiency of the mycoparasite *Sporidesmium sclerotivorum in vitro* and in soil. *Soil Biology and Biochemistry* 17, 155–158.

Ahmad, J.S. and Baker, R. (1987a) Rhizosphere competence of *Trichoderma harzianum. Phytopathology* 77, 182–189.

Ahmad, J.S. and Baker, R. (1987b) Competitive saprophytic ability and celluloytic activity of rhizosphere competent mutants of *Trichoderma harzianum. Phytopathology* 77, 358–362.

Ahmad, J.S. and Baker, R. (1988) Rhizosphere competence of benomyl-tolerant mutants of *Trichoderma* spp. *Canadian Journal of Microbiology* 34, 694–696.

Ainsworth, A.M. and Rayner, A.D.M. (1991) Ontogenetic stages from coenocyte to basidiome and their relation to phenoloxidase activity and colonization processes in *Phanerochaete magnoliae. Mycological Research* 12, 1414–1422.

Ainsworth, G.C. (1971) *Ainsworth and Bisby's Dictionary of the Fungi*, 6th Edn. CAB International, Wallingford, 663 pp.

Alfen, N.K., van (1982) Biology and potential for disease control of hypovirulence of *Endothia parasitica. Annual Review of Phytopathology* 20, 349–362.

Al-Hamdani, A.M. and Cooke, R.C. (1983) Effects of the mycoparasite *Pythium oligandrum* on cellulolysis and sclerotium production by *Rhizoctonia solani. Transactions of the British Mycological Society* 81, 619–621.

Al-Hamdani, A.M., Lutchmeah, R.S. and Cooke, R.C. (1983) Biological control of *Pythium ultimum*-induced damping-off by treating cress seed with the mycoparasite *Pythium oligandrum. Plant Pathology* 32, 449–454.

Allen, D.J. (1982) *Verticillium lecanii* on the bean rust fungus, *Uromyces appendiculatus. Transactions of the British Mycological Society* 79, 362–364.

Aluko, M.O. and Hering, T.F. (1970) The mechanisms associated with the antagonistic relationship between *Corticium solani* and *Gliocladium virens. Transactions of the British Mycological Society* 55, 173–179.

Anke, H. and Sterner, O. (1988) Transformation of isovelleral by the parasitic fungus *Calcarisporium arbuscula. Phytochemistry* 27, 2765–2767.

Anon. (1991) Biocontrol fungus registered for control of damping-off. *Phytopathology News* 25, 48.

Armentrout, V.N. and Wilson, C.L. (1969) Haustorium–host interaction during mycoparasitism of *Mycotypha microspora* by *Piptocephalis virginiana. Phytopathology* 59, 897–905.

Ayers, T.T. (1933) Growth of *Dispira cornuta* in artificial culture. *Mycologia* 25, 333–341.

Ayers, T.T. (1935) Parasitism of *Dispira cornuta. Mycologia* 27, 235–261.

Ayers, W.A. and Adams, P.B. (1978) Mycoparasitism of sclerotia of *Sclerotinia* and *Sclerotium* species by *Sclerotiorum sclerotiovorum. Canadian Journal of Microbiology* 25, 17–23.

Ayers, W.A. and Adams, P.B. (1979) Factors affecting germination, mycoparasitism, and survival of *Sporidesmium sclerotivorum. Canadian Journal of Microbiology* 25, 1021–1026.

Ayers, W.A. and Adams, P.B. (1981) Mycoparasitism of sclerotial fungi by *Teratosperma oligocladum. Canadian Journal of Microbiology* 27, 886–892.

Ayers, W.A. and Adams, P.B. (1983) Improved media for growth and sporulation of *Sporidesmium sclerotivorum. Canadian Journal of Microbiology* 29, 325–330.

Ayers, W.A. and Lumsden, R.D. (1977) Mycoparasitism of oospores of *Pythium* and *Aphanomyces* species by *Hyphochytrium catenoides. Canadian Journal of Microbiology* 23, 38–44.

Ayers, W.A., Lee, S.P., Tsuneda, A. and Hiratsuka, Y. (1980) The isolation, identification and bioassay of the antifungal metabolites produced by *Monicillium nordinii. Canadian Journal of Microbiology* 26, 766–773.

Aylmore, R.C. and Todd, N.K. (1984) Hyphal fusion in *Coriolus versicolor.* In: Jennings, D.H. and Rayner, A.D.M. (eds) *The Ecology and Physiology of the Fungal Mycelium.*

Cambridge University Press, Cambridge, pp. 103–125.

Baard, S.W. (1988) Necrotrophic mycoparasitism of *Chalara elegans* (*Thielaviopsis basicola*) by a sterile basidiomycete. *Journal of Phytopathology* 122, 166–173.

Backman, P.A. and Rodriguez-Kabana, R. (1975) A system for the growth and delivery of biological control agents to the soil. *Phytopathology* 65, 819–821.

Backus, M.P. and Stowell, E.A. (1953) A *Fusidium* disease of *Xylaria* in Wisconsin. *Mycologia* 45, 836–847.

Bahl, N. and Chowdhry, P.N. (1981) *Podospora faurelii*, a new competitor in the mushroom (*Volvariella volvacea*) cultivation. *Current Science* 50, 378.

Baker, K.F. and Cook, R.J. (1974) *Biological Control of Plant Pathogens.* W.H. Freeman, San Francisco, 433 pp.

Baker, K.L., Neneke, E.S., Hooper, G.R. and Fields, W.G. (1977) Host range and axenic culture of the mycoparasite *Syncephalis sphaerica* (Mucorales). *Mycologia* 69, 1008–1015.

Baker, R. (1988) *Trichoderma* spp. as plant-growth stimulants. *CRC Critical Reviews in Biotechnology* 7, 97–106.

Baker, R. (1989) Improved *Trichoderma* spp. for promoting crop productivity. *Trends in Biotechnology* 7, 34–38.

Balasubramanian, R. and Manocha, M.S. (1986) Proteinase, chitinase, and chitosanase activities in germinating spores of *Piptocephalis virginiana. Mycologia* 78, 157–163.

Bandoni, R.J. (1956) A preliminary survey of the genus *Platygloea. Mycologia* 48, 821–840.

Bandoni, R.J. (1961) The genus *Naematelia. American Midland Naturalist* 66, 319–328.

Bandoni, R.J. (1984) The Tremellales and Auriculariales: an alternative classification. *Transactions of the Mycological Society of Japan* 25, 489–530.

Bandoni, R.J. (1986) On an undescribed basidiomycetous anamorph. *Windahlia* 16, 53–58.

Bandoni, R.J. (1987) Taxonomic overview of the Tremellales. *Studies in Mycology* 30, 87–110.

Bandoni, R.J. and Oberwinkler, F. (1983) On some species of *Tremella* described by Alfred Möller. *Mycologia* 75, 854–863.

Bandoni, R.J. and Zang, M. (1990) On an undescribed *Tremella* from China. *Mycologia* 82, 270–273.

Barak, R. and Chet, I. (1986) Determination, by fluorescein diacetate staining, of fungal viability during mycoparasitism. *Soil Biology and Biochemistry* 18, 315–319.

Barak, R. and Chet, I. (1990) Lectin of *Sclerotium rolfsii*: its purification and possible function in fungal–fungal interaction. *Journal of Applied Bacteriology* 69, 101–112.

Barak, R., Elad, Y., Mirelman, D. and Chet, I. (1985) Lectins: a possible basis for specific recognition in the interaction between *Trichoderma* and *Sclerotium rolfsii. Phytopathology* 75, 458–462.

Barker, S.M. and Barnett, H.L. (1973) Nitrogen and vitamin requirements for axenic growth of the haustorial mycoparasite *Dispira cornuta. Mycologia* 65, 21–27.

Barnett, H.L. (1963) The nature of mycoparasitism by fungi. *Annual Reviews of Microbiology* 17, 1–14.

Barnett, H.L. (1968) The effects of light, pyridoxine and biotin on the development of the mycoparasite, *Gonatobotryum fuscum. Mycologia* 60, 244–251.

Barnett, H.L. (1970) Nutritional requirements for axenic growth of some haustorial mycoparasites. *Mycologia* 62, 750–760.

Barnett, H.L. and Ayers, W.A. (1981) Nutritional and environmental factors affecting growth and sporulation of *Sporidesmium sclerotivorum. Canadian Journal of Microbiology* 27, 685–691.

Barnett, H.L. and Binder, F.L. (1973) The fungal host–parasite relationship. *Annual Review of Phytopathology* 11, 273–292.

Barnett, H.L. and Lilly, V.G. (1958) Parasitism of *Calcarisporium parasiticum* on species of *Physalospora* and related fungi. *Bulletin of the West Virginia Agricultural Experiment Station* 420T, 1–36.

Barr, D.J.S. and Bandoni, R. (1979) A new species of *Rozella* on a basdiomycete. *Mycologia* 71, 1261–1264.

Barron, G.L. and Fletcher, J.T. (1972) *Rhopalomyces elegans* Corda, a host of *Mycogone perniciosa* Magn. *Mushroom Science* 8, 383–386.

Bartnicki-Garcia, S. (1968) Cell wall chemistry, morphogenesis, and taxonomy of fungi. *Annual Review of Microbiology* 22, 87–108.

Bary, A. de (1887) *Comparative Morphology and Biology of the Fungi Mycetozoa and Bacteria.* Clarendon Press, Oxford.

Bauer, R. and Oberwinkler, F. (1990a) Direct cytoplasm-cytoplasm connection: an unusual host–parasite interaction of the tremelloid mycoparasite *Tetragoniomyces ulginosus. Protoplasma* 154, 157–160.

Bauer, R. and Oberwinkler, F. (1990b) Haustoria of the mycoparasitic heterobasidiomycete *Christiansenia pallida. Cytologia* 55, 419–424.

Bauer, R. and Oberwinkler, F. (1991) The colacosomes: new structures at the host–parasite interface of a mycoparasitic basidiomycete. *Botanica Acta* 104, 53–57.

Bawcutt, R.A. (1983) Biology of *Syncephalis leadbeateri* sp. nov. (Mucorales) and observations on *Syncephalis. Transactions of the British Mycological Society* 80, 219–230.

Beagle-Ristaino, J.E. and Papavizas, G.C. (1985) Survival and proliferation of *Trichoderma* spp. and *Gliocladium virens* in soil and in plant rhizospheres. *Phytopathology* 75, 729–732.

Beaton, G., Pegler, D.N. and Young, T.W.K. (1985) Gasteroid Basidiomycota of Victoria State, Australia. 5, Boletales, *Kew Bulletin* 40, 573–586.

Benjamin, R.K. (1959) The merosporangiferous Mucorales. *Aliso* 4, 321–433.

Benjamin, R.K. (1961) Addenda to the merosporangiferous Mucorales, I. *Aliso* 5, 11–19.

Benjamin, R.K. (1979) Zygomycetes and their spores. In: Kendrick, B. (ed.) *The Whole Fungus*, volume 2. National Museum of Natural Sciences, National Museums of Canada and the Kananaskis Foundation, pp. 573–622.

Benjamin, R.K. (1985) A new genus of the Piptocephalidaceae (Zoopagales) from Japan. *Botanical Journal of the Linnean Society* 91, 117–133.

Benny, G.L. and Benjamin, R.K. (1976) Observations on Thamnidiaceae (Mucorales), II. *Chaetocladium, Cokeromyces, Mycotypha* and *Phascolomyces. Aliso* 8, 391–424.

Berry, C.R. (1958) Parasitism of *Piptocephalis virginiana.* PhD thesis, West Virginia University.

Berry, C.R. (1959) Factors affecting parasitism of *Piptocephalis virginiana* on other Mucorales. *Mycologia* 51, 824–832.

Berry, C.R. and Barnett, H.L. (1957) Mode of parasitism and host range of *Piptocephalis virginiana. Mycologia* 49, 374–386.

Bessey, E.A. (1964) *Morphology and Taxonomy of Fungi.* Hafner Publishing Company, New York, 791 pp.

Bezerra, J.L. and Kimbrough, J.W. (1978) A new species of *Tremella* parasitic on *Rhytidhysterium rufulum*. *Canadian Journal of Botany* 56, 3021–3033.

Bhattacharjee, M., Mukerji, K.G., Tewari, J.P. and Skoropad, W.P. (1982) Structure and hyperparasitism of a new species of *Gigaspora*. *Transactions of the British Mycological Society* 78, 184–188.

Binder, F.L. (1974) Differences in the uptake of glucose and glycerol by *Tieghemiomyces parasiticus* in axenic culture. *Canadian Journal of Botany* 52, 2571–2574.

Binder, F.L. and Barnett, H.L. (1973) Enzymes for carbohydrate catabolism in the mycoparasite *Tieghemiomyces parasiticus*. *Mycologia* 65, 99–1006.

Binder, F.L. and Barnett, H.L. (1974) Amino acid requirements for the axenic growth of *Tieghemiomyces parasiticus*. *Mycologia* 66, 265–271.

Binder, F.L. and Pierce, J.C. (1976) Glycerol uptake by the haustorial mycoparasite *Tieghemiomyces parasiticus* in axenic culture. *Canadian Journal of Botany* 54, 1403–1409.

Bliss, D.E. (1951) The destruction of *Armillaria* in citrus soils. *Phytopathology* 41, 665–683.

Bo, L. and Bau, Y.-S. (1980) Fungi and mycological literature of the Szechwan province in China. *Mycologia* 72, 1117–1126.

Boogert, van den, P.H.J.F. (1989a) Nutritional requirements of the mycoparasitic fungus *Verticillium biguttatum*. *Netherlands Journal of Plant Pathology*, 95, 149–156.

Boogert, van den, P.H.J.F. (1989b) Colonization of roots and stolons of potato by the mycoparasitic fungus *Verticillium biguttatum*. *Soil Biology and Biochemistry* 21, 255–262.

Boogert, van den, P.H.J.F. and Jager, G. (1983) Accumulation of hyperparasites of *Rhizoctonia solani* by addition of live mycelium of *R. solani* to soil. *Netherlands Journal of Plant Pathology* 89, 223–228.

Boogert, van den, P.H.J.F. and Jager, G. (1984) Biological control of *Rhizoctonia solani* on potatoes by antagonists. 3. Inoculation of seed potatoes with different fungi. *Netherlands Journal of Plant Pathology* 90, 117–126.

Boogert, van den, P.H.J.F. and Saat, T.A.W.M. (1991) Growth of the mycoparasitic fungus *Verticillium biguttatum* from different geographical origins at near-minimum temperatures. *Netherlands Journal of Plant Pathology* 97, 115–124.

Boogert, van den, P.H.J.F. and Velvis, H. (1992) Population dynamics of the mycoparasite *Verticillium biguttatum* and its host *Rhizoctonia solani*. *Soil Biology and Biochemistry* 24, 159–164.

Boogert, van den, P.H.J.F., Reinartz, H., Sjollema, K.A. and Veenhuis, M. (1989) Microscopic observations on the interaction of the mycoparasite *Verticillium biguttatum* with *Rhizoctonia solani* and other soil-borne fungi. *Antonie van Leeuwenhoek* 56, 161–174.

Boogert, van den, P.H.J.F., Jager, G. and Velvis, H. (1990) *Verticillium biguttatum*, an important mycoparasite for the control of *Rhizoctonia solani*. In: Hornby, D. (ed.) *Biological Control of Soil-borne Plant Pathogens*. CAB International, Wallingford, pp. 77–91.

Boosalis, M.G. (1956) Effect of soil temperature and green-manure amendment of unsterilized soil on parasitism of *Rhizoctonia solani* by *Penicillium vermiculatum* and *Trichoderma* sp. *Phytopathology* 46, 473–478.

Boosalis, M.G. (1964) Hyperparasitism. *Annual Review of Phytopathology* 2, 363–375.

Botha, W.J. and Eicker, A. (1986) Notes on the physiology and morphology of

Sepedonium niveum, a newly-recorded competitor mould in mushroom compost. *Crop Science* 10, 331–339.

Boudreau, M.A. and Andrews, J.H. (1987) Factors influencing antagonism of *Chaetomium globosum* to *Venturia inaequalis*: a case study in failed biocontrol. *Phytopathology* 77, 1470–1475.

Boyetchko, S.M. and Tewari, J.P. (1991) Parasitism of spores of the vesicular–arbuscular mycorrhizal fungus *Glomus dimorphicum*. *Phytoprotection* 72, 27–32.

Bracker, C.E. and Littlefield, L.J. (1973) Structural concepts of host–pathogen interfaces. In: Byrde, R.J.W. and Cutting, C.V. (eds) *Fungal Pathogenicity and the Plant's Response*. Academic Press, London, New York, pp. 159–317.

Brain, A.P.R., Jeffries, P. and Young, T.W.K. (1982) Ultrastructure of septa in *Tieghemiomyces parasiticus*. *Mycologia* 74, 173–181.

Brefeld, O. (1872) Botanisch untersuchchungen uber Schimmelpilze, Heft 1. Verlag Von Arthur Felix, Leipzig.

Brown, A. (1987) Activity of glucanases of *Zygorrhynchus moelleri* in relation to antagonism against some soil-borne plant pathogenic fungi. *Journal of Phytopathology* 120, 298–309.

Bruce, A. and Highley, T.L. (1989) Decay resistance of wood removed from poles treated with *Trichoderma*. The International Research Group on Wood Preservation. Document No. IRG/WP/1386.

Bruce, A. and King, B. (1986a) Biological control of decay in creosote treated distribution poles. I. Establishment of immunizing commensal fungi in poles. *Material und Organismen* 21, 1–13.

Bruce, A. and King, B. (1986b) Biological control of decay in creosote treated distribution poles. II. Control of decay in poles by immunizing commensal fungi. *Material und Organismen* 21, 165–179.

Bruce, A., Austin, W.J. and King, B. (1984) Control of growth of *Lentinus lepideus* by volatiles from *Trichoderma*. *Transactions of the British Mycological Society* 82, 423–428.

Bruce, A., Fairnington, A. and King, B. (1990) Biological control of decay in creosote treated distribution poles. III. Control of decay in poles by immunizing commensal fungi after extended incubation period. *Material und Organismen* 15, 15–28.

Brückner, H. and Przybylski, M. (1984) Isolation and structural characterization of polypeptide antibiotics of the peptaibol class by hplc with field desorption and fast atom bombardment mass spectrometry. *Journal of Chromatography* 296, 263–275.

Brunk, M.A. and Barnett, H.L. (1966) Mycoparasitism of *Dispira simplex* and *Dispira parvispora*. *Mycologia* 58, 518–532.

Budge, S.P. and Whipps, J.M. (1991) Glasshouse trials of *Coniothyrium minitans* and *Trichoderma* species for the biological control of *Sclerotinia sclerotiorum* in celery and lettuce. *Plant Pathology* 40, 59–66.

Buller, A.H.R. (1924) *Researches on Fungi, Vol. III*. Longmans, Green, New York, 611 pp.

Bullock, S., Adams, P.B., Willetts, H.J. and Ayers, W.A. (1986) Production of haustoria by *Sporidesmium sclerotivorum* in sclerotia of *Sclerotinia minor*. *Phytopathology* 76, 101–103.

Burdsall, H.H. Jr, Hoch, H.C., Boosalis, M.G. and Setiff, E.C. (1980) *Laetisaria arvalis* (Aphyllophorales, Corticiaceae): a possible biological control agent for *Rhizoctonia solani* and *Pythium* species. *Mycologia* 72, 728–736.

Burge, M.N. (ed.) (1988) *Fungi in Biological Control Systems.* Manchester University Press, Manchester, 269 pp.

Burgeff, H. (1924) Untersuchungen uber sexualitat und parasitismus bei Mucorineen. 1. *Botanische Abhandlungen* 4, 1–135.

Butin, H. (1981) *Keissleriella bavarica* sp.nov., a hyperparasite on *Ascodichaena rugosa* Butin. *Phytopathologische Zeitschrift* 100, 186–190.

Butler, E.E. (1957) *Rhizoctonia solani* as a parasite of fungi. *Mycologia* 49, 354–373.

Cal, A. de, M.-Sagasta, E. and Melgarejo, P. (1988) Antifungal substances produced by *Penicillium frequentans* and their relationship to the biocontrol of *Monilinia laxa. Phytopathology* 78, 888–893.

Cal, A. de, M.-Sagasta, E. and Melgarejo, P. (1990) Biological control of peach twig blight (*Monilinia laxa*) with *Penicillium frequentans. Plant Pathology* 39, 612–618.

Calderone, R.A. and Barnett, H.L. (1972) Axenic growth and nutrition of *Gonatobotryum fuscum. Mycologia* 64, 153–160.

Calpouzos, L., Thies, T. and Batlle, C.M.R. (1957) Culture of the rust parasite, *Darluca filum. Phytopathology* 47, 108–109.

Campbell, R. (1985) *Plant Microbiology.* Edward Arnold, London, 191 pp.

Campbell, R. (1986) The search for biological control agents against plant pathogens: a pragmatic approach. *Biological Agriculture and Horticulture* 3, 317–327.

Canter, H.M. and Ingold, C.T. (1984) A chytrid on *Dacrymyces. Transactions of the British Mycological Society* 82, 739–742.

Carling, D.E., Brown, M.F. and Millikan, D.F. (1976) Ultrastructural examination of the *Puccinia graminis–Darluca filum* host–parasite relationship. *Phytopathology* 66, 419–422.

Casulli, F. (1990) *Darluca filum*: un possibili agente per la lotta biologica contro le ruggini. *Phytopathologie Mediterranea*, 29, 196–201.

Chalutz, E. and Wilson, C.L. (1990) Postharvest biocontrol of green and blue mold and sour rot of citrus fruit by *Debaryomyces hansenii. Plant Disease* 74, 134–137.

Chand, T. and Logan, C. (1984) Antagonists and parasites of *Rhizoctonia solani* and their efficacy in reducing stem canker of potato under controlled conditions. *Transactions of the British Mycological Society* 83, 107–112.

Cherif, M. and Benhamou, N. (1990) Cytochemical aspects of chitin breakdown during the parasitic action of *Trichoderma* sp. on *Fusarium oxysporum* f. sp. *radicis-lycopersici. Phytopathology* 80, 1406–1414.

Chesters, C.G.C. and Bull, A.T. (1963) The enzymatic degradation of laminarin. I. The distribution of laminarinase amongst microorganisms. *Biochemical Journal* 86, 28–31.

Chet, I. and Baker, R. (1981) Isolation and biocontrol potential of *Trichoderma hamatum* from soil naturally suppressive to *Rhizoctonia solani. Phytopathology* 71, 286–290.

Chet, I., Harman, G.E. and Baker, R. (1981) *Trichoderma hamatum*: its hyphal interactions with *Rhizoctonia solani* and *Pythium* spp. *Microbial Ecology* 7, 29–38.

Chinn, S.H.F. (1971) Biological effects of Panogen PX in soil on common root rot and growth responses of wheat seedlings. *Phytopathology* 61, 98–101.

Christiansen, M.P. (1959) Danish resupinate fungi. Part I. *Dansk Botanisk Archiv* 19, 1–55.

Chu, F.F. and Wu, W.S. (1981) Antagonistic action of *Trichoderma* spp. and *Penicillium* spp. on *Rhizoctonia solani. Memoirs of the College of Agriculture, National Taiwan University* 21, 4–18.

Chung, Y.R. and Hoitink, H.A.J. (1990) Interactions between thermophilic fungi and *Trichoderma hamatum* in suppression of *Rhizoctonia* damping-off in a bark compost-amended medium. *Phytopathology* 80, 73–77.

Clauzade, G., Diederich, P. and Roux, C. (1989) Nelikenigintaj fungoj likenologaj. Ilustrita determindibro. *Bulletin de la Societe linneenne de Provence, numero* special 1, 1–142. [Keys to 682 lichenicolous fungi.]

Claydon, N., Allen, M., Hanson, J.R. and Avent, A.G. (1987) Antifungal alkyl pyrones of *Trichoderma harzianum. Transactions of the British Mycological Society* 88, 503–513.

Coffey, M.D. (1975) Ultrastructural features of the haustorial apparatus of the white blister fungus *Albugo candida. Canadian Journal of Botany* 53, 1285–1299.

Cole, J.S. and Zvenyika, Z. (1988) Integrated control of *Rhizoctonia solani* and *Fusarium solani* in tobacco transplants with *Trichoderma harzianum* and triadimenol. *Plant Pathology* 37, 271–277.

Coley-Smith, J.R., Ridout, C.J., Mitchell, C.M. and Lynch, J.M. (1991) Control of bottom rot disease of lettuce (*Rhizoctonia solani*) using preparations of *Trichoderma viride, T. harzianum* or tolclofos-methyl. *Plant Pathology* 40, 359–366.

Cooke, R.C. (1977) *Biology of Symbiotic Fungi.* Wiley, Chichester, 282 pp.

Cooke, R.C. and Rayner, A.D.M. (1984) *Ecology of Saprotrophic Fungi.* Longman, London and New York, 539 pp.

Cooke, R.C. and Whipps, J.M. (1980) The evolution of nutrition in fungi parasitic on terrestrial plants. *Biological Reviews* 55, 341–362.

Coppins, B.J. (1987) Two new lichenicolous species of *Opegrapha* from Western Scotland. *Notes from the Royal Botanic Garden Edinburgh* 44, 601–606.

Corke, A.T.K. (1978) Microbial antagonists affecting tree diseases. *Annals of Applied Biology* 89, 89–94.

Cullen, D. and Andrews, J.H. (1984) Evidence for the role of antibiosis in the antagonism of *Chaetomium globosum* to the apple scab pathogen, *Venturia inaequalis. Canadian Journal of Botany* 62, 1819–1823.

Cullen, D., Berbee, F. M. and Andrews, J.H. (1984) *Chaetomium globosum* antagonizes the apple scab pathogen, *Venturia inaequalis*, under field conditions. *Canadian Journal of Botany* 62, 1814–1818.

Cunningham, J.E. and Pickard, M.A. (1985) Maltol, a metabolite of *Scytalidium uredinicola* which inhibits spore germination of *Endocronartium harknessii*, the western gall rust. *Canadian Journal of Microbiology* 31, 1051–1055.

Curtis, F.C., Evans, G.H., Lillis, V., Lewis, D. and Cooke, R.C. (1978) Studies on mucoralean mycoparasites. I. Some effects of *Piptocephalis* species on host growth. *New Phytologist* 80, 157–165.

Cuthbert, A. (1984) Cell wall composition and resistance in haustorial mycoparasites. *Bulletin of the British Mycological Society* 17, suppl. 3, 1–2.

Cuthbert, A. and Jeffries, P. (1984) Mycoparasitism by *Piptocephalis unispora* within the Mortierellaceae. *Transactions of the British Mycological Society* 83, 700–702.

Dadwal, V.S. and Jamaluddin. (1984) A new disease of the white button mushroom (*Agaricus bisporus*). *Current Science* 53, 532–533.

Daft, G.C. and Tsao, P.H. (1983) Susceptibility of *Phytophthora cinnamomi* and *P. parasitica* to fungi known to parasitize other Oomycetes. *Transactions of the British Mycological Society* 81, 71–76.

Daft, G.C. and Tsao, P.H. (1984) Parasitism of *Phytophthora cinnamomi* and *P. parasitica* spores by *Catenaria anguillulae* in a soil environment. *Transactions of the British Mycological Society* 82, 485–490.

Daniels, B.A. and Menge, J.A. (1980) Hyperparasitization of vesicular–arbuscular mycorrhizal fungi. *Phytopathology* 70, 584–588.

Dayal, R. and Barron, G.L. (1970) *Verticillium psalliotae* as a parasite of *Rhopalomyces. Mycologia* 62, 826–830.

Deacon, J.W. (1976) Studies on *Pythium oligandrum*, an aggressive parasite of other fungi. *Transactions of the British Mycological Soiety* 66, 383–391.

Deacon, J.W. (1988a) Biocontrol of soil–borne plant pathogens with introduced inocula. *Philosophical Transactions of the Royal Society, London, B* 318, 249–264.

Deacon, J.W. (1988b) Behavioural responses of fungal zoospores. *Microbiological Sciences* 5, 249–252.

Deacon, J.W. and Henry, C.M. (1978) Mycoparasitism by *Pythium oligandrum* and *Pythium acanthicum. Soil Biology and Biochemistry* 10, 409–415.

Deacon, J.W., Laing, S.A.K. and Berry, L.A. (1991) *Pythium mycoparasiticum* sp. nov., an aggressive mycoparasite from British soils. *Mycotaxon* 42, 1–8.

Deighton, F.C. (1969) Microfungi IV: Some hyperparasitic hyphomycetes, and a note on *Cercosporella uredinophila* Sacc. *Mycological Papers* No 118, 1–41, CAB International, Wallingford.

Deighton, F.C. and Pirozynski, K.A. (1972) Microfungi V: More hyperparasitic hyphomycetes. *Mycological Papers* No 128 1–110, CAB International, Wallingford.

Dennis, C. and Webster, J. (1971a) Antagonistic properties of species groups of *Trichoderma.* I. Production of non-volatile antibiotics. *Transactions of the British Mycological Society* 57, 25–39.

Dennis, C. and Webster, J. (1971b) Antagonistic properties of species-groups of *Trichoderma.* II. Production of volatile antibiotics. *Transactions of the British Mycological Society* 57, 41–48.

Dennis, C. and Webster, J. (1971c) Antagonistic properties of species-groups of *Trichoderma.* III. Hyphal interaction. *Transactions of the British Mycological Society* 57, 363–369.

Dennis, R.W.G. (1977) *British Ascomycetes.* J. Cramer, London, 585 pp.

Dickinson, J.M., Hanson, J.R., Hitchcock, P.B. and Claydon, N. (1989) Structure and biosynthesis of harzianopyridone, an antifungal metabolite of *Trichoderma harzianum. Journal of the Chemical Society Perkin Transactions* I, 1885–1887.

Di Pietro, A., Gut-Rella,M., Pachlatko, J.P. and Schwinn, F.J. (1992) Role of antibiotics produced by *Chaetomium globosum* in biocontrol of *Pythium ultimum*, a causal agent of damping-off. *Phytopathology* 82, 131–135.

Dobbs, C.G. (1938) The life history and morphology of *Dicranophora fulva* Schrot. *Transactions of the British Mycological Society* 21, 167–192.

Dobbs, C.G. and English, M.P. (1954) *Piptocephalis xenophila* sp. nov. parasitic on non-mucorine hosts. *Transactions of the British Mycological Society* 37, 375–389.

Domsch, K.H., Gams, W. and Anderson, T-H. (1980) *Compendium of Soil Fungi*, Volume 1. Academic Press, London, 859 pp.

Doshi, A. and Singh, R.D. (1985) Control of weed fungi and their effect on yield of oyster mushroom (*Pleurotus sajor-caju*). *Indian Journal of Mycology and Plant Pathology* 15, 269–273.

Dos Santos, A.F. and Dhingra, O.D. (1982) Pathogenicity of *Trichoderma* spp. on the sclerotia of *Sclerotinia sclerotiorum. Canadian Journal of Botany* 60, 472–475.

Drechsler, C. (1938) Two Hyphomycetes parasitic on oospores of root–rotting Oomycetes. *Phytopathology* 2828, 81–103.

Drechsler, C. (1943) Another hyphomycetous fungus parasitic on *Pythium* oospores. *Phytopathology* 33, 227–233.

Drechsler, C. (1952) Another nematode strangulating *Dactyella* parasitic on *Pythium* oospores. *Mycologia* 44, 533–556.

Drechsler, C. (1961) Two additional species of *Dactyella* parasitic on *Pythium* oospores. *Sydowia* 15, 92–97.

Drechsler, C. (1963) A slender-spored *Dactylella* parasitic on *Pythium* oospores. *Phytopathology* 53, 1050–1053.

Droby, S., Chalutz, E., Wilson, C.L. and Wisniewski, M. (1989) Characterization of the biocontrol activity of *Debaryomyces hansenii* in the control of *Penicillium digitatum* on grapefruit. *Canadian Journal of Microbiology* 35, 794–800.

Dubos, B. (1987) Fungal antagonism in aerial agrobiocenoses. In: Chet, I. (ed.) *Innovative Approaches to Plant Disease Control.* John Wiley, New York, pp. 107–135.

Dutta, B.K. (1981) Studies on some fungi isolated from the rhizosphere of tomato plants and the consequent prospects for the control of *Verticillium* wilt. *Plant and Soil* 63, 209–216.

Ehrlich, M.A. and Ehrlich, H.G. (1971) Fine structure of the host–parasite interface in mycoparasitism. *Annual Review of Phytopathology* 9, 155–184.

Ekpo, E.J.A. and Young, T.W.K. (1979) Fine-structure of the dormant and germinating sporangiospore of *Syzygites megalocarpus* (Mucorales) with notes on *Sporodiniella umbellata. Microbios Letters* 101, 63–68.

Elad, Y., Chet, I. and Henis, Y. (1981) A selective medium for improving quantitative isolation of *Trichoderma* spp. from soil. *Phytoparasitica* 9, 59–67.

Elad, Y., Chet, I. and Henis, Y. (1982) Degradation of plant pathogenic fungi by *Trichoderma harzianum. Canadian Journal of Microbiology* 28, 719–725.

Elad, Y., Barak, R., Chet, I. and Henis, Y. (1983a) Ultrastructural studies of the interaction between *Trichoderma* spp. and plant pathogenic fungi. *Phytopathologische Zeitschrift* 107, 168–175.

Elad, Y., Chet, I., Boyle, P. and Henis, Y. (1983b) Parasitism of *Trichoderma* spp. on *Rhizoctonia solani* and *Sclerotium rolfsii* – scanning electron microscopy and fluorescence microscopy. *Phytopathology* 73, 85–88.

Elad, Y., Sadovski, Z. and Chet, I. (1983c) Detection of mycoparasitism by infrared photomicrography. *Microbial Ecology* 9, 185–187.

Elad, Y., Barak, R. and Chet, I. (1984) Parasitism of sclerotia of *Sclerotium rolfsii* by *Trichoderma harzianum. Soil Biology and Biochemistry* 16, 381–386.

Elad, Y., Lifshitz, R. and Baker, R. (1985) Enzymatic activity of the mycoparasite *Pythium nunn* during interaction with host and non-host fungi. *Physiological Plant Pathology* 27, 131–148.

Elad, U., Sadowsky, Z. and Chet, I. (1987) Scanning electron microscopical observations of early stages of interaction of *Trichoderma harzianum* and *Rhizoctonia solani. Transactions of the British Mycological Society* 88, 259–263.

Ellis, J.J. (1966) On growing *Syncephalis* in pure culture. *Mycologia* 58, 465–469.

Ellis, M.B. (1971) *Dematiaceous Hyphomycetes.* CAB International, Wallingford, 608 pp.

Ellis, M.B. and Ellis, J.P. (1985) *Microfungi on Land Plants.* Croom Helm, London, 818 pp.

Ellis, M.B. and Ellis, J.P. (1988) *Microfungi on Miscellaneous Substrates.* Croom Helm, London, 244 pp.

El-Nashaar, H.M., Moore, L.W. and George, R.A. (1986) Enzyme-linked immunosorbent assay quantification of initial infection of wheat by *Gaeumannomyces graminis* var. *tritici* as moderated by biocontrol agents. *Phytopathology* 76, 1319–1322.

El Shafie, A.E. and Webster, J. (1979) *Curvularia* species as parasites of *Rhizopus* and other fungi. *Transactions of the British Mycological Society* 73, 352–353.

England, W.H. (1969) Relation of age of two host fungi to development of the mycoparasite *Piptocephalis virginiana. Mycologia* 61, 586–592.

Epton, H.A.S. and Wynn A.R. (1984) A method for detecting and isolating parasites of oospores. *Abstracts of the British Society for Plant Pathology workshop on methods for studying biological control of diseases caused by soil-borne plant pathogens*, p. 17.

Erbeznik, M., Matavulj, M. and Stojilkovic, S. (1986) The effect of indole-3-acetic acid on cellulase production by *Gliocladium virens* CZ-RI. *Mikrobiologiya* 23, 39–47.

Evans G.H. and Cooke, R.C. (1981) *Piptocephalis*-induced changes in organization of marginal hyphae in *Mycotypha microspora* colonies. *Transactions of the British Mycological Society* 76, 343–345.

Evans, G.H. and Cooke, R.C. (1982) Studies on Mucoralean mycoparasites. III. Diffusible factors from *Mortierella vinacea* Dixon-Stewart that direct germ tube growth of *Piptocephalis fimbriata* Richardson & Leadbeater. *New Phytologist* 91, 245–253.

Evans, G.H., Lewis, D.H. and Cooke, R.C. (1978) Studies on mucoralean mycoparasites II. Persistent yeast phase growth of *Mycotypha microspora* Fenner when infected by *Piptocephalis fimbriata* Richardson & Leadbeater. *New Phytologist* 81, 629–635.

Evans, G.H., Curtis F.C. and Cooke, R.C. (1981) Host carbohydrate composition and trehalase activity in relation to mycoparasitism in *Piptocephalis* species. *Transactions of the British Mycological Society* 77, 21–26.

Evans, G.H., Cooke, R.C. and Parker, M.L. (1984) Penetration of yeast-phase cells of *Mycotypha microspora* by *Piptocephalis fimbriata. Transactions of the British Mycological Society* 82, 540–546.

Fernandez, M.R. (1992a) The effect of *Trichoderma harzianum* on fungal pathogens infesting soybean residues. *Soil Biology and Biochemistry* 24, 1027–1029.

Fernandez, M.R. (1992b) The effect of *Trichoderma harzianum* on fungal pathogens infesting wheat and black oat straw. *Soil Biology and Biochemistry* 24, 1031–1034.

Ferrata, M. and D'Ambra, V. (1985) Aspetti morphologici del parassitismo di *Trichoderma harzianum* Rifai su ife e sclerozi di *Sclerotium rolfsii* Sacc. *Rivista di Patologia Vegetale* 21, 53–60.

Filip, G.M. and Roth, L.R. (1977) Stump injections with soil fumigants to eradicate *Armillariella mellea* from young-growth ponderosa pine killed by root rot. *Canadian Journal of Forest Research* 7, 226–231.

Firman, I.D. and Waller, J.M. (1977) Coffee berry disease and other *Colletotrichum* diseases of coffee. *Phytopathological Paper No. 20.* Commonwealth Mycological Institute, Kew.

Fletcher, J.T. and Ganney, G.W. (1968) Experiments on the biology and control of *Mycogone perniciosa* Magn. *Mushroom Science* 7, 221–237.

Fletcher, J.T., Smewin, B.J. and O'Brien, A. (1990) *Pythium oligandrum* associated with a cropping disorder of *Agaricus bisporus. Plant Pathology* 39, 603–605.

Fletcher, J.T., White, P.F. and Gaze, R.H. (1986) *Mushrooms – Pest and Disease Control.* Intercept, Newcastle-upon-Tyne, 156 pp.

Fokkema, N.J. (1978) Fungal antagonists in the phyllosphere. *Annals of Applied Biology* 89, 115–119.

Foley, M.F. and Deacon, J.W. (1985) Isolation of *Pythium oligandrum* and other necrotrophic mycoparasites from soil. *Transactions of the British Mycological Society* 85, 631–639.

Foley, M.F. and Deacon, J.W. (1986) Physiological differences between mycoparasitic and plant-pathogenic *Pythium* spp. *Transactions of the British Mycological Society* 86, 225–231.

Fravel, D.R. and Marois, J.J. (1986) Edaphic parameters associated with establishment of the biocontrol agent *Talaromyces flavus. Phytopathology* 76, 643–646.

Fravel, D.R., Marois, J.J., Lumsden, R.D. and Connick, W.J. Jr (1985) Encapsulation of potential biocontrol agents in an alginate-clay matrix. *Phytopathology* 75, 774–777.

Fravel, D.R., Kim, K.K. and Papavizas, G.C. (1987) Viability of microsclerotia of *Verticillium dahliae* reduced by a metabolite produced by *Talaromyces flavus. Phytopathology* 77, 616–619.

Freitag, M., Morrell, J.J. and Bruce, A. (1991) Biological protection of wood: status and prospects. *Biodeterioration Abstracts* 5, 1–13

Fresenius, G. (1864) Ueber *Ascophora elegans* Corda. *Botanische Zeitschrift* 22, 154–155.

Fries, N. and Swedjemark, G. (1985) Sporophagy in Hymenomycetes. *Experimental Mycology* 9, 74–79.

Fuska, J., Nemec, P. and Kuhr, I. (1972) Vermiculine, a new antiprotozoal antibiotic from *Penicillium vermiculatum. Journal of Antibiotics* 25, 208–211.

Gain, R.E. and Barnett, H.L. (1970) Parasitism and axenic growth of the mycoparasite *Gonatorhodiella highlei. Mycologia* 62, 1122–1129.

Gams, W. (1977) A key to species of *Mortierella. Persoonia* 9, 381–391.

Gams, W. (1986) The fungicolous Hyphomycete *Zakatoshia erikssonii* n.sp. *in vivo* and *in vitro. Windahlia* 16, 59–64.

Gams, W. and Hoozemans, A.C.M. (1970) *Cladobotryum*-konidienformens von *Hypomyces*–arten. *Persoonia* 6, 99–110.

Gandy, D.G. (1979) Inhibition of *Mycogone perniciosa* growth by *Acremonium constrictum. Transactions of the British Mycological Society* 72, 151–155.

Gardner, D.E., Miller, T. and Kuhlman, E.G. (1979) *Tuberculina* and the life cycle of *Uromyces koae. Mycologia* 71, 848–852.

Garrett, S.D. (1956) *Biology of Root-infecting Fungi.* Cambridge University Press, Cambridge.

Garrett, S.D. (1963) *Soil Fungi and Soil Fertility.* Pergamon Press, Oxford, 150 pp.

Garrett, S.D. (1970) *Pathogenic Root-infecting Fungi.* Cambridge University Press, Cambridge, 242 pp.

Gerlagh, M. and Vos, I. (1991) Enrichment of soil with sclerotia to isolate antagonists of *Sclerotinia sclerotiorum.* In: Beemster, A.B.R., Bollen, G.J., Gerlagh, M., Ruissen, M.A., Schippers, B. and Tempel, A. (eds) *Biotic Interactions and Soil-borne Diseases.* Elsevier, Amsterdam, pp. 165–171.

Ghisalberti, E.L. and Sivasithamparam, K. (1991) Antifungal antibiotics produced by *Trichoderma* spp. *Soil Biology and Biochemistry* II, 1011–1020.

Ginns, J. (1986) The genus *Syzygospora* (Heterobasidiomycetes: Syzygosporaceae). *Mycologia* 78, 619–636.

Ginns, J. and Sunhede, S. (1978) Three species of *Christiansenia* (Corticiaceae) and the teratological galls on *Collybia dryophila. Botaniska Notiser* 131, 167–173.

Giralt, M. and Hawksworth, D.L. (1991) *Diplolaeviopsis ranula*, a new genus and species of lichenicolous coelomycetes growing on the *Lecanora strobilina* group in Spain. *Mycological Research* 95, 759–761.

Gladders, P. and Coley-Smith, J.R. (1980) Interactions between *Rhizoctonia tuliparum* sclerotia and soil micro-organisms. *Transactions of the British Mycological Society* 74, 579–586.

Gorodecki, B. and Hadar, Y. (1990) Suppression of *Rhizoctonia solani* and *Sclerotium rolfsii* diseases in container media containing composted separated cattle manure and composted grape marc. *Crop Protection* 9, 271–274.

Gourley, C.O. and MacNab, A.A. (1964) *Verticillium dahliae* and *Gliocladium roseum* isolation from strawberries in Nova Scotia. *Canadian Journal of Plant Science* 44, 544–549.

Graham, R.D. (1973) Preventing and stopping internal decay of Douglas-fir poles. *Holzforschung* 27, 168–173.

Gray, D.J. and Morgan-Jones, G. (1980) Notes on Hyphomycetes. XXXIV. Some mycoparasitic species. *Mycotaxon*, 10, 375–404.

Gray, D.J. and Morgan-Jones, G. (1981) Host–parasite relationships of *Agaricus brunnescens* and a number of mycoparasitic hyphomycetes. *Mycopathologia* 75, 55–59.

Griffith, N.T. and Barnett, H.L. (1967) Mycoparasitism by basidiomycetes in culture. *Mycologia* 59, 149–154.

Gwande, M.P. (1990) Biological control of groundnut (*Arachis hypogaea* L.) rust (*Puccinia arachidis* Speg.) in India. *Tropical Pest Management* 36, 17–20.

Hadar, Y. and Gorodecki, B. (1991) Suppression of germination of sclerotia of *Sclerotium rolfisii* in compost. *Soil Biology and Biochemistry* 23, 303–306.

Hajlaoui, M.R., Benhamou, N. and Belanger, R.R. (1992) Cytochemical study of the antagonistic activity of *Sporothrix flocculosa* on Rose Powdery Mildew, *Sphaerotheca pannosa* var. *rosae*. *Phytopathology* 82, 583–589.

Hall, R.A. (1981) The fungus *Verticillium lecanii* as a microbial insecticide against aphids and scales. In: Burges, H.D. (ed.) *Microbial Control of Pests and Plant Diseases.* Academic Press, London, pp. 483–498.

Hallett, S.G. and Ayres, P.G. (1992) Invasion of rust (*Puccinia lagenophorae*) aecia on groundsel (*Senecio vulgaris*) by secondary pathogens: death of the host. *Mycological Research*, 96, 142–144.

Harman, G.E. and Taylor, A.G. (1990) Development of an effective biological seed treatment system. In: Hornby, D. (ed.) *Biological Control of Soil-borne Plant Pathogens.* CAB International, Wallingford, pp. 415–426.

Harris, D.C. (1985a) *Microdochium fusarioides* sp. nov. from oospores of *Phytophthora syringae*. *Transactions of the British Mycological Society* 84, 358–361.

Harris, D.C. (1985b) Survival of *Phytophthora syringae* oospores in and on apple orchard soil. *Transactions of the British Mycological Society* 85, 153–155.

Hauerslev, K. (1969) *Christiansenia pallida* gen.nov., sp.nov., a new parasitic homobasidiomycete from Denmark. *Friesia* 9, 43–45.

Hawker, L.E. (1955) Hypogeous fungi. IV and V. *Transactions of the British Mycological Society* 38, 73–77.

Hawker, L.E. and Beckett, A. (1971) Fine structure and development of the zygospore of *Rhizopus sexualis* (Smith) Callen. *Philosophical Transactions of the Royal Society of London* B263, 71–100.

Hawksworth, D.L. (1977a) Taxonomic and biological observations on the genus

Lichenoconium (Sphaeropsidales). *Persoonia* 9, 159–198.

Hawksworth, D.L. (1977b) Three new genera of lichenicolous fungi. *Botanical Journal of the Linnean Society* 75, 195–209.

Hawksworth, D.L. (1979) The lichenicolous hyphomycetes. *Bulletin of the British Museum (Natural History), Botany Series* 6, 118.

Hawksworth, D.L. (1980) Notes on some fungi occurring on *Peltigera*, with a key to accepted species. *Transactions of the British Mycological Society* 74, 363–386.

Hawksworth, D.L. (1981) A survey of the fungicolous conidial fungi. In: Cole, G.T. (ed.) *Biology of Conidial Fungi*, Volume 1. Academic Press, New York, pp. 171–244.

Hawksworth, D.L. (1982a) Coevolution and the detection of ancestry in lichens. *Journal of the Hattori Botanical Laboratory* 52, 323–329.

Hawksworth, D.L. (1982b) Secondary fungi in lichen symbioses: parasites, saprophytes and parasymbionts. *Journal of the Hattori Botanical Laboratory* 52, 357–366.

Hawksworth, D.L. (1983) A key to the lichen-forming, parasitic, parasymbiotic and saprophytic fungi occurring on lichens in the British Isles. *Lichenologist* 15, 1–44.

Hawksworth, D.L. (1988) The variety of fungal-algal symbioses, their evolutionary significance and the nature of lichens. *Botanical Journal of the Linnean Society* 96, 3–20.

Hawksworth, D.L. (1991) The fungal dimension of biodiversity: magnitude, significance, conservation. *Mycological Research* 95, 641–655.

Hawksworth, D.L. and Hill, D.J. (1984) *The Lichen-forming Fungi.* Blackie, Edinburgh, 158 pp.

Hawksworth, D.L., James, P.W. and Coppins, B.J. (1980) Checklist of British lichen-forming, lichenicolous and allied fungi. *Lichenologist* 12, 1–115.

Hawksworth, D.L., Sutton, B.C. and Ainsworth, G.C. (1983) *Ainsworth and Bisby's Dictionary of the Fungi,* 7th edn. CAB International, Wallingford, 445 pp.

Hawksworth, D.L., Kirk, P.M., Pegler, D.N., Sutton, B.C. and Ainsworth, G.C. (1994) *Ainsworth and Bisby's Dictionary of the Fungi,* 8th edn. CAB International, Wallingford.

Heath, M.C. (1976) Ultrastructural and functional similarity of the haustorial neckband of rust fungi and the Casparian strip of vascular plants. *Canadian Journal of Botany* 54, 2484–2489.

Heim, R. (1948) Phylogeny and natural classification of macro-fungi. *Transactions of the British Mycological Society* 30, 161–178.

Heintz, C. and Blaich, R. (1990) *Verticillium lecanii* als Hyperparasit des Rebmehltaus (*Uncinula necator*). *Vitis* 29, 229–232.

Held, A.A. (1972) Host–parasite relations between *Allomyces* and *Rozella.* Parasite penetration depends on growth response of host cell-wall. *Archiv für Mikrobiologie* 82, 128–139.

Held, A.A. (1981) *Rozella* and *Rozellopsis*: naked endoparasitic fungi which dress-up as their hosts. *Botanical Review* 47, 451–515.

Helfer, W. (1991) Pije auf Pilzfruchtkorpen Untermulungen gur Okologie, Systematik und Chemie. *Libri Botanici* 1, 1–157.

Henis, Y. (1984) Biological control. Ecological principles of biological control of soil-borne plant pathogens: *Trichoderma* model. In King, U.Z. and Reddy, C.A. (eds) *Current Perspectives in Microbial Ecology.* Washington DC.

Hepperly, P.R. (1986) *Hansfordia* sp.: a parasitic pathogen of dematiaceous plant pathogenic fungi in Puerto Rico. *Journal of Agriculture of the University of Puerto Rico* 70, 113–119.

Herr, L.J. (1988) Biocontrol of *Rhizoctonia* crown and root rot of sugar beet by binucleate *Rhizoctonia* spp. and *Laetisaria arvalis*. *Annals of Applied Biology* 113, 107–118.

Hijwegen, T. (1992) Biological control of cucumber powdery mildew with *Tilletiopsis minor* under greenhouse conditions. *Netherlands Journal of Plant Pathology* 98, 221–225.

Hiratsuka, Y. and Powell, J.M. (1976) Pine Stem Rusts of Canada. *Forest Technical Report No 4, Department of the Environment Canada.*

Hoch, H.C. (1977a) Mycoparasitic relationships: *Gonatobotrys simplex* parasitic on *Alternaria tenuis*. *Phytopathology* 67, 309–314.

Hoch, H.C. (1977b) Mycoparasitic relationships. III. Parasitism of *Physalospora obtusa* by *Calcarisporium parasiticum*. *Canadian Journal of Botany* 55, 198–207.

Hoch, H.C. (1978) Mycoparasitic relationships. IV. *Stephanoma phaeospora* parasitic on a species of *Fusarium*. *Mycologia* 70, 370–379.

Hoch, H.C. and Abawi, G.S. (1979a) Mycoparasitism of oospores of *Pythium ultimum* by *Fusarium merismoides*. *Mycologia* 71, 621–625.

Hoch, H.C. and Abawi, G.S. (1979b) Biological control of *Pythium* root rot of table beet with *Corticium* sp. *Phytopathology* 69, 417–419.

Hoch, H.C. and Fuller, M.S. (1977) Mycoparasitic relationships. I. Morphological features of interactions between *Pythium acanthicum* and several fungal hosts. *Archives of Microbiology* 111, 207–224.

Hoch, H.C. and Provvidenti, R. (1979) Mycoparasitic relationships: cytology of the *Sphaerotheca fuliginea–Tilletiopsis* sp. interaction. *Phytopathology* 69, 359–362.

Holland, D.M. and Cooke, R.C. (1990) Activation of dormant conidia of the wet bubble pathogen *Mycogone perniciosa* by Basidomycotina. *Mycological Research* 94, 789–792.

Holland, D.M., Parker, M.L., Cooke, R.C. and Evans, G.H. (1985) Germination of bicellular conidia of *Mycogone perniciosa*, the wet bubble pathogen of the cultivated mushroom. *Transactions of the British Mycological Society* 85, 730–735.

Hoog, G.S. de, Hijwegen, T. and Batenburg-Van Der Vegte, W.H. (1991) A new species of *Dissoconium*. *Mycological Research* 95, 679–682.

Hornby, D. (ed.) (1990) *Biological Control of Soil-borne Plant Pathogens.* CAB International, Wallingford, 479 pp.

Hornok, L. and Walcz, I. (1983) *Fusarium heterosporum*, a highly specialised hyperparasite of *Claviceps purpurea*. *Transactions of the British Mycological Society* 80, 377–380.

Howell, C.R. (1982) Effect of *Gliocladium virens* on *Pythium ultimum*, *Rhizoctonia solani*, and damping-off of cotton seedlings. *Phytopathology* 72, 496–498.

Howell, C.R. (1987) Relevance of mycoparasitism in the biological control of *Rhizoctonia solani* by *Gliocladium virens*. *Phytopathology* 77, 992–994.

Howell, C.R. (1991) Biological control of *Pythium* damping-off of cotton with seed-coating preparations of *Gliocladium virens*. *Phytopathology* 81, 738–741.

Howell, C.R. and Stipanovic, R.D. (1983) Gliovirin, a new antibiotic from *Gliocladium virens*, and its role in the biological control of *Pythium ultimum*. *Canadian Journal of Microbiology* 29, 321–324.

Huang, H.C. (1977). Importance of *Coniothyrium minitans* in survival of sclerotia of *Sclerotinia sclerotiorum* in wilt of sunflower. *Canadian Journal of Botany* 56, 2243–2246.

Huang, H.C. (1978) *Gliocladium catenulatum*: hyperparasite of *Sclerotinia sclerotiorum* and *Fusarium* species. *Canadian Journal of Botany* 56, 2243–2246.

Huang, H.C. (1980) Control of sclerotinia wilt of sunflower by hyperparasites. *Canadian Journal of Plant Pathology* 2, 26–32.

Huang, H.C. and Hoes, J.A. (1976) Penetration and infection of *Sclerotinia sclerotiorum* by *Coniothyrium minitans. Canadian Journal of Botany* 54, 406–410.

Huang, H.C. and Kokko, E.G. (1988) Penetration of hyphae of *Sclerotinia sclerotiorum* by *Coniothyrium minitans* without the formation of appressoria. *Journal of Phytopathology* 124, 133–139.

Hubert, E.A. (1935) Observations on *Tuberculina maxima*, a parasite of *Cronartium ribicola*, *Phytopathology* 25, 253–261.

Hughes, S.J. (1979) Relocation of species of *Endophragmia* Auct. with notes on relevant generic names. *New Zealand Journal of Botany* 17, 139–188.

Humble, S.J. and Lockwood, J.L. (1981) Hyperparasitism of oospores of *Phytophthora megasperma* var. *sojae. Soil Biology and Biochemistry* 13, 355–360.

Hunter, W.E. and Butler, E.E. (1975) *Syncephalis californica*, a mycoparasite inducing giant hyphal swellings in species of Mucorales. *Mycologia* 67, 863–872.

Hunter, W.E., Duniway, J.M. and Butler, E.E. (1977) Influence of nutrition, temperature, moisture and gas composition on parasitism of *Rhizopus oryzae* by *Syncephalis californica. Phytopathology* 67, 664–669.

Hwang, K., Stelzig, D.A., Barnett H.L., Roller, P.P. and Kelsey, M.I. (1985) Partial purification of the growth factor mycotrophein. *Mycologia* 77, 109–113.

Ikediugwu, F.E.O. (1976a) The interface in hyphal interference by *Peniophora* against *Heterobasidion annosum. Transactions of the British Mycological Society* 66, 291–296.

Ikediugwu, F.E.O. (1976b) Ultrastructure of hyphal interference between *Coprinus heptemerus* and *Ascobolous crenulatus. Transactions of the British Mycological Society* 66, 281–290.

Ikediugwu, F.E.O. and Webster, J. (1970a) Antagonism between *Coprinus heptemerus* and other coprophilous fungi. *Transactions of the British Mycological Society* 54, 181–204.

Ikediugwu, F.E.O. and Webster, J. (1970b) Hyphal interference in a range of coprophilous fungi. *Transactions of the British Mycological Society* 54, 205–210.

Inbar, J. and Chet, I. (1992) Biomimics of fungal cell–cell recognition by use of lectin-coated nylon fibers. *Journal of Bacteriology* 174, 1055–1059.

Ing, B. (1974) Mouldy Myxomycetes. *Bulletin of the British Mycological Society* ʋ, 25–30.

Jackson, A.M., Whipps, J.M. and Lynch, J.M. (1991a) *In vitro* screening for the identification of potential biocontrol agents of *Allium* white rot. *Mycological Research* 95, 430–434.

Jackson, A.M., Whipps, J.M., Lynch, J.M. and Bazin, M.J. (1991b) Effects of some carbon and nitrogen sources on spore germination, production of biomass and antifungal metabolites by species of *Trichoderma* and *Gliocladium virens* antagonistic to *Sclerotium cepivorum. Biocontrol Science and Technology*, 1, 43–51.

Jager, G. and Velvis, H. (1988) Inactivation of sclerotia of *Rhizoctonia solani* on potato tubers by *Verticillium biguttatum*, a soil-borne mycoparasite. *Netherlands Journal of Plant Pathology* 94, 225–231.

Jager, G., Ten Hoopen, A. and Velvis, H. (1979) Hyperparasites of *Rhizoctonia solani* in Dutch potato fields. *Netherlands Journals of Plant Pathology* 85, 253–268.

Janisiewicz, W.J. (1987) Postharvest biological control of blue mold on apples. *Phytopathology* 77, 481–485.

Janisiewicz, W.J. (1988) Biocontrol of postharvest diseases of apples with antagonistic mixtures. *Phytopathology* 78, 194–198.

Jarvis, W.R. and Slingsby, K. (1977) The control of powdery mildew of greenhouse cucumber by water sprays and *Ampelomyces quisqualis. Plant Disease Reporter* 61, 728–730.

Jeffries, P. (1976) Mycoparasitism in *Piptocephalis unispora.* Ph.D. thesis, University of London.

Jeffries, P. (1984) Strategies of mycoparasitism within the Zygomycetes. *Bulletin of the British Mycological Society* 17, supplement 3, 1.

Jeffries, P. (1985) Mycoparasitism within the Zygomycetes. *Botanical Journal of the Linnean Society* 91, 135–150.

Jeffries, P. (1987) Pathways for the exchange of materials in mycoparasitic and plant–fungal interactions. In: Ayres, P.G. and Pegg, G.F. (eds) *Fungal Infection of Plants,* Cambridge University Press, Cambridge, Chap. 4, pp. 60–78.

Jeffries, P. and Cuthbert, A. (1984) Self–penetration by the mycoparasite *Dimargaris cristalligena. Protoplasma* 121, 129–131.

Jeffries, P. and Jeger, M.J. (1990) The biological control of postharvest diseases of fruits. *Biocontrol News and Information* 11, 333–336.

Jeffries, P. and Kirk, P.M. (1976) New technique for the isolation of mycoparasitic Mucorales. *Transactions of the British Mycological Society* 66, 541–543.

Jeffries, P. and Koomen, I. (1992) Strategies and prospects for biological control of diseases caused by *Colletotrichum.* In: Bailey, J.A. and Jeger, M.J. (eds) *Colletotrichum: Biology, Pathology and Control.* CAB International, Wallingford, pp. 337–357.

Jeffries, P. and Young, T.W.K. (1976a) Physiology and fine structure of sporangiospore germination in *Piptocephalis unispora* prior to infection. *Archives of Microbiology* 107, 99–107.

Jeffries, P. and Young, T.W.K. (1976b) Ultrastructure of infection of *Cokeromyces recurvatus* by *Piptocephalis unispora* (Mucorales). *Archives of Microbiology* 109, 277–288.

Jeffries, P. and Young, T.W.K. (1978) Mycoparasitism by *Piptocephalis unispora* (Mucorales): host range and reaction with *Phascolomyces articulosus. Canadian Journal of Botany* 56, 2449–2459.

Jeffries, P. and Young, T.W.K. (1981) Ultrastructure of the haustorial apparatus of *Dimargaris cristalligena. Annals of Botany* 47, 107–119.

Jeffries, P. and Young, T.W.K. (1983) Zygospore structure in *Cokeromyces recurvatus* with notes on the asexual apparatus. *Mycologia* 75, 509–517.

Jeffries, P., Spyropoulos, T. and Vardavarkis, E. (1988) Vesicular–arbuscular mycorrhizal status of various crops in different agricultural soils of northern Greece. *Biology and Fertility of Soils* 5, 333–337.

Jeger, M.J. and Jeffries, P. (1988) Alternatives to chemical usage for disease management in the post-harvest environment. *Aspects of Applied Biology* 17, 47–57.

Jones, D., Gordon, A.H. and Bacon, J.S.D. (1974) Co-operative action by endo- and exo-β-(1-3)-glucanases from parasitic fungi in the degradation of cell-wall glucans of *Sclerotinia sclerotiorum* (Lib.) de Bary. *Biochemical Journal* 140, 47–55.

Jones, R.W., Pettit, R.E. and Taber, R.A. (1984) Lignite and stillage: carrier and substrate for application of fungal biocontrol agents to soil. *Phytopathology* 74, 1167–1170.

Jordan, E.G. and Barnett, H.L. (1978) Nutrition and parasitism of *Melanospora zamiae. Mycologia* 70, 300–312.

Julich, W. (1984) Die Nichblatterpilze, Gallerpilze und Bauchpilze. In: Gams, H. (ed.) *Kleine Kryptogamenflora* IIb/I: G. Fischer Verlag, Stuttgart, 626 pp.

Karhuvaara, L. (1960) On the parasites of the sclerotia of some fungi. *Acta Agricultura Scandinavica.* 10, 127–134.

Karling, J.S. (1977) *Chytridiomycetarum Iconographia.* Vaduz, Lubrecht and Kramer, Monticello, New York.

Katan, T., Dunn, M.T. and Papavizas, G.L. (1984) Genetics of fungicide resistance in *Talaromyces flavus. Canadian Journal of Microbiology*, 30, 1079–1087.

Keinath, A.P., Fravel, D.R. and Papavizas, G.C. (1991) Potential of *Gliocladium roseum* for biocontrol of *Verticillium dahliae. Phytopathology* 81, 644–648.

Kellock, L.M. and Dix, N.J. (1984a) Antagonism by *Hypomyces aurantius.* I. Toxins and hyphal interactions. *Transactions of the British Mycological Society* 82, 327–334.

Kellock, L.M. and Dix, N.J. (1984b) Antagonism by *Hypomyces aurantius.* II. Ultrastructural studies of hyphal disruption. *Transactions of the British Mycological Society*, 82, 335–338.

Kemp, R.F.O. (1970) Inter-specific sterility in *Coprinus bisporus*, *C. congregatus* and other basidiomycetes. *Transactions of the British Mycological Society* 54, 488–489.

Kenerley, C.M. and Stack, J.P. (1987) Influence of assessment methods on selection of fungal antagonists of the sclerotium-forming fungus *Phymatotrichum omnivorum. Canadian Journal of Microbiology* 33, 632–635.

Kim, K.K., Fravel, D.R. and Papavizas, G.C. (1990a) Glucose oxidase as the antifungal principle of talaron from *Talaromyces flavus. Canadian Journal of Microbiology* 36, 760–764.

Kim, K.K., Fravel, D.R. and Papavizas, G.C. (1990b) Production, purification and properties of the glucose oxidase from the biocontrol fungus *Talaromyces flavus. Canadian Journal of Microbiology* 36, 199–205.

Klecan, A.L., Hippe, S. and Somerville, S.C. (1990) Reduced growth of *Erysiphe graminis* f.sp. *hordei* induced by *Tilletiopsis pallescens. Phytopathology* 80, 325–331.

Kligman A.M. (1944) Control of the truffle in beds of the cultivated mushroom. *Phytopathology* 34, 370–384.

Klingström, A.E. and Johansson, S.M. (1973) Antagonism of *Scytalidium* isolates against decay fungi. *Phytopathology* 63, 473–479.

Kobayashi, Y. (1941) The genus *Cordyceps* and its allies. Report No. 84. *Science Reports of the Tokyo Bunrika Daigaku, Section B* 5, 53–260.

Koc, N.K., Forrer, H.R. and Kern, H. (1981) Studies on the relationship between *Puccinia graminis* and the hyperparasite *Aphanocladium album. Phytopathologische Zeitschrift* 101, 131–135.

Koller, B. and Jahrmann, H.J. (1985) Life-cycle and physiological description of the yeast-form of the homobasidiomycete *Asterophora lycoperdoides* (Bull.: Fr) Ditm. *Antonie van Leeuwenhoek* 51, 255–261.

Kreisel, H. (1969) Grundzuge einer nabirlichen Systems der Pilze. Lehre, Cramer.

Kuhlman, E.G. and Matthews, F.R. (1976) Occurrence of *Darluca filum* on *Cronartium strobilinum* and *Cronartium fusiforme* on oak. *Phytopathology* 66, 1195–1197.

Kuhlman, E.G., Matthews, F.R. and Tillerson, H.P. (1980) Efficacy of *Darluca filum* for biological control of *Cronartium fusiforme* and *C. strobilinum*. *Phytopathology* 68, 507–511.

Kunz, C., Sellam, O. and Bertheau, Y. (1992) Purification and characterization of a chitinase from the hyperparasitic fungus *Aphanocladium album*. *Physiological and Molecular Plant Pathology* 40, 117–131.

Kurtzman, C.P. (1967) Parasitism, spore germination and axenic growth of *Dispira cornuta*. Ph.D. thesis, West Virginia University.

Kurtzman, C.P. (1968) Parasitism and axenic growth of *Dispira cornuta*. *Mycologia* 60, 915–923.

Kuter, G.A. (1984) Hyphal interactions between *Rhizoctonia solani* and some *Verticillium* species. *Mycologia* 76, 936–940.

Kuykendall, W.R., Hindal, D.F. and Barnett, H.L. (1983) Parasitism and nutrition of the contact mycoparasite *Acladium tenellum*. *Mycologia* 75, 656–665.

Kwee, L.T. and Teik, K.H. (1982) Effect of pesticides on mango leaf and flower microflora. *Zeitschrift für Pflanzenkrankheiten und Pflanzenschutz* 89, 125–131.

Kwon-Chung, K.J. (1976) Morphogenesis of *Filobasidiella neoformans*, the sexual state of *Cryptococcus neoformans*. *Mycologia* 68, 821–833.

Lagerheim, G. (1898) Mykologische Studien. I. Beitrage zur Kenntnis der parasitischen Pilze, 1–3. [3. *Iola (Cystobasidium) lasioboli* nov. spec.]. *Bihang Till K. Sv. Vet. -Akad Handl. Afd. III.* 24, 15–21.

Laing, S.A.K. and Deacon, J.W. (1990) Aggressiveness and fungal host ranges of mycoparasitic *Pythium* species. *Soil Biology and Biochemistry* 22, 905–911.

Laing, S.A.K. and Deacon, J.W. (1991) Video microscopical comparison of mycoparasitism by *Pythium oligandrum*, *P. nunn* and an unnamed *Pythium* species. *Mycological Research* 95, 469–479.

Leadbeater, G. and Mercer, C.K. (1957) *Piptocephalis virginiana* sp. nov. *Transactions of the British Mycological Society* 40, 461–471.

Le Picard, Y.T. and Trique, B. (1987) Antagonistes et hyperparasites du *Fulvia fulva* (Cooke) Ciferri. *Cryptogamie Mycologie* 8, 43–50.

Lewis, D.H. (1974) Microorganisms and plants: the evolution of parasitism and mutualism. *Symposium of the Society for General Microbiology* 24, 367–392.

Lewis, D.H., Whipps, J.M. and Cooke, R.C. (1989) Mechanisms of biological disease control with special reference to the case study of *Pythium oligandrum* as an antagonist. In: Whipps, J.M. and Lumsden, R.D. (eds) *Biotechnology of Fungi for Improving Plant Growth*. Cambridge University Press, Cambridge, pp. 191–217.

Lewis, J.A. and Papavizas, G.C. (1984) A new approach to stimulate population proliferation of *Trichoderma* species and other potential biocontrol fungi introduced into natural soils. *Phytopathology* 74, 1240–1244.

Lewis, J.A. and Papavizas, G.C. (1985) Characteristics of alginate pellets formulated with *Trichoderma* and *Gliocladium* and their effect on the proliferation of the fungus in soil. *Plant Pathology* 34, 571–577.

Lewis, J.A. and Papavizas, G.C. (1987) Application of *Trichoderma* and *Gliocladium* in alginate pellets for control of *Rhizoctonia* damping off. *Plant Pathology* 36, 438–446.

Lewis, J.A. and Papavizas, G.C. (1992) Potential of *Laetisaria arvalis* for the biocontrol of *Rhizoctonia solani*. *Soil Biology and Biochemistry* 24, 1075–1079.

Lewis, J.A., Roberts, D.P. and Hollenbeck, M.D. (1991) Induction of cytoplasmic leakage from *Rhizoctonia solani* hyphae by *Gliocladium virens* and partial characterization of a leakage factor. *Biocontrol Science and Technology* 1, 21–29.

Lifshitz, R., Dupler, M., Elad, Y. and Baker, R. (1984a) Hyphal interactions between a mycoparasite, *Pythium nunn*, and several soil fungi. *Canadian Journal of Microbiology* 30, 1482–1487.

Lifshitz, R., Sneh, B. and Baker, R. (1984b) Soil suppressiveness to a plant-pathogenic *Pythium* species. *Phytopathology* 74, 1054–1061.

Lim, T.K. and Rohrbach, K.G. (1980) Role of *Penicillium funiculosum* strains in the development of pineapple fruit diseases. *Phytopathology* 70, 663–665.

Lim, T.K. and Wan, Z.W.N. (1983) Mycoparasitism of the coffee rust pathogen, *Hemileia vastatrix*, by *Verticillium psalliotae* in Malaysia. *Pertanika* 6, 23–25.

Litkei, J. (1989) Characteristics of isolates of *Sporidesmium sclerotivorum* Uecker, Ayers and Adams in Hungary. *Acta Phytopathologica et Entomologica Hungarica* 24, 311–315.

Lloyd A.B. and Lockwood J.L. (1966) Lysis of fungal hyphae in soil and its possible relation to autolysis. *Phytopathology* 56, 595–602.

Lodha, B.C. and Webster, J. (1990) *Pythium acanthophoron*, a mycoparasite, rediscovered in India and Britain. *Mycological Research* 94, 1006–1008.

Lowen, R., Brady, B.L., Hawksworth, D.L. and Peterson, R.R.M. (1986) Two new lichenicolous species of *Hobsonia. Mycologia* 78, 842–846.

Lumsden, R.D. and Lewis, J.A. (1989) Selection, production, formation, and commercial use of plant disease biocontrol fungi: problems and progress. In: Whipps, J.M. and Lumsden, R.D. (eds) *Biotechnology of Fungi for Improving Plant Growth.* Cambridge University Press, Cambridge, pp. 171–190.

Lumsden, R.D. and Locke, J.C. (1989) Biological control of damping-off caused by *Pythium ultimum* and *Rhizoctonia solani* with *Gliocladium virens* in soilless mix. *Phytopathology* 79, 361–366.

Lumsden, R.D., Locke, J.C., Adkins, S.T., Walter, J.F. and Ridout, C.J. (1992) Isolation and localization of the antibiotic gliotoxin produced by *Gliocladium virens* from alginate prill in soil and soilless media. *Phytopathology* 82, 230–235.

Lundborg, A. and Unestam, T. (1980) Antagonism against *Fomes annosus*. Comparison between different test methods *in vitro* and *in vivo*. *Mycopathologia* 70, 107–115.

Lutchmeah, R.S. and Cooke, R.C. (1984) Aspects of antagonism by the mycoparasite *Pythium oligandrum*. *Transactions of the British Mycological Society* 83, 696–700.

Lutchmeah, R.S. and Cooke, R.C. (1985) Pelleting of seed with the antagonist *Pythium oligandrum* for biological control of damping off. *Plant Pathology* 34, 528–531.

Lynch, J.M. (1987) *In vitro* identification of *Trichoderma harzianum* as a potential antagonist of plant pathogens. *Current Microbiology* 16, 49–53.

Lynch, J.M., Wilson, K.L., Ousley, M.A., Whipps, J.P. (1991) Response of lettuce to *Trichoderma* treatment. *Letters in Applied Microbiology*, 12, 59–61.

Madelin, M.F. (1968) Fungi parasitic on other fungi and lichens. In: Ainsworth, G.C. and Sussman, A.S. *The Fungi*, Volume III. Academic Press, London, pp. 253–269.

Madelin, M.F. and Feest, A. (1982) *Dipodascus macrosporus* sp. nov. (Hemiascomycetes), associated with plasmodia of *Badhamia utricularis. Transactions of the British Mycological Society* 79, 331–335.

Madi, L., Katan, T. and Henis, Y. (1992) Inheritance of antagonistic properties and lytic enzyme activities in sexual crosses of *Talaromyces flavus. Annals of Applied Biology* 121, 565–576.

Magan, N. and Whipps, J.M. (1988) Growth of *Coniothyrium minitans*, *Gliocladium roseum*, *Trichoderma harzianum* and *T. viride* from alginate pellets and interactions with water availability. *EPPO Bulletin* 18, 37–45.

Magnani, G. (1970) A popular rust hyperparasite. *Pubblicazioni del Centro di Sperimentzaione Agricola e Forestale* 11, 27–35.

Makkonen, R. and Pohjakallio, O. (1960) On the parasites attacking the sclerotia of some fungi pathogenic to higher plants and on the resistance of these sclerotia to their parasites. *Acta Agricultura Scandinavica.* 10, 105–126.

Malençon, M.G. (1930) Études de parasitisme mycopathologique I. Sur une propriete mycetophage du *Claudopus byssisedus. Revue de Mycologie* (*Paris*) 7, 27–52.

Malloch, D. and Rogerson, C.T. (1978) Fungi of the Canadian boreal forest region: *Catalus aquilonius* gen. et sp.nov., a hyperparasite on *Seuratia millardetii. Canadian Journal of Botany* 56, 2344–2347.

Mandeel, Q. and Baker, R. (1991) Mechanisms involved in biological control of Fusarium wilt of cucumber with strains of nonpathogenic *Fusarium oxysporum. Phytopathology* 81, 462–469.

Manners, J.M. and Gay, J.L. (1983) The host–parasite interface and nutrient transfer in biotrophic parasitism. In: Callow, J.A. (ed.) *Biochemical Plant Pathology.* John Wiley, Chichester, pp. 163–169.

Manocha, M.S. (1975) Host–parasite relations in a mycoparasite III. Morphological and biochemical differences in the parasitic- and axenic-culture spores of *Piptocephalis virginiana. Mycologia* 67, 382–391.

Manocha, M.S. (1981) Host specificity and mechanism of resistance in a mycoparasitic system. *Physiological Plant Pathology* 18, 257–265.

Manocha, M.S. (1985) Specificity of mycoparasite attachment to the host cell surface. *Canadian Journal of Botany* 63, 772–778.

Manocha, M.S. (1987) Cellular and molecular aspects of fungal host–mycoparasite interaction. *Journal of Plant Diseases and Protection* 94, 431–444.

Manocha, M.S. (1988) Biotrophic mycoparasitism: a model system for investigations in host–parasite interactions. In: Agnihotri, V.P., Sarbhoy, A.K. and Kumar, D. (eds) *Perspectives in Mycology and Plant Pathology.* Malhotra Publishing House, New Delhi, India.

Manocha, M.S. and Begum, B. (1985) Properties of chitin synthase from mucoraceous hosts of a mycoparasite. *Canadian Journal of Microbiology* 31, 6–12.

Manocha, M.S. and Campbell, C.D. (1983) Host–parasite relations in a mycoparasite. VIII. Age-related host resistance in *Choanephora cucurbitarum. Mycologia* 75, 588–596.

Manocha, M.S. and Deven, J.M. (1975) Host–parasite relations in a mycoparasite. IV. A correlation between the levels of γ-linolenic acid and parasitism of *Piptocephalis virginiana. Mycologia* 67, 1148–1157.

Manocha, M.S. and Golesorkhi, R. (1979) Host–parasite relations in a mycoparasite. V. Electron microscopy of *Piptocephalis virginiana* infection in compatible and incompatible hosts. *Mycologia* 61, 565–576.

Manocha, M.S. and Graham, L.L. (1982) Host cell wall synthesis and its role in resistance to a mycoparasite. *Physiological Plant Pathology* 20, 157–164.

Manocha, M.S. and Lee, K.Y. (1971) Host–parasite relations in mycoparasite. I. Fine structure of host, parasite, and their interface. *Canadian Journal of Botany* 49, 1677–1681.

Manocha, M.S. and Lee, K.Y. (1972) Host–parasite relations in mycoparasite. II. Incorporation of tritiated *N*-acetyl-glucosamine into *Choanephora cucurbitarum* infected with *Piptocephalis virginiana. Canadian Journal of Botany* 50, 35–37.

Manocha, M.S. and Maharaj, R. (1981) Host–parasite relations in a mycoparasite. VI. Leakage of electrolytes, amino acids, and sugars after infection with *Piptocephalis virginiana. Canadian Journal of Botany* 59, 1271–1276.

Manocha, M.S., Chen, Y. and Rao, N. (1990) Involvement of cell surface sugars in recognition, attachment and appressorium formation by a mycoparasite. *Canadian Journal of Microbiology* 36, 771–778.

Maplestone, P.A., Whipps, J.M. and Lynch, J.M. (1991) Effect of peat–bran inoculum of *Trichoderma* species on biological control of *Rhizoctonia solani* in lettuce. *Plant and Soil* 136, 257–263.

Marlowe, A. and Romaine, C.P. (1982) Dry bubble of oyster mushrooms caused by *Verticillium fungicola. Plant Diseases* 66, 859–860.

Marois, J.J., Johnston, S.A., Dunn, M.T. and Papavizas, G.C. (1982) Biological control of *Verticillium* wilt of eggplant in the field. *Plant Disease* 66, 1166–1168.

Marois, J.J., Fravel, D.R. and Papavizas, G.C. (1984) Ability of *Talaromyces flavus* to occupy the rhizosphere and its interaction with *Verticillium dahliae. Soil Biology and Biochemistry* 16, 387–390.

Martin, F.N. and Hancock, J.G. (1986) Association of chemical and biological factors in soils suppressive to *Pythium ultimum. Phytopathology* 76, 1221–1231.

Martin, F.N. and Hancock, J.G. (1987) The use of *Pythium oligandrum* for biological control of preemergence damping-off caused by *P. ultimum. Phytopathology* 77, 1013–1020.

Martin, S.B., Hoch, H.C. and Abawi, G.S. (1983) Population dynamics of *Laetisaria arvalis* and low-temperature *Pythium* spp. in untreated and pasteurized beet field soils. *Phytopathology* 73, 1445–1449.

Marvanova, L. (1977) A contact biotrophic mycoparasite on aquatic hyphomycete conidia. *Transactions of the British Mycological Society* 68, 485–488.

Marvanova, L. and Bandoni, R.J. (1987) *Naiadella fluitans* gen. et sp. nov.: a conidial basidiomycete. *Mycologia* 79, 578–586.

Marvanova, L. and Suberkropp, K. (1990) *Camptobasidium hydrophilum* and its anamorph, *Crucella subtilis*: a new heterobasidiomycete from streams. *Mycologia* 82, 208–217.

McCredie, T.A. and Sivasithamparam, K. (1985) Fungi mycoparasitic on sclerotia of *Sclerotinia sclerotiorum* in some Western Australian soils. *Transactions of the British Mycological Society* 84, 736–739.

McDaniel, L.L. and Hindal, D.F. (1979) Host range and effect of host nutrition on parasitism by *Piptocephalis lepidula. Mycologia* 71, 996–1004.

McDaniel, L.L. and Hindal, D.F. (1982) Spore swelling, germination and germ tube formation in axenic culture among four species of *Piptocephalis. Mycologia* 74, 271–274.

McDougall, W.B. (1919) Development of *Stropharia epimyces. Botanical Gazette* 67, 258–263.

McKenzie, E.H.C. and Hudson, H.J. (1976) Mycoflora of rust-infected and non-infected plant material during decay. *Transactions of the British Mycological Society* 66, 223–238.

McLaren, D.L., Huang, H.C. and Rimmer, S.R. (1986) Hyperparasitism of *Sclerotinia sclerotiorum* by *Talaromyces flavus. Canadian Journal of Plant Pathology* 8, 43–48.

McLaren, D.L., Huang, H.C., Rimmer, S.R. and Kokko, E.G. (1989) Ultrastructural studies on infection of sclerotia of *Sclerotinia sclerotiorum* by *Talaromyces flavus*. *Canadian Journal of Botany* 67, 2199–2205.

McLaughlin, R.J., Wilson, C.L., Chalutz, E., Kurtzman, C.P., Fett, W.F. and Osman, S.F. (1990a) Characterization and reclassification of yeasts used for biological control of postharvest diseases of fruit and vegetables. *Applied and Environmental Microbiology* 56, 3583–3586.

McLaughlin, R.J., Wisniewski, M.E., Wilson, C.L. and Chalutz, E. (1990b) Effect of inoculum concentration and salt solutions on biological control of postharvest diseases of apples with *Candida* sp. *Phytopathology* 80, 456–461.

McMeekin, D. (1991) Basidiocarp formation in *Asterophora lycoperdioides*. *Mycologia*, 83, 220–223.

McNabb, R.F.R. (1965) Some auriculariaceous fungi from the British Isles. *Transactions of the British Mycological Society* 48, 187–192.

McQuilken, M.P., Whipps, J.M. and Cooke, R.C. (1990a) Control of damping-off in cress and sugar beet by commercial seed-coating with *Pythium oligandrum*. *Plant Pathology* 39, 452–462.

McQuilken, M.P., Whipps, J.M. and Cooke, R.C. (1990b) Oospores of the biocontrol agent *Pythium oligandrum* bulk-produced in liquid culture. *Mycological Research* 94, 613–616.

McQuilken, M.P., Whipps, J.M. and Cooke, R.C. (1992a) Effects of osmotic and matric potential on growth and oospore germination of the biocontrol agent *Pythium oligandrum*. *Mycological Research* 96, 588–591.

McQuilken, M.P., Whipps, J.M. and Cooke, R.C. (1992b) Use of oospore formulations of *Pythium oligandrum* for biological control of *Pythium* damping-off in cress. *Journal of Phytopathology* 135, 125–134.

Mendgen, K. and Casper, R. (1980) Detection of *Verticillium lecanii* in pustules of bean rust (*Uromyces phaseoli*) by immunofluorescence. *Phytopathologische Zeitschrift* 99, 362–364.

Mercer, P.C. and Kirk, S.A. (1984a) Biological treatments for the control of decay in tree wounds. I. Laboratory tests. *Annals of Applied Biology* 104, 211–219.

Mercer, P.C. and Kirk, S.A. (1984b) Biological treatments for the control of decay in tree wounds. II. Field tests. *Annals of Applied Biology*, 104, 221–229.

Metzler, B., Oberwinkler, F. and Petzold, H. (1989) *Rhynchogastrema* gen. nov. and Rhynchogastremaceae fam. nov. *Systematic and Applied Microbiology* 12, 280–287.

Mijuskovic, M. and Vucinic, Z. (1974) Nove pojave superparazita gliva u Crnoj Gori. *Zast, Bilja* 25, 128–129.

Milanez, A.I. (1967) Resting spores of *Phlyctochytrium planicorne* on Saprolegniaceae. *Transactions of the British Mycological Society* 50, 679–681.

Mirrington, R.N., Ritchie, E., Shoppee, C.W., Taylor, W.C. and Sternhell, S. (1964) The constitution of radicicol. *Tetrahedron Letters* 7, 365–370.

Mitchell, J.K., Taber, R.A. and Pettit, R.E. (1986) Establishment of *Dicyma pulvinata* in *Cercosporidium personatum* leaf spot of peanuts: effect of spray formulation, inoculation time, and hours of leaf wetness. *Phytopathology* 76, 1168–1171.

Mitchell, J.K., Smith, D.J. and Taber, R.A. (1987) Potential for biological control of *Cercosporidium personatum* leafspot of peanuts by *Dicyma pulvinata*. *Canadian Journal of Botany* 65, 2263–2269.

Montealegre, J.R. and Henriquez, J.L. (1990) Posibilidades de control integrado et

Sclerotium rolfsii Sacc. mediante hongos del genero *Trichoderma* y fungicidas. *Fitopatologia* 25, 68–74.

Morgan-Jones, G. and Gray, D.J. (1979) Notes on Hyphomycetes. XXVII. *Pycogone psilocybina* , a new mycoparasitic species. *Mycotaxon* 8, 144–148.

Morgan-Jones, G. and Gray, D.J. (1980) Notes on Hyphomycetes. XXVIII. *Sepedonium ampullosporum* and *Sibirina lutea* sp. nov. *Mycotaxon* 12, 241–248.

Morrell, J.J. and Sexton, C.M. (1990) Evaluation of a biocontrol agent for controlling basidiomycete attack of Douglas fir and Southern pine. *Wood and Fiber Science* 22, 10–21.

Morris, R.A.C., Coley-Smith, J.R. and Whipps, J.M. (1992) Isolation of the mycoparasite *Verticillium biguttatum* from sclerotia of *Rhizoctonia solani* in the United Kingdom. *Plant Pathology* 41, 513–516.

Morris, W. (1965) Taxonomy and physiology of a new and unusual species of *Piptocephalis.* PhD Thesis, University of London.

Moubasher, A.H. and Mazen, M.B. (1971) Selective effects of three fumigants on Egyptian soil fungi. *Transactions of the British Mycological Society* 57, 447–454.

Mueller, J.D., Cline, M.N., Sinclair, J.B. and Jacobsen, B.J. (1985) An *in vitro* test for evaluating efficacy of mycoparasites on sclerotia of *Sclerotinia sclerotiorum. Plant Disease* 69, 584–587.

Mukhopadhyay, A.N., Brahmbhatt, A. and Patel, G.J. (1986) *Trichoderma harzianum* a potential bio-control agent for tobacco damping-off. *Tobacco Research* 12, 26–35.

Mulligan, D.F.C. and Deacon, J.W. (1992) Detection of presumptive mycoparasites in soil placed on host-colonized agar plates. *Mycological Research* 96, 605–608.

Nagler, A. and Oberwinkler, F. (1989) Haustoria in *Urocystis* (Tilletiales). *Plant Systematics and Evolution* 165, 17–28.

Nelson, E.B., Harman, G.E. and Nash, G.T. (1988) Enhancement of *Trichoderma*-induced biological control of *Pythium* seed rot and pre-emergence damping-off of peas. *Soil Biology and Biochemistry* 20, 145–150.

Nooji, M.P. de and Paul, N.D. (1992) Invastion of rust (*Puccinia poarum*) pycnia and aecia on coltsfoot (*Tussilago farfara*) by secondary pathogens: death of host leaves. *Mycological Research* 96, 309–312.

Nordbring-Herz, B. and Mattiason, B. (1979) Action of a nematode-trapping fungus shows lectin-mediated host-microorganisms interaction. *Nature* 281, 477–479.

Oberwinkler, F. and Bandoni, R.J. (1982) Carcinomycetaceae: a new family in the Heterobasidiomycetes. *Nordic Journal of Botany* 2, 501–516.

Oberwinkler, F. and Bandoni, R.J. (1983) *Trimorphomyces*: a new genus in the Tremellaceae. *Systematic and Applied Microbiology* 4, 105–113.

Oberwinkler, F. and Bauer, R. (1990) *Cryptomycocolax*: a new mycoparasitic heterobasidiomycete. *Mycologia* 82, 671–692.

Oberwinkler, F. and Lowy, B. (1981) *Syzygospora alba* Martin, a mycoparasitic heterobasidiomycete. *Mycologia* 73, 1108–1115.

Oberwinkler, F., Bandoni, R.J., Bauer, R., Deml, G. and Kisimova-Horovitz, L. (1984) The life history of *Christiansenia pallida*, a dimorphic, mycoparasitic heterobasidiomycete. *Mycologia*, 76, 9–22.

Oberwinkler, F., Bauer, R. and Schneller, J. (1990) *Phragmoxenidium mycophilum*, an unusual mycoparasitic heterobasidiomycete. *Systematic and Applied Microbiology* 13, 186–191.

Olive, L.S. (1946) New or rare heterobasidiomycetes from North Carolina. *Journal of the*

Elisha Mitchell Scientific Society 62, 65–71.

Olive, L.S. (1951) New or noteworthy Tremellales from the southern Appalachians. *Bulletin of the Torrey Botantical Club* 78, 103–112.

Olive, L.S. (1968) An unusual heterobasidiomycete with *Tilletia*-like basidia. *Journal of the Elisha Mitchell Scientific Society* 84, 261–266.

Op den Camp, H.J.M., Stumm, C.K., Straatsma, G., Derikx, P.J.L. and van Griensven, L.J.L.D. (1990) Hyphal and mycelial interactions between *Agaricus bisporus* and *Scytalidium thermophilum* on agar media. *Microbial Ecology* 19, 303–309.

Ordentlich, A., Wiesman, Z., Gottlieb, H.E., Cojocaru, M. and Chet, I. (1992) Inhibitory furanone produced by the biocontrol agent *Trichoderma harzianum. Phytochemistry* 31, 485–486.

Orton, P.D. (1986) *British Fungus Flora. Agarics and Boleti. 4 Pluteaceae*: Pluteus *and* Volvariella. Royal Botanic Garden Edinburgh, 98 pp.

Ossana, N. and Mischke, S. (1990) Genetic transformation of the biocontrol fungus *Gliocladium virens* to benomyl resistance. *Applied and Environmental Microbiology* 56, 3052–3056.

Pachenari, A. and Dix, N.J. (1980) Production of toxins and wall degrading enzymes by *Gliocladium roseum. Transactions of the British Mycological Society* 74, 561–566.

Papavizas, G.C. (1981) Survival of *Trichoderma harzianum* in soil and in pea and bean rhizosphere. *Phytopathology* 71, 121–125.

Papavizas, G.C. (1985) *Trichoderma* and *Gliocladium*: biology, ecology, and potential for biocontrol. *Annual Review of Phytopathology* 23, 23–54.

Papavizas, G.C. and Collins, D.J. (1990) Influence of *Gliocladium virens* on germination and infectivity of sclerotia of *Sclerotium rolfsii. Phytopathology* 80, 627–630.

Papavizas, G.C. and Lewis, J.A. (1989) Effect of *Gliocladium* and *Trichoderma* on damping-off of snap bean caused by *Sclerotium rolfsii* in the greenhouse. *Plant Pathology* 38, 277–286.

Papavizas, G.C., Morris, B.B. and Marois, J.J. (1983) Selective isolation and enumeration of *Laetisaria arvalis* from soil. *Phytopathology* 73, 220–223.

Papavizas, G.C., Dunn, M.T., Lewis, J.A. and Beagle-Ristaino, J. (1984) Liquid fermentation technology for experimental production of biocontrol fungi. *Phytopathology* 74, 1171–1175.

Papavizas, G.C., Fravel, D.R. and Lewis, J.A. (1987) Proliferation of *Talaromyces flavus* in soil and survival in alginate pellets. *Phytopathology* 77, 131–136.

Papavizas, G.C., Roberts, D.P. and Kim, K.K. (1990) Develoment of mutants of *Gliocladium virens* tolerant to benomyl. *Canadian Journal of Microbiology* 36, 484–489.

Park, D. (1972) Methods of detecting fungi in organic detritus in water. *Transactions of the British Mycological Society* 58, 281–290.

Park, Y.-H., Stack, J.P. and Kenerley, C.M. (1992) Selective isolation and enumeration of *Gliocladium virens* and *G. roseum* from soil. *Plant Disease* 76, 230–235.

Parker, E.J. (1977) Viability tests for the biological control fungus *Peniophora gigantea* (Fr.) Mass. *European Journal of Forest Pathology* 7, 251–253.

Paulitz, T.C. and Baker, R. (1987a) Biological control of *Pythium* damping-off of cucumbers with *Pythium nunn*: Population dynamics and disease suppression. *Phytopathology* 77, 335–340.

Paulitz, T.C. and Baker, R. (1987b) Biological control of *Pythium* damping-off of cucumbers with *Pythium nunn*: Influence of soil environment and organic amendments. *Phytopathology* 77, 341–346.

Paulitz, T.C. and Baker, R. (1988) Interactions between *Pythium nunn* and *Pythium ultimum* on bean leaves. *Canadian Journal of Microbiology* 34, 947–951.

Paulitz, T.C. and Linderman, R.G. (1991) Lack of antagonism between the biocontrol agent *Gliocladium virens* and vesicular arbuscular mycorrhizal fungi. *New Phytologist* 117, 303–308.

Paulitz, T.C. and Menge, J.A. (1984) Is *Spizellomyces punctatum* a parasite or saprophyte of vesicular–arbuscular mycorrhizal fungi? *Mycologia* 76, 99–107.

Paulitz, T.C. and Menge, J.A. (1986) The effects of a mycoparasite on the mycorrhizal fungus, *Glomus deserticola. Phytopathology* 76, 351–354.

Paulitz, T.C., Ahmad, J.S. and Baker, R. (1990) Integration of *Pythium nunn* and *Trichoderma harzianum* isolate T-95 for the biological control of *Pythium* damping-off of cucumber. *Plant and Soil* 121, 243–250.

Pearce, M.H. (1990) *In vitro* interactions between *Armillaria luteobubalina* and other wood decay fungi. *Mycological Research* 94, 753–761.

Pegler, D.N., Spooner, B.M. and Young, T.W.K. (1993) *British Truffles. A Revision of British Hypogeous Fungi*. Royal Botanic Gardens, Kew, 216 pp.

Pemberton, C.M., Davey, R.M., Webster, J., Dick, M.W. and Clark, G. (1990) Infection of *Pythium* and *Phytophthora* species by *Olpidopis gracilis* (Oomycetes). *Mycological Research* 94, 1081–1085.

Peresse, M. and Le Picard, D. (1980) *Hansfordia pulvinata*, mycoparasite destructeur du *Cladosporium fulva. Mycopathologia* 71, 22–30.

Persson, Y. (1991) Mycoparasitism by the nematode-trapping fungus *Arthrobotrys oligospora.* PhD thesis, University of Lund.

Persson, Y. and Baath, E. (1992) Quantification of mycoparasitism by the nematode-trapping fungus *Arthrobotrys oligospora* on *Rhizoctonia solani* and the influence of nutrient levels. *FEMS Microbiology Ecology* 101, 11–16.

Persson, Y., Veenhuis, M. and Nordbring-Hertz, B. (1985) Morphogenesis and significance of hyphal coiling by nematode-trapping fungi in mycoparasitic relationships. *FEMS Microbiology Ecology* 31, 283–291.

Petrini, O., Hake, U. and Dreyfuss, M.M. (1990) An analysis of fungal communities isolated from fruticose lichens. *Mycologia* 82, 444–451.

Philipp, W.-D. (1985) Extracellular enzymes and nutritional physiology of *Ampelomyces quisqualis* Ces., hyperparasite of powdery mildew, *in vitro. Phytopathologische Zeitschrift* 114, 274–283.

Philipp, W.-D., Beuther, E., Hermann, D., Klinkert, F., Oberwalder, C., Schmidt, M. and Straub, B. (1990) Formulation of the powdery mildew hyperparasite *Ampelomyces quisqualis* Ces. *Journal of Plant Diseases and Protection* 97, 120–132.

Phillips, A.J.L. (1986) Factors affecting the parasitic activity of *Gliocladium virens* on sclerotia of *Sclerotinia sclerotiorum* and a note on its host range. *Journal of Phytopathology* 116, 212–220.

Phillips, A.J.L. (1987) Carpogenic germination of sclerotia of *Sclerotinia sclerotiorum*: a review. *Phytophylactica* 19, 279–283.

Phillips, A.J.L. and Price, K. (1983) Structural aspects of the parasitism of sclerotia of *Sclerotinia sclerotiorum* (Lib.) de Bary by *Coniothyrium minitans* Camp. *Phytopathologische Zeitschrift* 107, 193–203.

Phipps, P.M. and Barnett, H.L. (1975) Effect of nitrogen nutrition on the free amino acid pool of *Choanephora cucurbitarum* and parasitism by two haustorial mycoparasites. *Mycologia* 67, 1128–1142.

Pirozynski, K.A. and Hawksworth, D.L. (eds) (1988) *Coevolution of Fungi with Plants and Animals.* Academic Press, London, 285 pp.

Pollack, F.G. (1971) *Cercospora uromycestri*, hyperparasite of rust on *Cestrum diurnum. Mycologia* 63, 689–693.

Powell, M.J. (1981) Structure of the interface between the haustorium of *Caulochytrium protostelioides* and the hyphal cytoplasm of *Cladosporium cladosporioides. Journal of the Elisha Mitchell Scientific Society* 97, 171–182.

Powell, M.J. (1982) Ultrastructure of the host–parasite interface between *Allomyces javanicus* and its endoparasite *Catenaria allomyces. Botanical Gazette* 143, 176–187.

Powell, M.J. (1984) Fine structure of the unwalled thallus of *Rozella polyphagi* in its host *Polyphagus euglenae. Mycologia* 76, 1039–1048.

Prillinger, H. (1987) Yeasts and anastomoses: their occurrence and implications for the phylogeny of Eumycota. In: Rayner, A.D.M., Brasier, C.M. and Moore, D.M. (eds) *Evolutionary Biology of the Higher Fungi.* Cambridge University Press, Cambridge, pp. 355–377.

Prowse, G.A. (1954) *Sommerstorffia spinosa* and *Zoophagus insidians* predacious on rotifers, and *Rozellopsis inflata* the endoparasite of *Zoophagus. Transactions of the British Mycological Society* 37, 134–150.

Raghavendra Rao, N.N. and Pavgi, M.S. (1976) A mycoparasite on *Sclerospora graminicola. Canadian Journal of Botany* 54, 220–223.

Raghavendra Rao, N.N. and Pavgi, M.S. (1978) Two mycoparasites on powdery mildews. *Sydowia* 30, 145–147.

Rajarathnam, S., Singh, N.S. and Bano, Z. (1979) Efficacy of carboxin and heat treatment for controlling the growth of *Sclerotium rolfsii* during culture of the mushroom *Pleurotus flabellatus. Annals of Applied Biology* 92, 323–328.

Rakvidhyasastra, V. and Butler, E.E. (1973) Mycoparasitism by *Stephanoma phaeospora. Mycologia* 65, 580–583.

Ramsbottom, J. (1965) *A Handbook of the Larger British Fungi.* Alden Press, Oxford, 222 pp.

Rayner, A.D.M. (1991) The challenge of the individualistic mycelium. *Mycologia* 83, 48–71.

Rayner, A.D.M. and Webber, J.F. (1984) Interspecific mycelial interactions – an overview. In: Jennings, D.H. and Rayner, A.D.M. (eds) *The Ecology and Physiology of the Fungal Mycelium.* Cambridge University Press, Cambridge, pp. 383–417.

Rayner, A.D.M., Watling, R. and Frankland, J.C. (1985) Resource relations – an overview. In: Moore, D., Casselton, L.A., Wood, D.A. and Frankland, J.C. (eds) *Developmental Biology of Higher Fungi.* Cambridge University Press, Cambridge, pp. 1–40.

Rayner, A.D.M., Boddy, L. and Dowson, C.G. (1987) Temporary parasitism of *Coriolus* spp. by *Lenzites betulina*: a strategy for domain capture in wood decay fungi. *FEMS Microbiology Ecology* 45, 53–58.

Redhead, S.A. and Malloch, D.W. (1977) The Endomycetaceae: new concepts and new taxa. *Canadian Journal of Botany* 55, 1701–1711.

Renwick, A., Campbell, R. and Coe, S. (1991) Assessment of *in vivo* screening systems for potential biocontrol agents of *Gaeumannomyces graminis. Plant Pathology* 40, 524–552.

Reper, C. and Penninckx, M.J. (1987) Inhibition of *Pleurotus ostreatus* growth and fructification by a diffusible toxin from *Trichoderma hamatum. Lebensmittel-Wissenschaft u.-Technologie* 20, 291–292.

Rghei, N.A., Castle, A.J. and Manocha, M.S. (1992) Involvement of fimbriae in fungal host–mycoparasite interaction. *Physiological and Molecular Plant Pathology* 41, 139–148.

Ribeiro, W.R. and Butler, E.E. (1992) Isolation of mycoparasitic species of *Pythium* with spiny oogonia from soil in California. *Mycological Research* 96, 857–862.

Ricard, J.L. (1981) Commercialization of a *Trichoderma*-based mycofungicide: some problems and solutions. *Biocontrol News and Information.* 2, 95–98.

Ricard, J.L., Wilson, M.M. and Bollen, W.B. (1969) Biological control of decay in Douglas-fir poles. *Forest Products Journal* 19, 41–45.

Richardson, M.J. (1963) The biology of a new species of *Piptocephalis.* MSc thesis, Nottingham Regional College of Technology, University of Nottingham.

Richardson, M.J. and Leadbeater, G. (1972) *Piptocephalis fimbriata* sp. nov., and observations on the occurrence of *Piptocephalis* and *Syncephalis. Transactions of the British Mycological Society* 58, 205–215.

Ridout, C.J., Coley-Smith, J.R. and Lynch, J.M. (1988) Fractionation of extracellular enzymes from a mycoparasitic strain of *Trichoderma harzianum. Enzyme and Microbial Technology* 10, 180–187.

Rishbeth, J. (1963) Stump protection against *Fomes annosus.* III. Inoculation with *Peniophora gigantea. Annals of Applied Biology* 52, 63–77.

Rishbeth, J. (1988) Biological control of air-borne pathogens. *Philosophical Transactions of the Royal Society, London* B318, 265–281.

Ristaino, J.B., Perry, K.B. and Lumsden, R.D. (1991) Effect of solarization and *Gliocladium virens* on sclerotia of *Sclerotium rolfsii*, soil microbiota, and the incidence of southern blight of tomato. *Phytopathology* 81, 1117–1124.

Roberts, D.P. and Lumsden, R.D. (1990) Effect of extracellular metabolites from *Gliocladium virens* on germination of sporangia and mycelial growth of *Pythium ultimum. Phytopathology* 80, 461–465.

Roberts, P. (1992) Spiral-spored *Tulasnella* species from Devon and the New Forest. *Mycological Research* 96, 233–236.

Roberts, R.G. (1990) Postharvest biological control of gray mold of apple by *Cryptococcus laurentii. Phytopathology* 80, 526–530.

Robertson, R.W., Luttrell, E.S. and Fuller, M.S. (1990) Mycoparasitism of teliospores of *Ustilago bullata* by an oomycete. *Canadian Journal of Botany* 68, 2415–2421.

Rogerson, C.T. and Samuels, G.J. (1989) Boleticolous species of *Hypomyces. Mycologia* 81, 413–432.

Rogerson, C.T. and Stephenson, S.L. (1993) Myxomyceticolous fungi. *Mycologia* 85, 456–469.

Rombach, M.C. and Roberts, D.W. (1987) *Calcarisporium ovalisporum*, symbiotic with the insect pathogen *Hirsutella citriformis. Mycologia* 79, 153–155.

Ross, J.P. and Ruttencutter, R. (1977) Population dynamics of two vesicular–arbuscular endomycorrhizal fungi and the role of hyperparasitic fungi. *Phytopathology* 67, 490–496.

Roy, A.K. and Sayre, R.M. (1984) Electron microscopical studies of *Trichoderma harzianum* and *T. viride*, and mycoparasitic activity of the former on *Rhizoctonia solani* f. sp. *sasakii. Indian Phytopathology* 37, 710–712.

Rudakov, O.L. (1978) Physiological groups in mycophilic fungi. *Mycologia* 70, 150–159.

Saccardo, P.A. (1877) *Sylloge Fungorum* 6.

Saksirirat, W. and Hoppe, H.H. (1990) *Verticillium psalliotae*, an effective mycoparasite of the soybean rust fungus *Phakopsora pachyrhizi* Syd. *Zeitschrift für Pflanzenkrankheiten und Pflanzenschutz* 97, 622–633.

Samuels, G.J. (1988) Fungicolous, lichenicolous, and myxomyceticolous species of *Hypocreopsis, Nectriopsis, Nectria, Peristomialis,* and *Trichonectria. Memoirs of the New York Botanic Garden* 48, 1–78.

Satina, S. and Blakeslee, A.F. (1926) The *Mucor* parasite *Parasitella* in relation to sex. *Proceedings of the New York Academy of Sciences* 12, 202–207.

Savile, D.B.O. (1968) Possible interrelationships between fungal groups. In: Ainsworth, G.C. and Sussman, A.S. (eds) *The Fungi,* Volume III. Academic Press, London, pp. 649–676.

Savile, D.B.O. (1979) Fungi as aids to higher plant classification. *Botanical Review* 45, 377–503.

Schmitthenner, A.F. (1962) Isolation of *Pythium* from soil particles. *Phytopathology* 52, 1133–1138.

Schoen, J.F. (1983) Identification of seed-like structures: a taxonomic review of sclerotial-forming fungi. *Seed Science and Technology* 11, 639–650.

Schumacher, T. and Ryvarden, L. (1981) *Dipodascus polyporicola* nov. sp., a parasitic hemiascomycete on *Piptoporus soloniensis* (Fr.) Pil. *Mycotaxon* 12, 525–530.

Seaby, D. (1987) Infection of mushroom compost by *Trichoderma* species. *Bulletin of the Mushroom Growers Association,* 355–361.

Searle, T. and Tribe, H.T. (1984) *Coniothyrium minitans* as a biological control agent of sclerotial diseases. *Abstracts of the British Society for Plant Pathology Anglo French Workshop.*

Sharif, F.M., Okasha, A.M. and Kazem, K.T. (1988) *Penicillium stipitatum* and *Trichoderma harzianum* in the biological control of cucumber damping-off disease caused by *Pythium aphanidermatum. Journal of the University of Kuwait* 15, 107–113.

Sharma, J.K. and Heather, W.A. (1978) Parasitism of uredospores of *Melampsora larici-populina* Kleb. by *Cladosporium* sp. *European Journal of Forest Pathology* 8, 48–54.

Sharma, J.K. and Heather, W.A. (1983) Post-penetration antagonism by *Cladosporium tenuissimum* to uredinial induction by *Melampsora larici-populina. Transactions of the British Mycological Society* 80, 373–374.

Sharma, J.K. and Heather, W.A. (1988) Light and electron microscope studies on *Cladosporium tenuissimum,* mycoparasitic on poplar leaf rust, *Melampsora larici-populina. Transactions of the British Mycological Society* 90, 125–131.

Shaw, R. (1965) The occurrence of γ-linolenic acid in fungi. *Biochimica et Biophysica Acta* 98, 230–237.

Shaw, R. (1966) The fatty acids of phycomycete fungi, and the significance of the γ-linolenic acid component. *Comparative Biochemistry and Physiology* 18, 325–331.

Shigemitsu, H., Kunoh, H. and Akai, S. (1978) Scanning electron microscopic studies of *Corticium rolfsii* sclerotia parasitized by *Aspergillus terreus. Mycologia* 70, 935–943.

Shigo, A.L. (1960a) Parasitism of *Gonatobotryum fuscum* on species of *Ceratocystis. Mycologia* 52, 584–598.

Shigo, A.L. (1960b) Mycoparasitism of *Gonatobotryum fuscum* and *Piptocephalis xenophila. Transactions of the New York Academy of Science* 22, 365–372.

Shigo, A.L., Anderson, C.D. and Barnett, H.L. (1961) Effects of concentration of host nutrients on parasitism of *Piptocephalis xenophila* and *P. virginiana. Phytopathology* 51, 616–620.

Sideris, C.P. (1932) Taxonomic studies in the family Pythiaceae. 2. *Pythium. Mycologia* 24, 14–61.

Simon, A. (1989) Biological control of take-all of wheat by *Trichoderma koningii* under controlled environmental conditions. *Soil Biology and Biochemistry* 21, 323–326.

Simon, A. and Sivasithamparam, K. (1989) Pathogen suppression: a case study in biological suppression of *Gaeumannomyces graminis* var. *tritici* in soil. *Soil Biology and Biochemistry* 21, 331–337.

Sinden, J.W. (1971) Ecological control of pathogens and weed-molds in mushroom culture. *Annual Review of Phytopathology* 9, 411–432.

Sinden, J.W. and Hauser, E. (1954) Nature and control of three mildew diseases of mushrooms in America. *Mushroom Science* 2, 177–180

Singh, N. (1975) Nutritional requirements for growth of *Dispira cornuta* in axenic culture. *Journal of General Microbiology* 88, 372–376.

Singh, N. and Plunkett, B.E. (1967) Cyanide tolerance and growth stimulation by the carbon of organic nitrogen compounds in hymenomycete-inhabiting fungi. *Transactions of the British Mycological Society* 50, 359–376.

Singh, U.P., Vishwakarma, S.N. and Basuchaudhury, K.C. (1978) *Acremonium sordidulum* mycoparasitic on *Colletotrichum dematium* f. *truncata* in India. *Mycologia* 70, 453–455.

Siqueira, J.O., Hubbell, D.H., Kimbrough, J.M. and Schenck, N.C. (1984) *Stachybotrys chartarum* antagonistic to azygospores of *Gigaspora margarita in vitro. Soil Biology and Biochemistry* 16, 679–681.

Sivakami Sundari, S.S. and Manocha, M.S. (1991) Interaction between biotrophic haustorial mycoparasite and protoplasts of host and nonhost fungi. *Botanica Helvetica* 101, 141–148.

Sivan, A. and Chet, I. (1989) The possible role of competition between *Trichoderma harzianum* and *Fusarium oxysporum* in rhizosphere colonization. *Phytopathology* 79, 198–203.

Sivan, A. and Harman, G.E. (1991) Improved rhizosphere competence in a protoplast fusion progeny of *Trichoderma harzianum. Journal of General Microbiology* 137, 23–29.

Sivan, A., Ucko, O. and Chet, I. (1987) Biological control of *Fusarium* crown rot of tomato by *Trichoderma harzianum* under field conditions. *Plant Disease* 71, 587–592.

Sivan, A., Stasz, T.E., Hemmat, M., Hayes, C.K. and Harman, G.E. (1992) Transformation of *Trichoderma* spp. with plasmids conferring hygromycin B resistance. *Mycologia* 84, 687–694.

Sivakami Sundari, S.S. and Manocha, M.S. (1991) Interaction between biotrophic haustorial mycoparasite and protoplasts of host and nonhost fungi. *Botanica Helvetica* 101, 141–148.

Skidmore, A.M. and Dickinson, C.H. (1976) Colony interactions and hyphal interference between *Septoria nodorum* and phylloplane fungi. *Transactions of the British Mycological Society* 66, 57–64.

Slifkin, M.K. (1963) Parasitism of *Olpidiopsis incrassata* on members of the Saprolegniaceae. II. Effect of pH and host nutrition. *Mycologia* 55, 172–181.

Smith, F.E. (1924) Three diseases of cultivated mushrooms. *Transactions of the British Mycological Society* 10, 81–97.

Smith, V.L., Wilcox, W.F. and Harman, G.E. (1990) Potential for biological control of

Phytophthora root and crown rots of apple by *Trichoderma* and *Gliocladium* spp. *Phytopathology* 80, 880–885.

Sneh, B. (1977) A method for observation and study of living fungal propagules incubated in soil. *Soil Biology and Biochemistry* 9, 65–66.

Sneh, B., Humble, S.J. and Lockwood, J.L. (1977) Parasitism of oospores of *Phytophthora megasperma* var *sojae*, *P. cactorum*, *Pythium* sp. and *Aphanomyces euteiches* in soil by Oomycetes, Chytridiomycetes, Hyphomycetes, Actinomycetes and bacteria. *Phytopathology* 67, 622–628.

Spencer, D.M. (1980) Parasitism of carnation rust (*Uromyces dianthi*) by *Verticillium lecanii*. *Transactions of the British Mycological Society* 74, 191–194.

Spencer, D.M. and Atkey, P.T. (1981) Parasitic effects of *Verticillium lecanii* on two rust fungi. *Transactions of the British Mycological Society* 77, 535–542.

Spink, D.S. and Rowe, R.C. (1989) Evaluation of *Talaromyces flavus* as a biological control agent against *Verticillium dahliae* in potato. *Plant Disease* 73, 230–236.

Srivastava, A.K., Defago, G. and Boller, T. (1985a) Secretion of chitinase by *Aphanocladium album*, a hyperparasite of wheat rust. *Experientia* 41, 1612–1613.

Srivastava, A.K., Defago, G. and Kern, H. (1985b) Hyperparasitism of *Puccinia horiana* and other microcyclic rusts. *Phytopathologische Zeitschrift* 114, 73–78.

Stahle, U. and Kranz, J. (1984) Interactions between *Puccinia recondita* and *Eudarluca caricis* during germination. *Transactions of the British Mycological Society* 82, 562–563.

Staples, R.C. and Macko, V. (1980) Formation of infection structures as a recognition response in fungi. *Experimental Mycology* 4, 2–16.

Straatsma, G., Gerrits, J.P.G., Augustijn, M.P.A.M., Op Den Camp, H.J.M., Vogels, G.D. and Van Griensven, L.J.L.D. (1989) Population dynamics of *Scytalidium thermophilum* in mushroom compost and stimulatory effects on growth rates and yield of *Agaricus bisporus*. *Journal of General Microbiology* 135, 751–759.

Strashnow, Y., Elad, Y., Sivan, A. and Chet, I. (1985) Integrated control of *Rhizoctonia solani* by methyl bromide and *Trichoderma harzianum*. *Plant Pathology* 34, 146–151.

Strunz, G.M., Kakushima, M. and Stilwell, M.A. (1972) Scytalidin: a new fungitoxic metabolite produced by *Scytalidium* sp. *Journal of the Chemical Society* 18, 2280–2283.

Studer, M., Fluck, K. and Zimmerman, W. (1992) Production of chitinases by *Aphanocladium album* grown on crystalline and colloidal chitin. *FEMS Microbiology Letters* 99, 213–216.

Sundheim, L. (1982) Control of cucumber powdery mildew by the hyperparasite *Ampelomyces quisqualis* and fungicides. *Plant Pathology* 31, 209–214.

Sundheim, L. (1986) Use of hyperparasites in biological control of biotrophic plant pathogens. In: Fokkema, N.J. and van den Heuvel, J. (eds) *The Microbiology of the Phyllosphere*. Cambridge University Press, Cambridge, pp. 333–347.

Sundheim, L. and Amundsen, T. (1982) Fungicide tolerance in the hyperparasite *Ampelomyces quisqualis* and integrated control of cucumber powdery mildew. *Acta Agriculturae Scandinavica* 32, 349–355.

Sundheim, L. and Krekling, T. (1982) Host–parasite relationships of the hyperparasite *Ampelomyces quisqualis* and its powdery mildew host *Sphaerotheca fuliginea*. *Phytopathologische Zeitschrift* 104, 202–210.

Sutherland, E.D., Baker, K. and Lockwood, J.L. (1984) Ultrastructure of *Phytophthora megasperma* f. sp. *glycinea* oospores parasitized by *Actinoplanes missouriensis* and *Humicola fuscoatra*. *Transactions of the British Mycological Society* 82, 726–729.

Sutton, B.C. (1981) *The Coelomycetes*. CAB International, Farnham Royal, Slough, 696 pp.

Sy, A.A., Sarr, A., Albertini, L. and Moletti, M. (1990) Biological control of *Pyricularia oryzae*: parameters of stability of antagonistic activity. *Phytopathologia Mediterranea* 29, 175–183.

Sykes, E.E. and Porter, D. (1980) Infection and development of the obligate parasite *Caternaria allomycis* on *Allomyces arbuscula*. *Mycologia* 72, 288–300.

Sylvia, D.M. and Schenck, N.C. (1983) Germination of chlamydospores of three *Glomus* species as affected by soil matric potential and fungal contamination. *Mycologia* 74, 30–35.

Sztejnberg, A., Freeman, S., Chet, I. and Katan, J. (1987) Control of *Rosellinia necatrix* in soil and in apple orchard by solarization and *Trichoderma harzianum*. *Plant Disease* 71, 365–369.

Sztejnberg, A., Galper, S., Mazar, S. and Lisker, N. (1989) *Ampelomyces quisqualis* for biological and integrated control of powdery mildews in Israel. *Journal of Phytopathology* 124, 285–295.

Sztejnberg, A., Galper, S. and Lisker, N. (1990) Conditions for pycnidial production and spore formation by *Ampelomyces quisqualis*. *Canadian Journal of Microbiology* 36, 193–198.

Taber, R.A. (1982) *Gigaspora* spores and associated hyperparasites in weed seeds in soil. *Mycologia* 74, 1026–1031.

Taiwo, F.O., Chant, S.R. and Young, T.W.K. (1987a) Hydrolytic activity of *Zygorhynchus* species. *Microbios* 51, 23–28.

Taiwo, F.O., Chant, S.R. and Young, T.W.K. (1987b) Hydrolytic activity of *Zygorhynchus* species in relation to nitrogen source. *Microbios* 52, 183–189.

Templeton, G.E., Smith, R.J. and Klomparens, W. (1980) Commercialization of fungi and bacteria for biological control. *Biocontrol News and Information* 1, 291–294.

Thompson, G.E. (1936) *Nyctalis parasitica* and *N. asterophora* in culture. *Mycologia* 28, 222–227.

Tirilly, Y. (1990) The role of fosetyl-Al in the potential integrated control of *Fulvia fulvia*. *Canadian Journal of Botany* 69, 306–310.

Tirilly, Y., Lambert, F. and Thouvenot, D. (1991) Bioproduction of [^{14}C]desoxyphomenone, a fungistatic metabolite of the hyperparasite *Dicyma pulvinata*. *Phytochemistry* 30, 3963–3965.

Traquair, J.A. and McKeen, W.E. (1977) Hyphal interference by *Trametes hispida*. *Canadian Journal of Microbiology* 23, 1675–1682.

Traquair, J.A. and McKeen, W.E. (1978) Necrotrophic mycoparasitism of *Ceratocystis fimbriata* by *Hirschioporus pargamenus* (Polyporaceae). *Canadian Journal of Microbiology* 24, 869–874.

Traquair, J.A., Meloche, R.B., Jarvis, W.R., and Baker, K.W. (1984) Hyperparasitism of *Puccinia violae* by *Cladosporium uredinicola*. *Canadian Journal of Botany* 62, 181–184.

Tribe, H.T. (1966) Interactions of soil fungi on cellulose film. *Transactions of the British Mycological Society* 49, 457–466.

Trogoff, H. de and Ricard, J.L. (1976) Biological control of *Verticillium malthousei* by *Trichoderma viride* spray on casing soil in commercial mushroom production. *Plant Disease Reporter* 60, 677–680.

Tronsmo, A. (1986) Use of *Trichoderma* spp. in biological control of necrotrophic pathogens. In: Fokkema, N.J. and van den Heuvel, J. (eds) *Microbiology of the Phyllosphere.* Cambridge University Press, Cambridge, pp. 348–362.

Tronsmo, A. and Dennis, C. (1977) The use of *Trichoderma* species to control strawberry fruit rots. *Netherlands Journal of Plant Pathology* 83 (suppl. 1), 449–455.

Tronsmo, A. and Raa, J. (1977) Antagonistic action of *Trichoderma pseudokoningii* against the apple pathogen *Botrytis cinerea. Phytopathologische Zeitschrift* 89, 216–220.

Tronsmo, A. and Ystaas, J. (1980) Biological control of *Botrytis cinerea* on apple. *Plant Disease* 64, 1009.

Trutmann, P. and Keane, P.J. (1990) *Trichoderma koningii* as a biological control agent for *Sclerotinia sclerotiorum* in southern Australia. *Soil Biology and Biochemistry* 22, 43–50.

Trutmann, P., Keane, P.J. and Merriman, P.R. (1982) Biological control of *Sclerotinia sclerotiorum* on aerial parts of plants by the hyperparasite *Coniothyrium minitans. Transactions of the British Mycological Society* 78, 521–529.

Tsuneda, A. and Hiratsuka, Y. (1979) Mode of parasitism of a mycoparasite, *Cladosporium gallicola,* on western gall rust *Endocronartium harknessii. Canadian Journal of Plant Pathology* 1, 31–36.

Tsuneda, A. and Hiratsuka, Y. (1980) Parasitization of pine stem rust fungi by *Monocillium nordinii. Phytopathology* 70, 1101–1103.

Tsuneda, A. and Hiratsuka, Y. (1981) *Scopinella gallicola,* a new species from rust galls of *Endocronartium harknessii* on *Pinus contorta. Canadian Journal of Botany* 59, 1192–1195.

Tsuneda, A. and Skoropad, W.P. (1977) The *Alternaria brassicae–Nectria inventa* host–parasite interface. *Canadian Journal of Botany* 55, 448–454.

Tsuneda, A. and Skoropad, W.P. (1978a) Behaviour of *Alternaria brassicae* and its mycoparasite *Nectria inventa* on intact and on excised leaves of rapeseed. *Canadian Journal of Botany* 56, 1333–1340.

Tsuneda, A. and Skoropad, W.P. (1978b) Nutrient leakage from dried and rewetted conidia of *Alternaria brassicae* and its effect on the mycoparasite *Nectria inventa. Canadian Journal of Botany* 56, 1341–1345.

Tsuneda, A. and Skoropad, W.P. (1980) Interactions between *Nectria inventa,* a destructive mycoparasite, and fourteen fungi associated with rapeseed. *Transactions of the British Mycological Society* 74, 501–507.

Tsuneda, A., Hiratsuka, Y. and Maruyama, P.J. (1980) Hyperparasitism of *Scytalidium uredinicola* on western gall rust, *Endocronartium harknessii. Canadian Journal of Botany* 58, 1154–1159.

Tu, J.C. (1980) *Gliocladium virens,* a destructive mycoparasite of *Sclerotinia sclerotiorum. Phytopathology* 70, 670–674.

Tu, J.C. (1984) Mycoparasitism by *Coniothyrium minitans* on *Sclerotinia sclerotiorum* and its effect on sclerotial germination. *Phytopathologische Zeitschrift* 109, 261–268.

Tu, J.C. and Vaartaja, O. (1981) The effect of the hyperparasite (*Gliocladium virens*) on *Rhizoctonia solani* and on *Rhizoctonia* root rot of white beans. *Canadian Journal of Botany* 59, 22–27.

Turhan, G. (1990) Further hyperparasites of *Rhizoctonia solani* Kühn as promising candidates for biological control. *Zeitschrift für Pflanzenkrankheiten und Pflanzenschutz* 97, 208–215.

Turner, G.J. and Tribe, H.T. (1975) Preliminary field plot trials on the biological control of *Sclerotinia trifoliorum* by *Coniothyrium minitans*. *Plant Pathology* 24, 109–113.

Turner, G.J. and Tribe, H.T. (1976) On *Coniothyrium minitans* and its parasitism of *Sclerotinia* species. *Transactions of the British Mycological Society* 66, 97–105.

Tzean, S.S. and Estey, R.H. (1978a) Nematode-trapping fungi as mycopathogens. *Phytopathology* 68, 1266–1270.

Tzean, S.S. and Estey, R.H. (1978b) *Schizophyllum commune* Fr. as a destructive mycoparasite. *Canadian Journal of Microbiology* 24, 780–784.

Tzean, S.S. and Estey, R.H. (1991) *Geotrichopsis mycoparasitica* gen. et sp. nov; (Hyphomycetes), a new mycoparasite. *Mycological Research* 95, 1350–1354.

Tzean, S.S. and Estey, R.H. (1992) *Geotrichopsis mycoparasitica* as a destructive mycoparasite. *Mycological Research* 96, 263–269.

Uma, N.U. and Taylor, G.S. (1987) Parasitism of leek rust urediniospore by four fungi. *Transactions of the British Mycological Society* 88, 335–340.

Upadhyay, J.P. and Mukhopadhyay, A.N. (1983) Effect of non-volatile and volatile antibiotics of *Trichoderma harzianum* on the growth of *Sclerotium rolfsii*. *Indian Journal of Mycology and Plant Pathology* 13, 232–233.

Upadhyay, R.S., Bharat, R. and Gupta, R.C. (1981) *Fusarium udum* parasitic on *Aspergillus luchuensis* and *Syncephalastrum racemosum*. *Plant and Soil* 63, 407–413.

Urbasch, I. (1986) *Harzia acremonioides* as biotrophic contact mycoparasite on *Stemphyllium botryosum*. *Journal of Plant Diseases and Protection* 93, 392–396.

Vakili, N.G. (1985) Mycoparasitic fungi associated with potential stalk rot pathogens of corn. *Phytopathology* 75, 1201–1207.

Vakili, N.G. (1989) *Gonatobotrys simplex* and its teleomorph, *Melanospora damnosa*. *Mycological Research* 93, 67–74.

Vakili, N.G. and Bailey, T.B., Jr (1989) Yield response of corn hybrids and inbred lines to phylloplane treatment with mycopathogenic fungi. *Crop Science* 29, 183–190.

Velvis, H. and Jager, G. (1983) Biological control of *Rhizoctonia solani* on potatoes by antagonists. 1. Preliminary experiments with *Verticillium biguttatum*, a sclerotium-inhabiting fungus. *Netherlands Journal of Plant Pathology* 89, 113–123.

Velvis, H., Boogert, P.H.J.F. and Jager, G. (1989) Role of antagonism in the decline of *Rhizoctonia solani* inoculum in soil. *Soil Biology and Biochemistry* 21, 125–129.

Vesely, D. (1977) Potential biological control of damping-off pathogens in emerging sugar beet by *Pythium oligandrum* Drechsler. *Phytopathologische Zeitschrift* 90, 113–115.

Vesely, D. (1978) Biological protection of emerging sugar-beet against damping-off established by mycoparasitism in non-sterilized soil. *Zentraalblatt für Backteriologie*, 133, S. 436–443.

Vesely, D. (1989) Biological control of damping-off pathogens by treating seed with a powdery preparation of the mycoparasite *Pythium oligandrum* in large-scale field trials. In: Vancura, V. and Kinc, F. (eds) *Interrelationships Between Microorganisms and Plants in Soil*. Academia, Czechoslovak Academy of Sciences, Praha, pp. 445–449.

Von Arx, J.A. (1977) Notes on *Dipodascus*, *Endomyces* and *Geotrichum* with the description of two new species. *Antonie van Leeuwenhoek* 43, 333–340.

Walker, J.A. and Maude, R.B. (1975) Natural occurrence and growth of *Gliocladium roseum* on the mycelium and sclerotia of *Botrytis allii*. *Transactions of the British Mycological Society* 65, 335–339.

Walker, J.C., Jones, T.G. and Hedger, J.N. (1982) Preliminary studies on the contact mycoparasite *Nematogonium ferrugineum*. *Transactions of the British Mycological Society*, 78, 374–378.

Walker, J.C. and Minter, D.W. (1981) Taxonomy of *Nematogonum*, *Gonatobotrys*, *Gonatobotryum* and *Gonatorrhodiella*. *Transactions of the British Mycological Society* 77, 299–319.

Walther, D. and Gindrat, D. (1988) Biological control of damping-off of sugar-beet and cotton with *Chaetomium globosum* and a fluorescent *Pseudomonas* sp. *Canadian Journal of Microbiology* 34, 631–637.

Warcup, J.H. (1976) Studies on soil fumigation. IV. Effects on fungi. *Soil Biology and Biochemistry* 8, 261–266.

Warren, J.R. (1948) An undescribed species of *Papulospora* parasitic on *Rhizoctonia solani* Kuhn. *Mycologia* 40, 391–401.

Watling, R. (1974) Dimorphism in *Entoloma abortivum*. *Bulletin de la Société Linneene de Lyon*. Numero special, 43, 449–470.

Watson, P. (1962) Culture of *Spinellus*. *Nature* 195, 1018.

Webster, J. and Lomas, N. (1964) Does *Trichoderma viride* produce gliotoxin and viridin? *Transactions of the British Mycological Society* 47, 535–540.

Weete, J.D. (1980) *Lipid Biochemistry of Fungi and Other Organisms*. Plenum Press, New York, 388 pp.

Weindling, R. (1932) *Trichoderma lignorum* as a parasite of other fungi. *Phytopathology* 22, 837–845.

Weindling, R. (1934) Studies on a lethal principle effective in the parasitic action of *Trichoderma lignorum* on *Rhizoctonia solani* and other soil fungi. *Phytopathology* 24, 1153–1179.

Whaley, J.W. and Barnett, H.L. (1963) Parasitism and nutrition of *Gonatobotrys simplex*. *Mycologia* 55, 199–210.

Whipps, J.M. (1987a) Effect of media on growth and interactions between a range of soil–borne glasshouse pathogens and antagonistic fungi. *New Phytologist* 107, 127–142.

Whipps, J.M. (1987b) Behaviour of fungi antagonistic to *Sclerotinia sclerotiorum* on plant tissue segments. *Journal of General Microbiology* 133, 1495–1501.

Whipps, J.M. and Budge, S.P. (1990) Screening for sclerotial mycoparasites of *Sclerotinia sclerotiorum*. *Mycological Research* 94, 607–612.

Whipps, J.M. and Gerlagh, M. (1992) Biology of *Coniothyrium minitans* and its potential for use in disease biocontrol. *Mycological Research* 96, 897–907.

Whipps, J.M. and Lumsden, R.D. (eds) (1989) *Biotechnology of Fungi for Improving Plant Growth*. Cambridge University Press, Cambridge, 303 pp.

Whipps, J.M. and Lumsden, R.D. (1991) Biological control of *Pythium* species. *Biocontrol Science and Technology* 1, 75–90.

Whipps, J.M., Lewis, D.H. and Cooke, R.C. (1988) Mycoparasitism and plant disease control. In: Burge, M.N. (ed.) *Fungi in Biological Control Systems*. Manchester University Press, Manchester, pp. 161–187.

Whipps, J.M., Budge, S.P. and Ebben, M.H. (1989). Effect of *Coniothyrium minitans* and *Trichoderma harzianum* on *Sclerotinia* disease of celery and lettuce in the glasshouse at a range of humidities. In: Cavalloro, R. and Pelerents, C. (eds) *Integrated Pest Management in Protected Vegetable Crops, Proceedings of CEC IOBC Group Meeting Cabrils, 27–29 May, 1987*. A.A. Balkema, Rotterdam, The Netherlands, pp. 233–243.

Whipps, J.M., Grewal, S.K. and Van der Goes, P. (1991) Interactions between *Coniothyrium minitans* and sclerotia. *Mycological Research* 95, 295–299.

White, P.F. (1981) Spread of the mushroom disease *Verticillium fungicola* by *Megaselia halterata* (Diptera: Phoridae). *Protection Ecology* 3, 17–24.

White, J.G., Wakeham, A.J. and Petch, G.M. (1992) Deleterious effect of soil-applied Metalaxyl and Mancozeb on the mycoparasite *Pythium oligandrum. Biocontrol Science and Technology* 2, 335–340.

White, W.L. (1942) A new hemiascomycete. *Canadian Journal of Research* Sect. C 20, 389–395.

Wicker, E.F. (1981) Natural control of white pine blister rust by *Tuberculina maxima. Phytopathology* 71, 997–1000.

Wicker, E.F. and Woo, J.Y. (1973) Histology of blister rust cankers parasitized by *Tuberculina maxima. Phytopathologische Zeitschrift* 76, 356–366.

Wicklow, D.T. and Wilson, D.M. (1990) *Paecilomyces lilacinus*, a colonist of *Aspergillus flavus* sclerotia buried in soil in Illinois and Georgia. *Mycologia* 82, 393–395.

Williams, P.G., Scott, K.J. and Kuhl, J.L. (1966) Vegetative growth of *Puccinia graminis tritici* in axenic culture. *Nature* 229, 181–182.

Williams, P.G., Scott, K.J., Kuhl, J.L. and Maclean, D.J. (1967) Sporulation and pathogenicity of *Puccinia graminis* f.sp. *tritici* grown on an artificial medium. *Phytopathology* 56, 1418–1419.

Williamson, M.A. and Fokkema, N.J. (1985) Phyllosphere yeasts antagonize penetration from appressoria and subsequent infection of maize leaves by *Colletotrichum graminicola. Netherlands Journal of Plant Pathology* 91, 265–276.

Willoughby, L.G. (1988) *Saprolegnia* parasitized by *Mortierella alpina. Transactions of the British Mycological Society* 90, 496–499.

Wilson, C.L. and Chalutz, E. (1989) Postharvest biological control of *Penicillium* rots of citrus with antagonistic yeasts and bacteria. *Scientia Horticulturae* 40, 105–112.

Wood, S.N. and Cooke, R.C. (1986) Effect of *Piptocephalis* species on growth and sporulation of *Pilaira anomala. Transactions of the British Mycological Society* 86, 672–674.

Woodbridge, B., Coley-Smith, J.R. and Reid, D.A. (1988) A new species of *Cylindrobasidium* parasitic on sclerotia of *Typhula incarnata. Transactions of the British Mycological Society* 91, 166–169.

Wynn, A.R. and Epton, H.A.S. (1979) Parasitism of oospores of *Phytophthora erythroseptica* in soil. *Transactions of the British Mycological Society* 73, 255–259.

Yaniv, Z., Kenneth, R.G. and Miura, J. (1979) Teliospore formation in *Puccinia graminis* f.sp. *tritici* grown in axenic culture, induced by the fungus *Aphanocladium album. Physiological Plant Pathology* 14, 153–156.

Young, C.S. and Andrews, J.H. (1990) Recovery of *Athelia bombacina* from apple leaf litter. *Phytopathology* 80, 530–535.

Young, T.W.K. (1968) Electron microscopic study of the asexual structures in Mucorales. *Proceedings of the Linnean Society* 179, 1–9.

Young, T.W.K. (1970) Ultrastructure of the spore wall of *Linderina. Transactions of the British Mycological Society* 54, 15–25.

Young, T.W.K. (1973a) Ultrastructure of the sporangiospore of *Coemansia reversa* (Mucorales). *Transactions of the British Mycological Society* 60, 57–63.

Young, T.W.K. (1973b) Ultrastructure of the sporangiospore of *Coemansia aciculifera* (Mucorales). *Annals of Botany* 37, 973–979.

Young, T.W.K. (1974) Ultrastructure of the sporangiospore of *Kickxella alabastrina* (Mucorales). *Annals of Botany* 38, 873–876.

Young, T.W.K. (1985) Ultrastructure of mucoralean sporangiospores. *Botanical Journal of the Linnean Society* 91, 151–165.

Zaayen, A. van and Van der Pol-Luiten, B. (1977) Heat resistance, biology and prevention of *Diehliomyces microsporus* in crops of *Agaricus* species. *Netherlands Journal of Plant Pathology* 83, 221–240.

Zaayen, A. van and Gams, W. (1982) Contribution to the taxonomy and pathogenicity of fungicolous *Verticillium* species. II. Pathogenicity. *Netherlands Journal of Plant Pathology* 88, 57–78.

Zazzerini, A. and Tosi, L. (1985) Antagonistic activity of fungi isolated from sclerotia of *Sclerotinia sclerotiorum. Plant Pathology* 34, 415–421.

Zhou, T. and Reeleder, R.D. (1990) Selection of strains of *Epicoccum purpurascens* for tolerance to fungicides and improved biocontrol of *Sclerotinia sclerotiorum. Canadian Journal of Microbiology*, 36, 754–759.

Species Index

Subject Index